# Emerging Zoonotic and Wildlife Pathogens

# Emerging Zoonotic and Wildlife Pathogens

Disease Ecology, Epidemiology, and Conservation

**Dan Salkeld**

*Research Scientist, Colorado State University*

**Skylar Hopkins**

*Assistant Professor, Department of Applied Ecology, North Carolina State University*

and

**David Hayman**

*Professor, Massey University, New Zealand*

OXFORD
UNIVERSITY PRESS

# OXFORD
UNIVERSITY PRESS

Great Clarendon Street, Oxford, OX2 6DP,
United Kingdom

Oxford University Press is a department of the University of Oxford.
It furthers the University's objective of excellence in research, scholarship,
and education by publishing worldwide. Oxford is a registered trade mark of
Oxford University Press in the UK and in certain other countries

Published in the United States of America by Oxford University Press
198 Madison Avenue, New York, NY 10016, United States of America

British Library Cataloguing in Publication Data
Data available

Library of Congress Control Number: 2023942278

ISBN 9780198825920
ISBN 9780198825937 (pbk.)

DOI: 10.1093/oso/9780198825920.001.0001

Printed and bound by
CPI Group (UK) Ltd, Croydon, CR0 4YY

# Authors' Notes

This looks like a boring book. I would only read it for the pictures...

—Linnea Salkeld, top-notch person

In a way, this is a textbook written by us—now curmudgeons with PhDs and cumulative decades of experience, successes, mistakes, and misunderstandings—for our former selves, when we knew little about disease ecology and epidemiology but were keen to learn. The resultant book will now hopefully work for students as an introduction to the science of emerging infectious diseases, from outbreak epidemiology to ecosystem impacts.

Though all three of us hail from northern climes, we have attempted to include a diversity of zoonoses and wildlife disease from across the globe. We did our best to make sure that everything that we included in this book was accurate, but we included research on case studies that are relatively new, or poorly studied, or studied so often that it would be impossible for us to read everything about them. So, we will have made mistakes. You may also notice glaring omissions—and so we encourage you to use this book as a motivation to learn and research more. At the very least, we hope that you find the book interesting, clear, and nice to look at.

We wrote this book during challenging times, and we would not have finished it without the support of our families and friends. If not for their guidance, there would be fewer good jokes in this book. Truly, we stole some of them.

We would like to thank a (parasitic) host of scientific experts, colleagues, and friends for sharing their expertise. When acknowledging someone by name, there is no implicit implication that the person endorses or approves of this book. Most likely they will be oblivious to its existence, and dubious of its content. Nonetheless, thank you to Mike Antolin, APOPO and their HeroRATs, Tomas Aragon, Carol Boggs, Danielle Buttke, Scott Carver, Dave Civitello, Christopher Cleveland, Meggan Craft, Paul Cross, Jake DeBow, Giulio De Leo, Terence Farrell, Matt Ferrari, Brian Foy, Lynne Gaffikin, Tony Goldberg, Gabe Hamer, Madison Harman, Carl Herzog, Richard Hoffman, James Holland Jones, Anne Kjemtrup, Kevin Lafferty, Bob Lane, Edward S. Lino, Andrea Lund, Parag Mahale, Alynn Martin, Stephen Schneider, Paul Stapp, Jean Tsao, and J.D. Willson.

We are especially grateful to Hanna Kiryluk, who made so many figures, wrangled reference lists, and confirmed that content was understandable.

We are also thrilled that so many people shared their stunning photos, data, and graphics. Thank you!

At OUP, thank you to Ian Sherman for instigating and investing in this book. Thank you to Bethany Kershaw, whose original response to Chapter 1 helped set the tone for the rest of the book (though we the authors accept the blame). To Charles Bath for patiently encouraging rich content and avoiding any mention of deadlines whilst on UK lockdown. And to Katie Lakina for shepherding us through the final stages.

Dan Salkeld would like to thank his beloved parents; a wonderful, scenic-road-taking Virginian redhead; two brilliant girls who've already racked up miles of expert tick-collecting; and the much-missed Nate Nieto. And also, whisper it, thank you Skylar and Dave; after fifteen chapters I still think that you're funny and interesting.

S. Hopkins would like to thank Carrot the Dog for loudly voicing his opinions during every Zoom meeting and Stuart Robertson for providing artistic feedback and helping to translate British co-author communications.

Dave Hayman would like to thank his two co-authors for doing most of the work, being exceptionally patient, and humouring him. He should probably also thank his family, but they will never read this, and possibly his team, but they will also probably never read this. He would like to thank anyone who does actually read it.

A quick mention of stylistic oddities. We have deliberately tried to restrain our use of citations to let the writing flow as much as possible—references might only be cited once within an entire paragraph. Latin names have been repeated in full, instead of abbreviating the genus name, for readability and to help lodge them in the reader's brain. Like an errant *Umingmakstrongylus*.

With all that in mind, you may now turn the page, and start learning about some of the greatest challenges of our time: emerging zoonoses and wildlife pathogens.

# Contents

**1 Spillover and emerging infectious diseases**     **1**

1.1  Introduction     1
1.2  From spillover to pandemic     2
1.3  Zoonoses     4
1.4  Zoonotic origins of the 'Big Three'     7
1.5  Barriers to emergence     9
1.6  Ebola virus as a case study of spillover     9
1.7  Improved diagnostics and increasing rate of pathogen discovery     11
1.8  Epidemiology meets disease ecology     13
1.9  Why study the ecology and epidemiology of infectious disease?     13
1.10  Notes on sources     14
1.11  References     14

**SECTION 1  Describing Outbreaks**     **17**

**2  The anatomy of disease**     **19**

2.1  Modes of pathogen transmission—direct contact     19
2.2  Airborne transmission     19
2.3  Environmental transmission     20
2.4  Vehicle-borne transmission     21
2.5  Vector-borne transmission     24
2.6  Vertical or congenital transmission     25
2.7  Portals of host entry     28
2.8  Host exits     30
2.9  Infectious, latent, incubation, and symptomatic periods     30
2.10  Disease     34
2.11  Disease agent groups     40
2.12  Summary     42
2.13  Notes on sources     44
2.14  References     44

**3  Descriptive epidemiology of disease outbreaks**     **47**

3.1  Primary and index cases     47
3.2  Epidemic curves     47

3.3    Interpreting epidemic curve patterns                                                49
3.4    Common source outbreaks                                                            51
3.5    Incubation periods and outbreak exposures                                          53
3.6    Propagated transmission                                                            54
3.7    Test validity                                                                      57
3.8    Test validity, within-host pathogen dynamics, and test type                       60
3.9    Test validity and local pathogen prevalence                                       63
3.10   Test validity and repeat tests                                                    63
3.11   Pooled samples                                                                    63
3.12   Summary                                                                           64
3.13   Notes on sources                                                                  64
3.14   References                                                                         64

4   Surveillance                                                                          67

4.1    Surveillance approaches                                                            67
4.2    Aggregating data                                                                   74
4.3    Aggregating data: ecologic fallacy and Simpson's paradox                          74
4.4    Surveillance and zoonotic outbreaks                                               78
4.5    Outbreak surveillance                                                             79
4.6    An outbreak case study                                                            83
4.7    Summary                                                                           85
4.8    References                                                                         85

5   Making simple predictions using models                                               89

5.1    Introduction                                                                       89
5.2    Mathematical models are simplifications of disease systems                         89
5.3    Basic compartmental models                                                         92
5.4    How does host density affect pathogen transmission?                                95
5.5    Using simple models to make predictions                                            97
5.6    The basic reproductive number, $R_0$                                               99
5.7    Deterministic vaccination thresholds                                              102
5.8    Deterministic invasion thresholds                                                 104
5.9    When do pathogens drive host species to extinction?                               105
5.10   Predicting long-term dynamics while ignoring random chance                        106
5.11   Incorporating random chance when predicting long-term dynamics                    107
5.12   When models are wrong                                                             109
5.13   Summary                                                                           110
5.14   References                                                                         111

SECTION 2  Pathogen Sources                                                             113

6   The environment as a pathogen reservoir                                             115

6.1    Introduction                                                                      115
6.2    Five questions to define 'environmental reservoirs'                               120
6.3    Soil and plants as environmental reservoirs                                       121

6.4   Faeces as an environmental reservoir                                              123
6.5   Water as an environmental reservoir                                               126
6.6   Summary                                                                           129
6.7   References                                                                        129

**7   Reservoir hosts**                                                                 **133**

7.1   Introduction                                                                      133
7.2   Spillover from a single host reservoir: armadillos and leprosy                    133
7.3   Multiple reservoir hosts: rabies, dogs, and wildlife                              136
7.4   Interactions between domestic and wildlife reservoirs                             138
7.5   Spillover from multiple reservoirs: Lyme disease                                  143
7.6   Reservoir hosts and the dilution effect                                           145
7.7   Multiple reservoir hosts and multiple pathogens: tick-borne diseases             148
7.8   Idiosyncrasies of human behaviour and exposure to tick-borne pathogens           149
7.9   Contributions of non-reservoir hosts to local disease ecology                    150
7.10  Summary                                                                           151
7.11  References                                                                        152

**8   Identifying animal reservoirs during an epidemic**                                **155**

8.1   Evidence of infection                                                             155
8.2   Evidence of exposure is not evidence of reservoir competence                      160
8.3   Genomic analyses to identify reservoir sources                                    161
8.4   Causal association                                                                163
8.5   Finding the reservoir for Hantavirus Pulmonary Syndrome                           164
8.6   Outbreaks are not always caused by spillover from reservoirs                      165
8.7   Notes on sources                                                                  170
8.8   References                                                                        170

**SECTION 3  Drivers of Infectious Disease Emergence**                                  **173**

**9   Emerging infectious diseases and globalization—travel, trade,
and invasive species**                                                                  **175**

9.1   Travel brings zoonotic infections to non-endemic areas                            175
9.2   Travel drives stuttering outbreaks in non-endemic areas                           177
9.3   Travel drives pandemics                                                           180
9.4   Wildlife trade drives the spread of infectious diseases                           182
9.5   Invasive species drive disease emergence                                          188
9.6   Summary                                                                           194
9.7   Notes on sources                                                                  194
9.8   References                                                                        196

**10  Climate change and emerging infectious diseases**                                 **199**

10.1  Climate affects host–parasite interactions                                        199
10.2  Thermal performance curves                                                        204
10.3  Climate change and shifting distributions                                         206

10.4   Extreme weather events                                          215
10.5   Summary: interpreting complex climate–disease patterns          218
10.6   Notes on sources                                                218
10.7   References                                                      219

**11   Land use change and emerging infectious diseases**              **221**

11.1   Introduction                                                    221
11.2   What is land use change?                                        223
11.3   Deforestation and forest fragmentation                          225
11.4   Wild meat                                                       226
11.5   Wildlife farming                                                229
11.6   Livestock farming                                               231
11.7   Agriculture and water use                                       235
11.8   Urbanization                                                    241
11.9   Poverty traps                                                   242
11.10  Summary                                                         243
11.11  Notes on sources                                                244
11.12  References                                                      244


**SECTION 4   Conservation, Ecology, and Control**                     **249**

**12   Impacts of emerging infectious disease on wildlife populations** **251**

12.1   Introduction—Arctic foxes and otodectic mange                   251
12.2   Small carnivore populations threatened by pathogens from
       domestic dog reservoirs                                         256
12.3   Small carnivore populations threatened by pathogens from
       wildlife reservoirs                                             259
12.4   Environmental reservoirs and resistant reservoir hosts drive
       amphibian species to extinction                                 260
12.5   Infectious diseases can make common species rare                264
12.6   Summary                                                         266
12.7   References                                                      267

**13   Infectious diseases in ecosystems**                             **271**

13.1   Communities and ecosystems                                      271
13.2   Bottom-up effects                                               271
13.3   Top-down effects: mesopredator release                          273
13.4   Top-down effects: trophic cascades                              276
13.5   Parasites in food webs                                          282
13.6   Ecosystem functions performed by parasites                      283
13.7   Co-infection                                                    285
13.8   Summary                                                         287
13.9   References                                                      289

**14   Infectious disease control** **291**

14.1   Treating infected wildlife: a tale of scabid wombats   291
14.2   Vector control and vaccination: conserving the plagued
        black-footed ferret   294
14.3   Control interventions   299
14.4   Culling wildlife to prevent wildlife–livestock disease
        transmission: the case of badgers and bovine tuberculosis   306
14.5   Culling and wildlife disease reservoirs, more generally   311
14.6   Unintended consequences—bison, elk, cattle, and brucellosis   313
14.7   Pathogen invasion and disease control   316
14.8   A final case study: Guinea worm disease   318
14.9   Summary   320
14.10 References   321

**15   COVID-19, One Health, and pandemic prevention** **325**

15.1   Rapid emergence of a novel pathogen   325
15.2   From outbreak to pandemic   328
15.3   The source of SARS-CoV-2 spillover   328
15.4   Controlling the spread of a pandemic virus … or not   331
15.5   Drivers of the COVID-19 pandemic   336
15.6   Complexity and wicked problems   337
15.7   The One Health approach and interdisciplinary collaboration   338
15.8   Preventing pandemics   344
15.9   Conclusion   344
15.10 Notes on sources   344
15.11 References   345

Index   349

# CHAPTER 1

# Spillover and emerging infectious diseases

Lamb castration is not always the idyllic pastime that some people may think it is.

—**Marc Abrahams**

## 1.1 Introduction

*Wyoming, USA, June 2011*—In a world that is increasingly urbanized, and where farming is practised by fewer and fewer people, you may not be familiar with any of the time-honoured techniques used to castrate lambs. A web search will suggest that you can use your teeth, and that this has the advantage of reducing the likelihood of a consequent infection in the newly emasculated lamb. However, there are risks for the castrator.

At a multiday event to castrate and dock the tails of 1600 lambs on a Wyoming sheep ranch, two men used their teeth to castrate some of the lambs. Ten other people did not adopt this technique. Subsequently, the two men were stricken by diarrhoea, and one also suffered abdominal cramps, fever, nausea, and vomiting. One patient was hospitalized for a day, but both men recovered from their bouts of illness (Van Houten et al. 2011).

The Wyoming Department of Health investigated these cases and managed to isolate indistinguishable bacteria—*Campylobacter jejuni*—from both men's stool samples. Although *Campylobacter* causes an estimated 1.5 million illnesses each year in the United States, the *Campylobacter jejuni* strain identified in the Wyoming shepherds was extremely rare (8 from 8817 strains), suggesting a common source of infection. The source turned out to be the lambs, some of which were experiencing their own bouts of diarrhoea.

This is a mildly amusing case of a **zoonosis—a pathogen transmitted from non-human animals to humans** (Fig. 1.1). *Campylobacter jejuni* was circulating in the sheep flock and the unusual contact between lamb scrotum and human teeth and mouths enabled zoonotic disease transmission.

It could also be considered an example of an **emerging infectious disease** (EID)—**caused by a pathogen spreading to new species, spreading to new geographic areas and populations, or spreading at faster rates in places or populations where it has previously been endemic** (Jones et al. 2008). In this case, *Campylobacter jejuni* appeared to have entered the local human population (Wyoming shepherds) for the first time.

Scientists pride themselves on being objective and systematic, but when it's you with an arachnid in your nasal cavity, that sense of distance goes right out of the window. When you first realize you have a tick up your nose, it takes all your willpower not to claw your face off.

—Tony Goldberg

*Kibale, Uganda and Wisconsin, USA, June 2012*—Exposure to wildlife zoonoses and/or their vectors is not an unusual circumstance affecting only luckless shepherds. In June 2012, Dr Goldberg, a veterinary epidemiologist with a research background in primatology and emerging infectious diseases, reluctantly acknowledged a severe pain in his nose (Goldberg 2013). Several days previously, Goldberg

*Emerging Zoonotic and Wildlife Pathogens.* Dan Salkeld, Skylar Hopkins, and David Hayman, Oxford University Press.
© Oxford University Press (2023). DOI: 10.1093/oso/9780198825920.003.0001

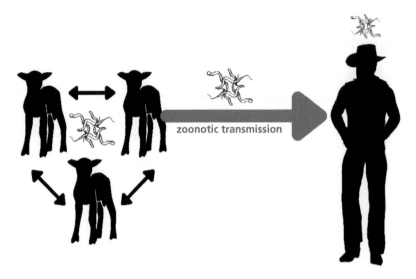

**Figure 1.1**  Zoonotic (non-human animal to human) transmission of *Campylobacter jejuni* from Wyoming lambs to humans. Created by Hanna D. Kiryluk using Canva.com.

had been in a tropical forest in Kibale, Uganda, studying how anthropogenic changes to ecosystems alter health-related outcomes and infectious disease dynamics in people, wildlife, and domestic animals. Now back at his Wisconsin home, Goldberg used an angled mirror to examine his nostril, and confirmed his suspicions by seeing the 'smooth, rounded backside of a fully engorged tick'. While rare, these 'nostril ticks' are not unheard of in visitors to Kibale National Park. This was Dr Goldberg's third infestation (and yet he kept returning!). Fortunately, he suffered no lasting harm, besides the mental trauma of having a blood-sucking arachnid inside his nose.

Collaborating with medical entomologists and primatologists, Goldberg and colleagues were able to describe this tick as likely a new *Amblyomma* species (Fig. 1.2). Using high resolution photographs of chimpanzee noses (Fig. 1.2), they showed that roughly 20% of young chimpanzees had tick-infested nostrils (Hamer et al. 2013). Presumably, the ticks infest the nostril because chimpanzees love to groom each other, so ticks need a place to hide from chimps' stubby fingers. Further research suggests that sporadic cases of nostril ticks have occurred in people visiting Africa's equatorial forests for several decades. Can the nostril ticks transmit pathogens, and could unwary

travellers inadvertently carry disease agents from tropical Africa to their far-flung homes? That question remains unanswered.

This incident has several pertinent lessons. First, human–wildlife interactions occur at individual levels, and often go unrecognized. Second, although contact is at an individual level, the impacts are no longer limited to small geographical locales. Air travel means that pathogens and vectors now have opportunities to travel globally and within infectious periods (Chapter 9). Third, human–wildlife interactions have not always been assiduously described or catalogued. *Amblyomma* ticks have probably been exploring primate nostrils for a long time, but it took the chance infestation of a qualified veterinarian to result in its scientific identification and further research.

## 1.2 From spillover to pandemic

**Spillover events—the occasions when a pathogen jumps to new species and thereby infects humans (zoonotic) or animals that are not the usual host species—**occur at the scale of the individual, but the impacts of pathogens in a new host species are unpredictable. Dr Goldberg did not report any pathogen transmission after being bitten by an African *Amblyomma* tick—perhaps because the tick

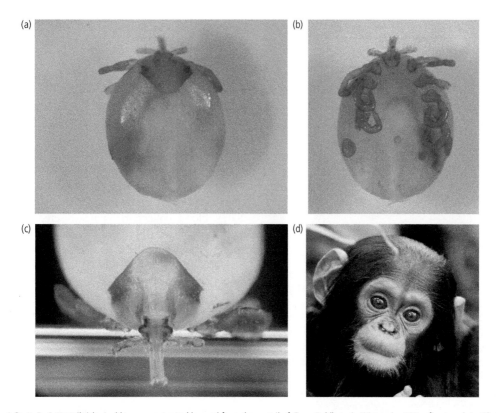

**Figure 1.2**  A, B, C. Nostril tick, *Amblyomma*, extracted by, and from the nostril of, Tony Goldberg in Wisconsin, USA, after travels in Kibale National Park, Uganda. Photos by Gabriel L. Hamer. D. Nostril tick in a young chimpanzee, Thatcher, in Kibale National Park, Uganda, 2012. Photo by Andrew B. Bernard/Kibale Chimpanzee Project.

Originally published in Hamer et al. (2013). Coincident tick infestations in the nostrils of wild chimpanzees and a human in Uganda. American Journal of Tropical Medicine and Hygiene, 89, 924–7. Used with the permission of American Journal of Tropical Medicine and Hygiene.

was not carrying a potentially pathogenic infection, or because any potential pathogen it carried was not able to infect a human host, or was unable to infect Dr Goldberg in particular because he was not *susceptible* (see Chapter 2). The bacterium infecting shepherds in Wyoming did cause illness, but as far as we know, there was no subsequent transmission from those two human cases to other humans or lambs.

Terms to describe a disease's impact include clusters, outbreaks, epidemics and pandemics.

- A **cluster** is an **observation of an above-normal number of cases**, normally aggregated in time at a local scale. Often this is simply the term for the beginning of an outbreak, but before connections and transmission chains have been officially identified.

- An **outbreak** or **epidemic** (also called an **epizootic** when referring to wildlife or domestic animal disease) is an **increase of disease cases**, often with continued transmission and with established causative links between cases.
- A **pandemic** is an **outbreak/epidemic at the international or global scale**.

These descriptions blur and bleed into each other; there is not normally an official threshold for when a cluster becomes an outbreak, or for when an epidemic becomes a pandemic.

How does one define an 'above-normal number of cases?' The normal number of cases refers to observed baseline infection rates of a persistent pathogen in a population or community. This is referred to as the **endemic** level of the disease (called **enzootic** when referring to wildlife or domestic animal disease) (Box 1.1). Where transmission in a population is persistent and at relatively high levels, the disease pattern can be referred to as **hyperen-demic**. In contrast, diseases that occur infrequently and irregularly are often termed **sporadic**.

## 1.3 Zoonoses

Zoonotic pathogen transmission can be classi-fied depending upon the transmission patterns in human populations (Wolfe et al. 2007, Lloyd-Smith et al. 2009, Woolhouse et al. 2013). (Similar

classifications can be used for any infection entering any new host species, such as giraffes or newts instead of humans.) Importantly, pathogens could fall into different categories in different contexts.

**Dead-end zoonoses** can be transmitted from ani-mals to humans, but do not subsequently spread from human-to-human. For example, West Nile virus circulates in the US in birds and mosquitoes and can cause illness in people after a mosquito bite. However, human-to-human transmission does not occur (Fig. 1.3). (Although infections acquired from blood transfusions are possible (Iwamoto et al. 2003), blood collection agencies screen donated blood for West Nile virus.) Another dead-end zoonosis is rabies virus, which is transmitted to people through bites of rabid wildlife or domestic species. Despite being vaccine-preventable, rabies kills approximately 59,000 people every year (the equivalent of one person every nine minutes), 40% of whom are children living in Asia and Africa. Humans are capable of shedding rabies virus dur-ing the clinical phase of disease but there has not been a documented case of human-to-human trans-mission, except from unusual cases involving organ or tissue transplantation (Blackburn et al. 2022, Chapter 2).

**Stuttering zoonoses** can jump from animals into humans, and then, with some limited success, into other people. In these outbreaks, **human-to-human transmission is limited and the pathogen *stutters* or *fades out* before a full-blown epidemic occurs**. For example, since 2001, in Bangladesh, there have been recurrent spillovers of Nipah virus to humans from Indian fruit bats (*Pteropus medius*). Often, the spillover results in an isolated human case, but human-to-human transmission can infect a mean of 7 persons (range 1–22) resulting in a stuttering outbreak (Luby et al. 2009) (Fig. 1.4). In another example from South Texas, USA, there are suit-able mosquito vectors for Zika virus, but Zika out-breaks in humans have tended to fade out, perhaps because the mosquitoes are predominantly biting non-human hosts (dogs, domestic animals) (Olson et al. 2020).

**Sustainable zoonoses** originate and persist in animals, **but once in human populations, they can spread through human-to-human transmission to cause outbreaks that persist through multiple generations of transmission and human cohorts**.

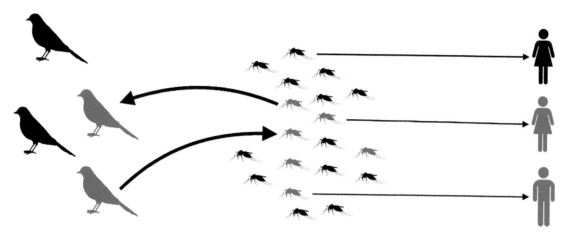

**Figure 1.3** West Nile virus is transmitted among wild birds by mosquitoes in the United States. Infected (red) mosquitoes that feed on humans may cause a zoonotic spillover by infecting people (red), but there is no further transmission from person-to-person; it is a dead-end zoonosis.
Created by Hanna D. Kiryluk with BioRender.com

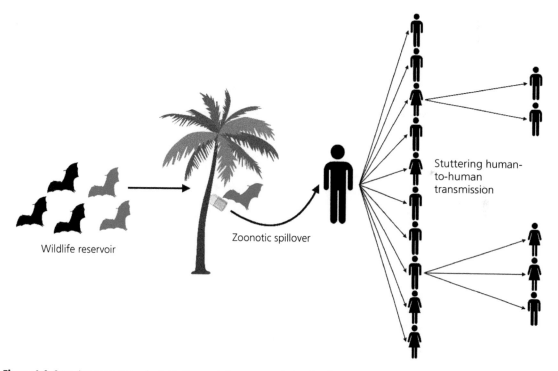

**Figure 1.4** Stuttering zoonoses involve limited human-to-human transmission after the initial zoonotic spillover and end up fading out. For example, in Bangladesh, Nipah virus persists in populations of greater Indian fruit bats (*Pteropus medius*) but occasionally makes the jump into humans through raw date palm sap contaminated by bat waste. The primary human case can then infect other people. A stuttering human-to-human Nipah virus outbreak occurred in 2004, when the index patient infected four of the people caring for him (his mother, son, aunt and a neighbour; the incubation period was 15–27 days). Whilst in hospital, the aunt was cared for by a popular local religious leader, who became ill 13 days later, and was visited by many of his relatives and members of his religious community at his home. Twenty-two people were infected after contact with the religious leader. Further transmissions occurred and, in the end, the chain of transmission involved five generations and affected 34 people (Luby et al. 2009).
Created by Hanna D. Kiryluk using BioRender.com and a sketch on an old pizza box.

A classic example of a sustainable zoonosis would be plague, caused by the bacterium *Yersinia pestis*, famous for the Black Death in medieval Europe. In the modern era, *Yersinia pestis* still sporadically jumps from animal populations and subsequently spreads from person-to-person, e.g. in 2017 a plague outbreak occurred in Madagascar when a man, presumably infected from an animal source, travelled by taxi through the nation's capital, Antananarivo, and transmitted *Yersinia pestis* to dozens of people. Person-to-person transmission continued and caused pneumonic disease (infection of the lungs), resulting in 2348 confirmed, probable, and suspected cases, including 202 deaths.

Sustainable zoonoses also include pathogens now regarded as human diseases, but historically or evolutionarily the disease had a zoonotic origin, whether that was from domesticated animals, livestock, or wildlife (Fig. 1.5). Recurrent introduction from animal to human populations is no longer a concern; the pathogens are adapted to human transmission. During the slow writing of this book,

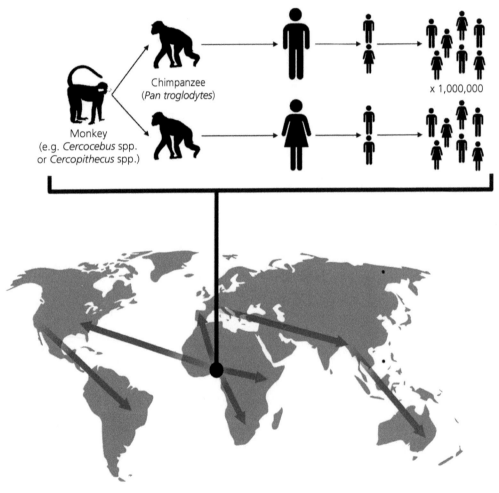

**Figure 1.5** Sustainable zoonoses can emerge from animal hosts into human populations, possibly via a 'bridge' reservoir host, and then adapt to persistent transmission within humans. Human travel can then elevate the spillover incident to an outbreak and then to pandemic transmission. For example, human immunodeficiency virus-1 (HIV-1) probably jumped from monkey populations to chimpanzees to humans, and the chimpanzee-to-human spillover occurred more than once. Human-to-human transmission is now responsible for maintaining the virus, which is now present throughout global populations (arrows simply represent the concept of global travel). Similarly, SARS-CoV-2 jumped into human populations in 2019 and is now a sustainable pandemic pathogen.
Created by Hanna D. Kiryluk with BioRender.com.

SARS-CoV-2 has provided a prime example of a novel virus (source still unknown) that has spilled over into human populations and continued evolution and circulation in its new human host (Chapter 15).

## 1.4 Zoonotic origins of the 'Big Three'

Malaria, tuberculosis, and AIDS are considered the 'Big Three' infectious diseases, with catastrophic impacts upon human populations, especially in low-income countries (Roche et al. 2018). These three diseases illustrate varied evolutionary and zoonotic histories.

Human malaria is caused by the protist *Plasmodium*, which is transmitted by female *Anopheles* mosquitoes. Estimates of the public health burden created by malaria suffer from high uncertainty, but in 2013, there were between 95 million and 284 million new cases globally, incurring 703,000–1,032,000 deaths (Murray et al. 2014). *Plasmodium falciparum* is the deadliest of the various malaria protozoa. By sampling for *Plasmodium* in faecal samples from wild apes that were living in remote forested areas away from humans, researchers discovered that *Plasmodium falciparum* probably originated in gorillas. Though the timing of this event in evolutionary history is unknown, the spillover to human populations may have been a single cross-species transmission occurrence (Liu et al. 2010, Prugnolle et al. 2011). There is little evidence to suggest that there are recurrent introductions of malaria from apes to humans in Africa (Sundararaman et al. 2013) (Fig. 1.6).

However, 'zoonotic' malaria cases of *Plasmodium* do occur: *Plasmodium knowlesi* typically infects monkeys in south-east Asia but was recently discovered as a human pathogen (Singh et al. 2004, Singh and Daneshvar 2013, Raja et al. 2020).

The sooty mangabey (*Cercocebus atys*) is a smoky-gray creature with a dark face and hands, white eyebrows, and flaring white muttonchops, not nearly so decorative as many monkeys on the continent but arresting in its way, like an elderly chimney sweep of dapper tonsorial habits.
—David Quammen

Today, the human immunodeficiency virus (HIV) that causes acquired immunodeficiency syndrome (AIDS) infects approximately 30 million people and is responsible for over a million deaths a year (data for 2013; Murray et al. 2014). The virus jumped from non-human primates to human populations multiple times, and now is firmly established as a human pathogen. For example, HIV-2 likely emerged in human populations from a sooty mangabey (*Cercocebus atys*) sometime in the latter half of the 20th century. This interpretation is based on three pieces of evidence: the common ancestor, simian immunodeficiency virus ($SIV_{sm}$), is prevalent in West African populations of sooty mangabeys; HIV-2 is endemic in West Africa; and the genomes of $SIV_{sm}$ and HIV-2 are closely related (Hirsch et al. 1989). HIV-1 has a different origin, and phylogenetic studies suggest that chimpanzees in south-eastern Cameroon were the original reservoirs of at least two HIV-1 strains. These ape populations probably acquired the virus from red-capped mangabeys or *Cercopithecus* species at a time when chimpanzee populations were evolving into subspecies (Keele et al. 2006). The spillover to human populations probably occurred early in the 20th century (Keele et al. 2006, Worobey et al. 2008).

Tuberculosis (TB) is one of the world's deadliest infectious diseases—the World Health Organization (WHO) estimates that the bacterium kills between two and three million people annually, 98% of whom live in the developing world—even though a cure is available (Arnold 2007). A third of people infected with HIV have TB, and the evolution of multidrug resistant tuberculosis (MDR TB) strains is a growing public health concern. Tuberculosis is usually caused by *Mycobacterium tuberculosis*, but the World Health Organization (WHO) estimates that >10,000 deaths were due to 'bovine' TB globally in 2010 (Olea-Popelka et al. 2017), which is caused by the related *Mycobacterium bovis*. *Mycobacterium bovis* has a broad host range including cattle, European badgers (*Meles meles*), Australian brushtail possums (*Trichosurus vulpecula*), deer, and so on (Brites and Gagneux 2015) (see Chapter 14). Consequently, it has been assumed that *Mycobacterium bovis* was ancestral to *Mycobacterium tuberculosis*, which would make human TB (*Mycobacterium tuberculosis*) another example of a sustainable zoonosis. However, genetic analyses have discovered that certain human *Mycobacterium tuberculosis* lineages are actually more ancestral (phylogenetically more

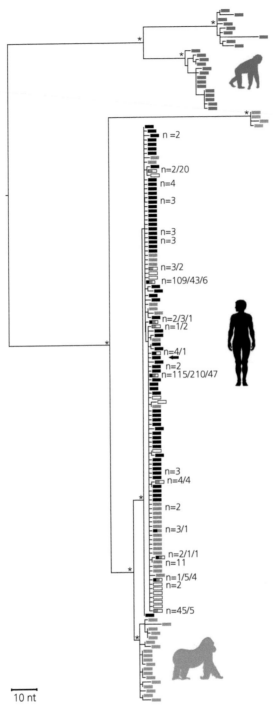

**Figure 1.6** Phylogeny of *Plasmodium falciparum* strains from rural Cameroon. *Plasmodium falciparum* is the mosquito-borne disease agent that causes falciparum malaria. Sequences from humans living in Cameroon (black) are compared to sequences from Genbank (white) and the Sanger Institute (grey), *Plasmodium praefalciparum* from gorillas (green), and *Plasmodium reichenowi* from chimpanzees (red).

From: Sundararaman et al. (2013). *Plasmodium falciparum*-like parasites infecting wild apes in southern Cameroon do not represent a recurrent source of human malaria. Proceedings of the National Academy of Sciences USA, 110, 7020–5.

basal) than the animal-adapted *Mycobacterium bovis* (Arnold 2007, Brites and Gagneeux 2015, Brosch et al. 2002). This evolutionary story appears more complex that those of malaria and HIV/AIDS!

## 1.5 Barriers to emergence

Many pathogens with zoonotic origin that became adapted to human populations are termed 'crowd diseases', because the pathogens could only become established in human societies that had reached high enough densities to allow sustained transmission and replacement of susceptible individuals. Crowd diseases tend to be very infectious and cause some antibody-derived resistance. Historically, some of these pathogen spillovers occurred synchronously with the development of agriculture and domestication, implying a pathogen spillover double jeopardy: increased exposure to livestock species, and increased human densities (Wolfe et al. 2007, Woolhouse et al. 2013).

However, not all animal pathogens spillover and become human-specialized pathogens (Wolfe et al. 2007), even if there are many humans living in one place. Humans might not be susceptible to the pathogen, or opportunities for human exposure may be limited, even if a pathogen is potentially capable of infecting people. Behaviour can reduce or prevent exposures. For example, instead of using your teeth, using alternative castration techniques to reduce the risk of *Campylobacter jejuni* transmission from lambs; or avoiding potential exposure to prions like bovine spongiform encephalopathy (BSE), also known as mad cow disease or chronic wasting disease, by not eating parts of the animal's central nervous system (CNS) and lymphoid tissue.

## 1.6 Ebola virus as a case study of spillover

*Guinea, Liberia, Sierra Leone, Spain, United States, etc., 2013–2016*—First described in 1976 from an outbreak in the remote village of Yambuku, Democratic Republic of the Congo, and named after the Ebola River, the Ebola virus (or *Zaire ebolavirus*) is a zoonotic pathogen that can cause high mortality among human cases (i.e. a high case-fatality risk or, more commonly known as, case fatality rate).

Spillover events for Ebola virus in human populations have been linked to exposure from diverse animal sources, e.g. carcasses of duiker (a small forest antelope), gorillas, and chimpanzees (Rouquet et al. 2005). Typically, Ebola virus outbreaks have occurred in remote locations in Central Africa: Gabon, the Republic of Congo, and the Democratic Republic of Congo. Between description in 1976 and early 2014, the largest known Ebola virus disease outbreaks infected approximately 300 people, and 47–89% of cases were fatal. After the outbreaks ended, several years normally passed before another Ebola virus outbreak. This boom-and-bust pattern is often exhibited by zoonotic diseases.

Illness strikes men when they are exposed to change.

—Herodotus

More recently, though, outbreaks have followed very different trajectories (Fig. 1.7). In December 2013, an 8-month-old boy from a small village in south-eastern Guinea was believed to be the index case for an Ebola virus epidemic (Baize et al. 2014, Marí Saéz et al. 2015). Though the cause of the boy's death went unrecognized at the time, the Ebola virus caused further cases of 'fatal diarrhoea' and an official medical alert was issued to district health officials in late January 2014. The virus spread to Conakry, Guinea's capital city, and was confirmed as *Zaire ebolavirus* in March 2014. Guinea is about 2500 km from the traditional range of Ebola virus in Central Africa. In this case, although it has been recognized in Central Africa for four decades, Ebola virus also constitutes an **emerging infectious disease**, because it is a **pathogen that has recently increased in incidence or geographic range**, even if it has infected humans historically (Jones et al. 2008).

Ebola virus had never infected people anywhere close to West Africa before, and the region was blindsided. On top of that, this was the first time that Ebola virus was active in heavily populated and connected urban centres rather than isolated, rural areas. A combination of dense populations, inadequate public health infrastructure and response, and a lack of prior experience with the disease, meant that the Ebola virus spread among an unprecedented number of people. By July 2014, the virus was also present in Monrovia and Freetown, the

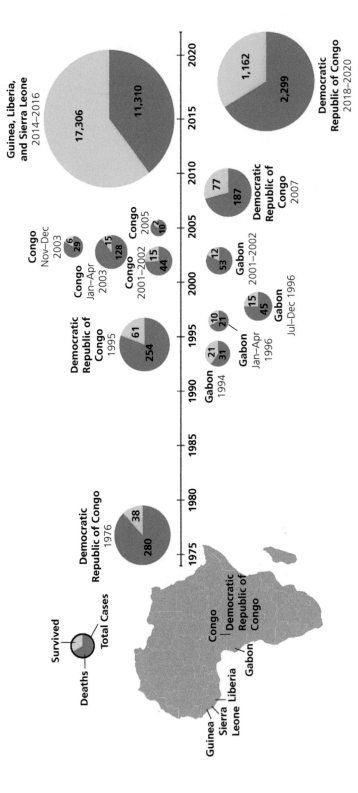

**Figure 1.7** Ebola virus outbreaks in Africa from 1976 to 2020, scaled by the outbreak size (number of cases; blue = deaths, yellow = survived). Recent outbreaks in West Africa and the Democratic Republic of Congo have been much larger than previous outbreaks and affected more urbanized populations.

Created by Hanna D. Kiryluk.

capitals of Guinea's neighbouring countries Liberia and Sierra Leone. Infected travellers or medical personnel were reported in Italy, Mali, Nigeria, Senegal, Spain, and the United States. The official tallies for the pandemic in Guinea, Liberia, and Sierra Leone were 28,610 reported cases, with 11,308 fatal cases.

During the West African Ebola virus outbreak, speculation was rife that the virus would evolve to become more deadly and transmissible. **Newly evolving strains of pathogen** are another class of **emerging infectious disease** (Jones et al. 2008). Normally circulating in wildlife species, Ebola virus certainly mutated during transmission in this naive human population (e.g. Gire et al. 2014). However, the mutations that accrued do not seem to have affected disease progression, at least in *in vivo* laboratory animal models (mice and macaques) (Marzi et al. 2018). Instead, the spread of Ebola virus in the West Africa epidemic can be better explained by the social context of closely connected human populations, and not as the result of changes in the virus's intrinsic biological properties (Marzi et al. 2018).

## 1.7 Improved diagnostics and increasing rate of pathogen discovery

Many emerging infectious diseases are probably instances of recent scientific description, rather than novel introductions to human populations (Gire et al. 2012). Illnesses often display 'flu-like symptoms' such as fever, headache, and nausea. Without proper laboratory diagnostics, it is difficult to discern whether these symptoms are a result of malaria, typhoid, shigella, or even mild cases of viral haemorrhagic fever. Thus, 'rare' spillovers of zoonoses into human populations are perhaps more frequent than recognized, and only considered rare because surveillance and diagnostics are lacking or not applied.

Diagnostic tests and our ability to describe pathogens are constantly improving. Consider that when plague emerged across the globe around the turn of the 20th century—newly invading the United States, Australia, Brazil, South Africa—bacteriology was in its infancy, germ theory (the recognition that disease might be caused by microorganisms rather than bad air) was an idea not even half a century old, and, in the absence of antibiotics, investigative public health agents might be as afeared of infected cuts as the threat of flea-borne bubonic plague (Chase 2004). Now, molecular techniques can discern previously unknown pathogens within days and even hours of the observation of suspicious new illnesses using genomic technologies, e.g. the discovery of the coronaviruses that caused Middle East respiratory syndrome (MERS) in 2012 (Zaki et al. 2012) and the related SARS-CoV-2 in China in 2020 (Zhou et al. 2020).

It is often suggested that the rate of emergence of infectious zoonotic disease is higher in the early 21st century than it has been in the more distant past; a result of combined drivers of emergence such as human population growth, population density, industrialization and changing farming methods, globalization of goods and human transport, and so on (see Chapters 9–11). This hypothesis is a difficult one to test (Woolhouse et al. 2013)! Arguably, there have been just a handful of globally devastating EID events in the past century, e.g. 'Spanish' H1N1 influenza (1918–1920) and other influenza outbreaks; HIV-1 entering human populations several times in the 20th century; the current SARS-CoV-2 pandemic. Other events, like the SARS coronavirus outbreak in 2002–2003 (see Chapter 8), had substantial impacts upon economies, government policies, and public perception, but its impacts pale in comparison to HIV, H1N1 influenza, or SARS-CoV-2. Simultaneously, other devastating zoonoses like plague, which devastated medieval Europe, represent less of a current threat because of antibiotics, sanitation, and urbanization. Have better public health responses prevented increased mortalities from emerging pathogens? Undoubtedly. The impression, then, of a modern world undergoing unprecedented threats of zoonotic epidemics is less clear cut when examined in a longer historical context.

Perhaps less debatable is the pattern of increased emergence of zoonoses in new populations and new geographies as a result of globalization and new dense human populations (Fig. 1.8). Still, though, advances in diagnostic techniques and increased interest in EIDs mean that patterns of disease emergence should be considered with nuance and thoughtfulness.

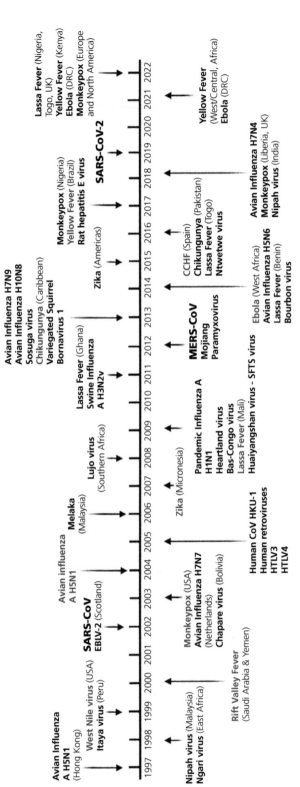

**Figure 1.8** A timeline (1997 to present) of zoonotic emergence and spillover of some RNA viruses and monkeypox virus. Repeat spillovers are indicated in red; the countries involved are in parentheses. Three recent emerging coronavirus pathogens are labelled with large font (i.e. SARS-CoV, MERS-CoV, and SARS-CoV-2). Abbreviations include: EBLV-2, European bat lyssavirus type 2; DRC, Democratic Republic of Congo; HKU-1, HKU-1 coronavirus; HTLV3, human T-lymphotropic virus type 3; HTLV4, human T-lymphotropic virus type 4; SFTS, severe fever with thrombocytopenia syndrome virus; CCHF, Crimean-Congo haemorrhagic fever virus. From Keusch et al. (2022).

## 1.8 Epidemiology meets disease ecology

When infecting humans, emerging infectious diseases are mainly the concern of **epidemiologists** who study **the distribution and determinants of health-related states or events in specified (human) populations, and the application of this study to control health problems**. Traditionally, epidemiologists have not often been experts in the animal component of cross-species pathogen transmission.

Instead, study of disease transmission in animal populations has been the purview of **disease ecologists**. Broadly, **ecology** is the study of the **distribution of organisms and their interactions with each other and their environment**, and disease ecology pays special attention to the ecology of infectious pathogens. In essence, **disease ecology is the multidisciplinary study of the interactions and causal relationships between hosts, pathogens, and vectors within the context of their environment and evolution, embracing molecular to population scales, and optimistically providing insights into forecasting disease risk and transmission** (Stapp 2007, Waller 2008, Kilpatrick and Altizer 2010, Koprivnikar and Johnson 2016) (Box 1.1). Disease ecology often broaches expertise from fields as diverse as parasitology, mathematics, engineering, conservation, climate science, and public health policy.

In this book, we mostly focus on the epidemiology and ecology of zoonotic infectious diseases and wildlife pathogens, but we also illustrate mechanisms and phenomena using examples of infectious diseases in populations of domestic animals. Admittedly, there is a bias towards systems that we are familiar with or excited by, e.g. plague, histoplasmosis, the Serengeti, tick-borne diseases, mange, Ebola virus, and northern hairy mammals. We also suffer biases based on our backgrounds: we are all white academics with roots in the northern hemisphere. We have tried to include cases outside of our expertise, but we will have missed things and we will have made mistakes. We will welcome feedback to broaden the scope of the book's topics for the second and third editions.

## 1.9 Why study the ecology and epidemiology of infectious disease?

We would argue that nostril ticks, lamb castrations, sick cowboys, and fatal haemorrhagic viruses spreading from unknown sources across the world should be answers enough! Why would you study anything else?

However, there are other reasons that this field, and hopefully this book, should excite your interest. David Quammen, the author of *Spillover* (which will be much quoted in this book), suggests that *zoonosis* is 'a word of the future, destined for heavy use in the twenty-first century' (Quammen, 2012). Zoonotic diseases and/or their vectors are certainly emerging and shifting in geographical scope and incidence (Jones et al. 2008). Changes in distribution may be due to habitat change, climate change, globalization, shifts in host community ecology, happenstance, or a combination of these factors (e.g. Chapters 9, 10 and 11). It is an exciting time to be monitoring and attempting to explain the fluctuating patterns. For the more mercenary among us, the field of disease ecology also offers job opportunities. Understanding processes influencing zoonoses and disease incidence in human populations is an important area of study, as is the ability to innovate and implement interventions that can combat and control the diseases.

The applied nature of the phenomena can also be rewarding. Devoting time and effort to understanding, for example, interactions between a harmless and obscure parasite in charismatic-but-obscure lizards may be inherently interesting to some (Salkeld and Schwarzkopf 2005), but others may revel in their work having some relevance to conservation science, or public health impacts, or environmental policy. Indeed, this added edge of potential implications of the research can add heat and passion to debated hypotheses within disease ecology, which lends the topic motivation and interest, even to those outside the field.

Like all good science, disease ecology *should* be **hypothesis driven**. Because new pathogens are emerging, there is the thrill of developing hypotheses at the most fundamental level: what might the cause of this new disease be? Where did it come from? How is it transmitted? Why is it emerging now? And because facts and data begin to roll in, one is forced to revisit or reject hypotheses at an invigorating pace. And, thus, the field of disease ecology is constantly undergoing 'Bayesian updating'—the constant revising of a theory, procedure, or opinion as new data become available (Schneider 2005). Given the potential stakes—sticking to a particular pet hypothesis that is false can have deleterious effects—disease ecology must be evidence-based, and the evidence must be translatable across disciplines, which ideally results in a science that embraces humility and transparency. How very noble!

## 1.10 Notes on sources

Information on lamb castration techniques: http://www.infovets.com/books/smrm/c/c104.htm (accessed 5/7/2018).

Information on *Campylobacter jejuni* infections in the US: https://www.cdc.gov/campylobacter/index.html (accessed 3/3/2021).

Information on Ebola virus outbreaks: https://www.cdc.gov/vhf/ebola/history/chronology.html (accessed 1/6/2018).

Information on rabies case numbers: https://www.who.int/health-topics/rabies#tab=tab_1 (accessed 1/24/2022).

Information on the 2017 plague outbreak in Madagascar: www.who.int/csr/don/27-november-2017-plague-madagascar/en/ (accessed 6/7/2018).

## 1.11 References

Arnold, C. (2007). Molecular evolution of *Mycobacterim tuberculosis*. Clinical Microbiology and Infection, 13, 120–28.

Baize, S., Pannetier, D., Oestereich, L., et al. (2014). Emergence of Zaire Ebola virus disease in Guinea. New England Journal of Medicine, 371, 1418–25.

Blackburn, D., Minhaj, F.S., Al Hammoud, R., et al. (2022). Human rabies—Texas, 2021. Morbidity and Mortality Weekly Report (MMWR), 71, 1547–9.

Brites, D., Gagneux, S. (2015). Co-evolution of *Mycobacterium tuberculosis* and *Homo sapiens*. Immunological Reviews, 264, 6–24.

Brosch, R., Gordon, S.V., Marmiesse, M., et al. (2002). A new evolutionary scenario for the *Mycobacterium tuberculosis* complex. Proceedings of the National Academy of Sciences USA, 99, 3684–9.

Chase, M. (2004). The Barbary Plague: The Black Death in Victorian San Francisco. Penguin Random House.

Gire, S.K., Goba, A., Andersen, K.G., et al. (2014). Genomic surveillance elucidates Ebola virus origin and transmission during the 2014 outbreak. Science, 345, 1369–72.

Gire, S.K., Stremlau, M., Andersen, K.G., et al. (2012). Emerging disease or diagnosis? Science, 338, 750–52.

Goldberg, T. (2013). Experience: I discovered a new species up my nose. The Guardian: https://www.theguardian.com/lifeandstyle/2013/dec/07/experience-new-species-up-my-nose, accessed 5/7/2018.

Hamer, S.A., Bernard, A.B., Donovan, R.M., et al. (2013). Coincident tick infestations in the nostrils of wild chimpanzees and a human in Uganda. American Journal of Tropical Medicine and Hygiene, 89, 924–7.

Hirsch, V.M., Olmsted, R.A., Murphey-Corb, M., et al. (1989). An African primate lentivirus (SIV_{sm}) closely related to HIV-2. Nature, 339, 389–92.

Iwamoto, M., Jernigan, D.B., Guasch, A., et al. (2003). Transmission of West Nile virus from an organ donor to four transplant recipients. New England Journal of Medicine, 348, 2196–203.

Jones, K.E., Patel, N.G., Levy, M.A., et al. (2008). Global trends in emerging infectious diseases. Nature, 451, 990–94.

Keele, B.F., Van Heuverswyn, F., Li, Y., et al. (2006). Chimpanzee reservoirs of pandemic and nonpandemic HIV-1. Science, 313, 523–6.

Keusch, G.T., Amuasi, J.H., Anderson, D.E., et al. (2022). Pandemic origins and a One Health approach to preparedness and prevention: solutions based on SARS-CoV-2 and other RNA viruses. Proceedings of the National Academy of Sciences USA, 119, e2202871119.

Kilpatrick, A.M., Altizer, S. (2010). Disease ecology. Nature Education Knowledge, 3, 55.

Koprivnikar, J., Johnson, P.T.J. (2016). The rise of disease ecology and its implications for parasitology—a review. Journal of Parasitology, 102, 397–409.

Liu, W., Li, Y., Learn, G.H., et al. (2010). Origin of the human malaria parasite *Plasmodium falciparum* in gorillas. Nature, 467, 420–25.

Lloyd-Smith, J.O., George, D., Pepin, K.M., et al. (2009). Epidemic dynamics at the human-animal interface. Science, 326, 1362–7.

Luby, S.P., Gurley, E.S., Hossain, M.J. (2009). Transmission of human infection with Nipah Virus. Clinical Infectious Diseases, 49, 1743–8.

Marí Saéz, A., Weiss, S., Nowak, K., et al. (2015). Investigating the zoonotic origin of the West African Ebola epidemic. EMBO Molecular Medicine, 7, 17–23.

Marzi, A., Chadinah, S., Haddock, E., et al. (2018). Recently identified mutations in the Ebola Virus-Makona genome do not alter pathogenicity in animal models. Cell Reports, 23, 1806–16.

Murray, C.J.L, Ortblad, K.F., Guinovart, C., et al. (2014). Global, regional, and national incidence and mortality for HIV, tuberculosis, and malaria during 1990–2013: a systematic analysis for the Global Burden of Disease Study 2013. The Lancet, 384, 1005–70.

Olea-Popelka, F., Muwonge, A., Perera, A., et al. (2017). Zoonotic tuberculosis in human beings caused by *Mycobacterium bovis*—a call for action. Lancet Infectious Disease, 17, e21–e25.

Olson, M.F., Ndeffo-Mbah, M.L., Juarez, J.G., et al. (2020). High rate of non-human feeding by *Aedes aegypti* reduces Zika virus transmission in South Texas. Viruses, 12, 453.

Prugnolle, F., Durand, P., Ollomo, B., et al. (2011). A fresh look at the origin of *Plasmodium falciparum*, the most malignant malaria agent. PLOS Pathogens, 7(2), e1001283.

Quammen, D. (2012). Spillover: Animal Infections and the Next Human Pandemic. W.W. Norton.

Raja, T.N., Hu, T.H., Kadir, K.A., et al. (2020). Naturally acquired human *Plasmodium cynomolgi* and *P. knowlesi* infections, Malaysian Borneo. Emerging Infectious Diseases, 26, 1801–09.

Roche, B., Baldet, T., Simard, F. (2018). Infectious diseases in low-income countries: where are we now? 1–16 in B. Roche, T. Baldet, F. Simard (Eds.), Ecology and Evolution of Infectious Diseases: Pathogen Control and Public Health Management in Low-Income Countries, Oxford University Press.

Rouquet, P., Froment, J-M., Bermejo, M., et al. (2005). Wild animal mortality monitoring and human Ebola outbreaks, Gabon and Republic of Congo, 2001–2003. Emerging Infectious Diseases, 11, 283–90.

Salkeld, D.J., Schwarzkopf, L. (2005). Epizootiology of blood parasites in an Australian lizard: a mark-recapture study of a natural population. International Journal for Parasitology, 35, 11–18.

Schneider, S.H. (2005). The Patient from Hell. Da Capo Press.

Singh, B., Daneshvar, C. (2013). Human infections and detection of *Plasmodium knowlesi*. Clinical Microbiology Reviews, 26, 165–84.

Singh, B., Kim Sung, L., Matusop, A., et al. (2004). A large focus of naturally acquired *Plasmodium knowlesi* infections in human beings. Lancet, 363, 1017–24.

Stapp, P. (2007). Trophic cascades and disease ecology. EcoHealth, 4, 121–4.

Sundararaman, S.A., Liu, W., Keele, B.F., et al. (2013). *Plasmodium falciparum*-like parasites infecting wild apes in southern Cameroon do not represent a recurrent source of human malaria. Proceedings of the National Academy of Sciences USA, 110, 7020–25.

Van Houten, C., Musgrave, K., Weidenbach, K., et al. (2011). Notes from the field: *Campylobacter jejuni* infections associated with sheep castration—Wyoming, 2011. Morbidity and Mortality Weekly Report (MMWR), 60, 1654.

Waller, L.A. (2008). Statistics in disease ecology: introduction to a special issue. Environmental and Ecological Statistics, 15, 259–63.

Wolfe, N.D., Dunavan, C.P., Diamond, J. (2007). Origins of major human infectious diseases. Nature, 447, 279–83.

Woolhouse, M.E.J., Adair, K., Brierley, L. (2013). RNA viruses: a case study of the biology of emerging infectious diseases. Microbiology Spectrum, 1, OH-0001-2012.

Worobey, M., Gemmel, M., Teuwen, D.E., et al. (2008). Direct evidence of extensive diversity of HIV-1 in Kinshasa by 1960. Nature, 455, 661–4.

Zaki, A.M., van Boheemen, S., Bestebroer, T.M., et al. (2012). Isolation of a novel coronavirus from a man with pneumonia in Saudi Arabia. New England Journal of Medicine, 8, 1814–20.

Zhou, P., Yang, X.L., Wang, X.G., et al. (2020). A pneumonia outbreak associated with a new coronavirus of probable bat origin. Nature, 579, 270–73.

# SECTION 1

# Describing Outbreaks

This section (Chapters 2–5) delves into the basic tools and concepts for understanding **infectious disease agents** and the **diseases** that they cause—what we call the 'anatomy of disease'. We illustrate how to describe and interpret an outbreak's patterns to provide clues to determine a pathogen's identity and transmission mechanisms, and how to translate diagnostic tests and design effective disease surveillance programmes. Finally, we introduce ways to use (and abuse) mathematical models to understand observed outbreak patterns and make predictions about disease dynamics in places or times where no data exist yet. Furthermore, giant worms emerge from anuses; Mr Charles Darwin makes a brief appearance; we alert you to the potential perils of eating chicken livers and the potential health benefits of eating fried chicken; and we try to persuade you to mask up the next time you perform a necropsy in your garage.

A researcher releasing a flying fox (*Pteropus medius*) after sampling for Nipah virus and other zoonoses in Bangladesh.
Photo by Rajib Ausraful Islam from icddr, b.

# CHAPTER 2

# The anatomy of disease

## 2.1 Modes of pathogen transmission—direct contact

Pathogens and parasites must be transmitted from one host to another to be successful; otherwise, individual parasites/pathogens cannot pass on their genes and parasite/pathogen species cannot persist in evolutionary time. In this chapter, we will describe the different modes of transmission, the different types of parasites and pathogens, and some factors that influence whether hosts become infected and diseased after being exposed. We refer to these distinguishing characteristics of parasites/pathogens and the diseases that they cause as the 'anatomy of disease'. These details help determine the most efficient transmission control methods for any given pathogen, which we will discuss further in Chapter 14.

*Missouri, USA, 2008*—Sometime shortly before Halloween, a family noticed a bat loitering in the rafters of their porch (Pue et al. 2009). After a few days, the bat flew inside the house. Unworried, an experienced outdoorsman in his fifties, who had kept a variety of wild animals as pets, managed to catch the animal and allowed it to crawl up his arm and neck. When the bat petulantly bit the man on the left ear, the victim remarked on the potential of rabies transmission from bats. Yet they allowed the bat to remain in the house for two days, unrestrained and apparently healthy, before releasing it. The man did not seek medical evaluation or report the incident to public health authorities, which proved to be a fatal mistake.

The unfortunate man visited two emergency departments in late November displaying symptoms consistent with rabies—complaints of dehydration, generalized shaking, tingling of the left face, and hydrophobia—and was hospitalized. (**Hydrophobia**—an abnormal or unnatural dread of water—was the historical name for rabies; it occurs at later stages of rabies infection because of painful spasms when trying to swallow, causing panic when presented with water to drink.) Infection with a rabies virus variant associated with silver-haired bats (*Lasionycteris noctivagans*) was confirmed just four days before the patient died from the disease (Pue et al. 2009).

Rabies is an archetypal example of **direct contact transmission** (also called **direct transmission**), where a pathogen spreads from an infectious to a susceptible host through close contact. In the case of rabies, the 'close contact' is usually an infectious host biting a susceptible host; the rabies virus is transmitted through the infectious host's saliva.

Other examples of direct transmission include the *Campylobacter jejuni* cases transmitted from lambs to shepherds in Wyoming during lamb castration (Chapter 1.1) and cases of *Salmonella* Typhimurium (54 people across 23 US states) associated with close contact between people and their pet hedgehogs. There is also **sexual transmission**, a specific kind of direct transmission that leads to the spread of sexually transmitted diseases (STDs) or infections (STIs) like acquired immunodeficiency syndrome (AIDS), caused by human immunodeficiency virus (HIV); gonorrhoea, caused by the bacterium *Neisseria gonorrhoeae*; and syphilis, caused by the spirochete bacterium *Treponema pallidum*.

## 2.2 Airborne transmission

During **airborne transmission**, pathogens are carried by respiratory **droplets** and **aerosols** that

*Emerging Zoonotic and Wildlife Pathogens*. Dan Salkeld, Skylar Hopkins, and David Hayman, Oxford University Press.
© Oxford University Press (2023). DOI: 10.1093/oso/9780198825920.003.0002

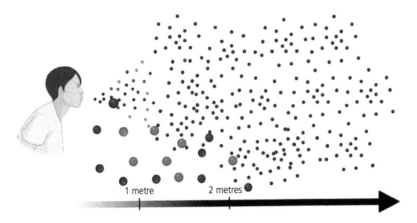

**Figure 2.1** Pathogens can be transmitted through the air by large respiratory droplets that fall relatively quickly to the ground (like rain) or by dust or droplet nuclei that drift through the air (like fog).
Created by Hanna D. Kiryluk using BioRender.com.

are generated during breathing, talking, coughing, sneezing, or disturbance of substrates. Once inhaled, pathogens can invade a new host from the respiratory tract—larger particles tend to be deposited in the upper airway, but smaller aerosols can also penetrate the alveolar region of the lungs (Wang et al. 2021) (Fig. 2.1).

Particle size is important because smaller aerosols can travel further and stay suspended in the air longer than droplets, and these details determine how effectively masks and indoor ventilation can prevent transmission (Tang et al. 2021). When emitted from a height of 1.5 metres, 100 μm is the largest particle size that can remain suspended in still air for more than 5 seconds and travel horizontally 1–2 metres. (For comparison, exhaled particles are mostly smaller than 5 μm, and a large fraction are smaller than 1 μm.) Because larger respiratory droplets (>100 μm) or secretions (i.e. mucus) fall to the ground faster, they are analogous to rain, whereas small aerosols (<100 μm) float in the air in a process analogous to fog. Of course, the distance travelled by the airborne pathogens is influenced by whether the infectious person is talking, coughing, or sneezing; whether the hosts are outside or inside or in poorly ventilated crowded places; and the heat, humidity, and airspeed of the local environment.

Let's look at a few examples of pathogens carried by aerosols versus droplets. Measles virus, which likely evolved from cattle rinderpest virus in the 6th century (Düx et al. 2020), tends to be transmitted by aerosols; if someone with measles coughs, the measles virus can drift in a room and infect someone who walks in later (Bloch et al. 1985). Hantavirus can also be aerosolized; for example, cleaning a storage shed or cabin frequented by deer mice and contaminated by mouse urine and faeces causes one to sweep dust containing hantavirus into the air and then to become infected by inhaling the virus (Sinclair et al. 2007). Some aerosolized pathogens can travel long distances when attached to dust particles, such as Q fever, *Coxiella burnetii*, a bacterium that can float 5 kilometres from infectious sheep or goats (or up to 18 km on gale force winds) before being inhaled by susceptible human hosts (Clark and Soares Magalhães 2018). In contrast, respiratory **droplets** (>100 μm) can only travel 1–2 metres (3–6 feet) (Prather et al. 2020), and this is how pathogens like pneumonic plague and smallpox are transmitted (Institute of Medicine 2002).

## 2.3 Environmental transmission

During **indirect transmission**, the infectious and susceptible hosts usually occupy the same space, but at different times; after the pathogen leaves the infectious host, it must persist for some time before a susceptible host comes along. Indirectly transmitted pathogens can persist outside the host on inanimate objects or surfaces, called **fomites**.

There are several ways that fomites can become contaminated with pathogens. Pathogens can be deposited during contact with body secretions or fluids, such as when house finches (*Carpodacus mexicanus*) touch their swollen, goopy eyes to bird feeders and thus contaminate the feeders with *Mycoplasma gallisepticum*, a bacterium that causes conjunctivitis (Dhondt et al. 2007). Fomites can also be contaminated by soiled hands or contact with previously airborne pathogens generated via talking, sneezing, coughing, or vomiting, etc.

When pathogens can remain viable on fomites, only small numbers of infectious organisms may be needed to infect a host upon contact (Boone and Gerba 2007). For instance, small amounts of nasal mucus containing rhinovirus—the agent responsible for the common cold—can spread from fingertips to doorknobs, faucet handles, or other environmental surfaces and remain infectious for many hours. For rhinovirus, quantities as small as one plaque-forming unit can contaminate a susceptible person's fingers and be transferred to their nasal and conjunctival mucosa via autoinoculation (Pancic et al. 1980).

High-touch surfaces are especially likely to be fomites that serve as transmission hot spots. For example, in aircraft cabins, the backs of seats along the aisles and the toilets are high-touch surfaces. It only takes 2–3 hours for most of these high-touch surfaces within a cabin to become contaminated! Therefore, even on short haul flights, aisle passengers have a higher risk of exposure to contaminated fomites (Lei et al. 2017).

Fomites are just one type of **environmental transmission**, a form of indirect transmission where susceptible hosts are exposed to pathogens in their environments rather than direct exposure to infectious hosts or vectors. Other examples include hookworms (Hotez et al. 2005) and schistosome trematodes that persist in soil or water before finding hosts. Parasites and pathogens that persist for some period in the environment can spread among hosts even when those hosts rarely contact, such as solitary host species or host populations with low density. Similarly, environmental transmission may often provide opportunities for interspecific transmission (transmission between two host species) that would rarely arise via close contact; in fact,

overlapping space use between host species can be a good indicator that the host species likely share parasites and pathogens (Woodroffe et al. 2016). We will cover environmental transmission and environmental reservoirs in detail in Chapter 6.

However, it is worth remembering that definitions and interpretations of transmission modes should be flexible. Is airborne transmission direct or indirect? It depends! Does transmission occur when the infectious and susceptible host are in close proximity at the same time? Airborne transmission can be considered a form of direct transmission if the pathogens are suspended for a short time and distance. But if pathogens spend longer periods in the air and can travel further when carried by wind, airborne transmission could be considered as indirect.

## 2.4 Vehicle-borne transmission

Oh, my goodness, my guts are coming out of me!
—Anonymous

*California, 2018*—A man was sitting on the toilet when he was struck by the sensation that his entrails were leaking from his body. He began to fret that death was upon him, but then realized it was a tapeworm leaving his anus, sliding out inch-by-inch. For 66 inches. The worm was 5½ feet long (1.7 m). The culprit was a Japanese broad tapeworm, *Diphyllobothrium nihonkaiense*. As the name suggests, this parasite is relatively uncommon in the United States, so both physician and patient were perplexed as to how the man had become infected; the patient had not travelled abroad or consumed questionable water. But then the man mentioned his passion for, and almost daily indulgence of, salmon sashimi (Romo 2018).

This is an example of **vehicle-borne transmission**, where the pathogen enters the host by way of a vehicle, e.g. ingestion of food—**food-borne transmission**—or water—**water-borne transmission**. The vehicle for the tapeworm in this example was presumably salmon sushi. *Diphyllobothrium nihonkaiense* is increasing in incidence in human populations due to the rising popularity of eating raw or undercooked fish in dishes such as sushi, sashimi, and ceviche; if you are worried about that, cooking or sufficiently freezing the fish will kill the tapeworm (Kuchta et al. 2017). Other food-borne illnesses, including salmonella, norovirus, and cholera, are transmitted by the **oral-faecal route** where contamination of food or water by faeces allows a parasite or pathogen to infect new hosts (Table 2.1).

Like other tapeworms, *Diphyllobothrium nihonkaiense* has a **complex life cycle** (also called a **multi-host life cycle**): the parasite must use multiple host species, in a specific sequence, to successfully develop and reproduce, and thus the parasite spends specific life stages in different host species or the environment (Fig. 2.2). As with many other parasites with complex life cycles, the life cycle of *Diphyllobothrium nihonkaiense* relies on a predator or other consumer to ingest it along with its prey. In ecological circles, this is called **trophic transmission**.

Vehicle-borne transmission also includes transmission via **needles**, e.g. during blood transfusions or intravenous injections. Thus, blood banks screen blood samples for pathogens like HIV, West Nile virus, hepatitis viruses, and syphilis to prevent accidental transmission. When infection via needles occurs during medical treatment, it is termed **iatrogenic** (no such elegant term is available for when intravenous drug users are infected, even though both are unintentional).

In most cases, vehicle-borne transmission describes pathogen transmission that is associated with a behaviour actively performed by the host, e.g. drinking, eating, or using needles. This can allow subsequent clues as to the infection source, as well as offering control options, e.g. stop consuming particular foods/liquids or cook them appropriately, or use sterile needles, or screen blood banks.

*North Carolina, United States, 2011*—In August 2011, after returning from a fishing trip, a previously healthy man sought medical help complaining of nausea, vomiting, and upper extremity paraesthesias ('pins and needles'), and then suffering fever and a seizure (Vora et al. 2013). While hospitalized, he had difficulty swallowing liquids and altered mental status, his body crashed, and he was declared brain dead 17 days after symptom onset. The presumed cause of death was ciguatera poisoning: a food-borne illness caused by eating fish whose flesh is contaminated with ciguatoxins produced by algal plankton. Further investigations attributed the cause of death to complications of severe gastroenteritis. No increased risk for infectious disease transmission was suspected or identified, and kidneys, heart, and liver were transplanted into four recipients in September 2011.

**Table 2.1** Examples of uncommon infections associated with ingestion of unusual uncooked foods, eaten either purposely or inadvertently (Rosenthal 2019).

| Disease agent | Case patient and presentation | Likely exposure |
|---|---|---|
| *Angiostrongylus cantonensis*, the rat lungworm | Hawaiian cases, including headache, myalgia, fever, vomiting, death in 2/82 cases | Ingestion of larvae associated with slugs in unwashed produce |
| *Angiostrongylus cantonensis*, the rat lungworm | Korean woman and her adult son, eosinophilic meningitis | Ingestion of live centipedes |
| *Anisakis* spp., nematode worm | American man, with recurrent abdominal pain | Ingestion of home-cured salmon gravlax |
| *Gongylonema pulchrum*, nematode | American man, 7-month history of recurrent 'zig-zagging blisters' in his mouth | Likely acquired from inadvertent ingestion of cockroaches, which had infested the man's grain stores |
| *Paragonimus westermani*, lung fluke | Nepali woman, with dyspnoea (difficulty breathing), chest pain, fever, sweats | Ingestion of live slugs as traditional medicine to accelerate bone healing |
| *Spirometra* canine or feline tapeworms | Chinese woman, requiring surgery for brain mass | Frequent ingestion of potentially undercooked stir-fried frogs |
| *Spirometra* canine or feline tapeworms | Thai resident of Switzerland with chronic history of subcutaneous nodules | Ingestion of untreated water containing infected intermediate host copepods |
| *Toxocara canis*, canine roundworm | Korean woman, with malaise and preprandial epigastric discomfort | Patient reported a history of yearly consumption of a cup of raw roe deer (*Capreolus pygargus*) blood for health benefits |

Practical and proactive advice included in Rosenthal (2019): 'Ingestion of raw centipedes is best avoided'; 'Travelers to Hawaii. . . should avoid ingesting uncooked slugs or snails'; 'Ingestion of raw animal blood is best avoided'.

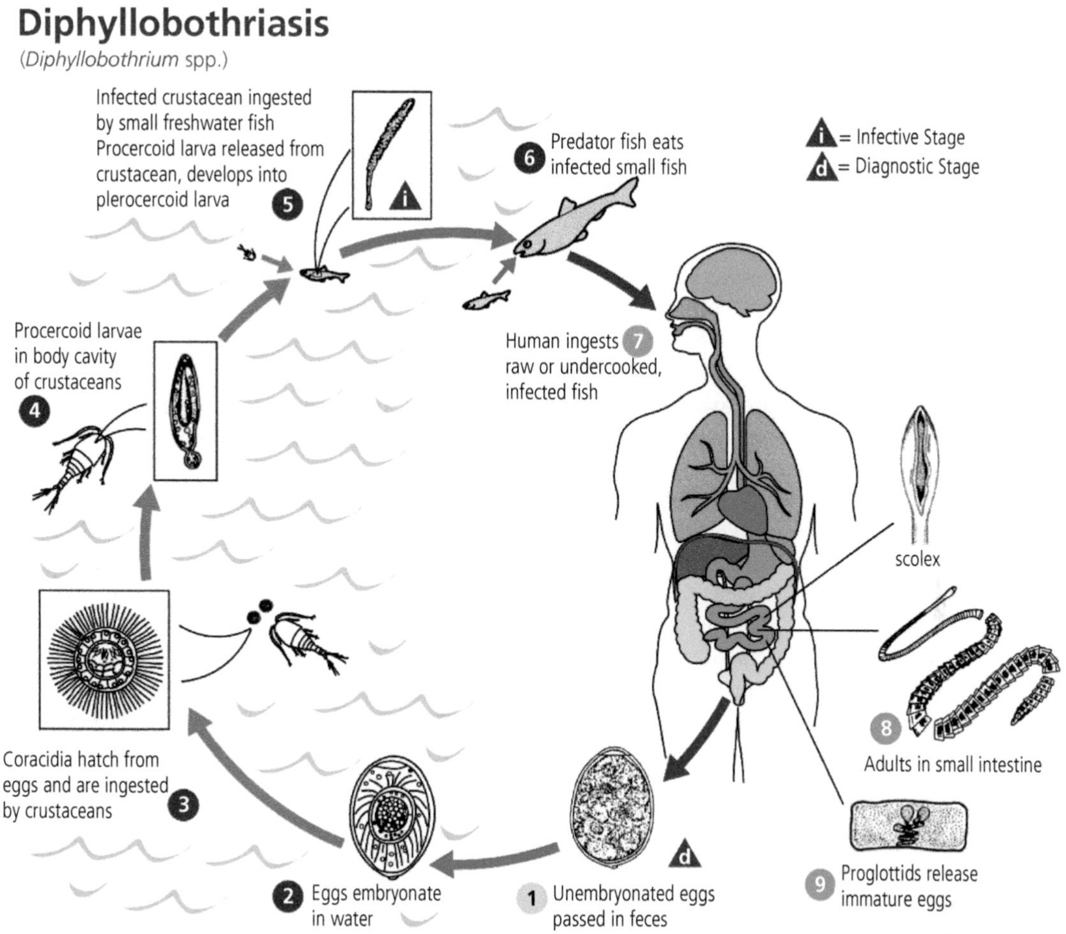

# Diphyllobothriasis

(*Diphyllobothrium* spp.)

**Figure 2.2** The life cycle of *Diphyllobothrium* spp. parasitic cestodes (also called tapeworms), the causal agents of the disease diphyllobothriasis. *Credit:* CDC/Alexander J. da Silva/Melanie Moser. This image is in the public domain and free of any copyright restrictions: https://phil.cdc.gov/Details.aspx?pid=3388

In February 2013, the recipient of the left kidney presented to a hospital emergency department complaining of right hip pain. He was diagnosed with sciatica and discharged but was admitted four days later with fever, abnormal and excessive sweating, nausea, and weakness and pain near the site of his transplanted kidney. He developed encephalopathy, excessive salivation, and blood flow instability, and died 22 days after admission. Post-mortem examinations discovered abnormalities throughout the brain and spinal cord. Serum collected five days before death and saliva and central nervous system samples revealed evidence of rabies virus infection.

The rabies virus was associated with raccoons, and the organ donor had had significant exposure to raccoons in North Carolina, US, '... whilst trapping and keeping them in captivity, using them as live bait during dog training exercises, and preparing pelts for display. During these activities, the donor sustained at least two raccoon bites, 18 and 7 months prior to symptom onset, for which he did not seek medical care. The captive raccoon responsible for the latter bite was healthy up to 4 weeks after the bite. Neither raccoon was available for testing' (Vora et al. 2013). Re-examination of the organ donor's serum confirmed the presence of rabies virus antibodies. Fortuitously, the other three organ recipients did not develop rabies, and received postexposure prophylaxis (PEP) as a precaution.

In a separate but similar case, in May 2005, four organ transplant recipients became severely ill with lymphocytic choriomeningitis virus (LCMV), an Old-World arenavirus normally found in rodents (CDC 2005). No infection had been suspected in the donor, who had died following a stroke, but three of the four persons who received the liver, lungs, and two kidneys declined in health and died within a month of the transplant operation. Two cornea recipients were asymptomatic.

An investigation identified the most likely source as an infected hamster living in the organ donor's home; a family member who cared for the hamster had specific antibodies to LCMV, and the pet hamster was determined positive for LCMV. Normally, lymphocytic choriomeningitis virus is asymptomatic or causes mild self-limited illness in humans, but immunosuppressed people (e.g. organ recipients) may be more likely to develop serious disease symptoms. Acute LCMV infection in an organ donor is nonetheless a rare event.

As these examples illustrate, organ transplants can provide opportunities for vehicle-borne transmission, which is why extensive health screening is required for organ donors (see also Box 2.1, Table 2.2).

## 2.5 Vector-borne transmission

At night I experienced an attack (for it deserves no less a name) of the Benchuca (a species of Reduvius), the great black bug of the Pampas. It is most disgusting to feel soft wingless insects, about an inch long, crawling over one's body. Before sucking they are quite thin, but afterwards they become round and bloated with blood, and in this state are easily crushed. They are also found in the northern parts of Chile and in Peru. One which I caught at Iquique, was very empty. When placed on the table, and though surrounded by people, if a finger was presented, the bold insect would immediately draw its sucker, make a charge, and if allowed, draw blood. No pain was caused by the wound. It was curious to watch its body during the act of sucking, as it changed in less than ten minutes, from being as flat as a wafer to a globular form. This one feast, for which the Benchuca was indebted to one of the officers, kept it fat during four whole months; but, after the first fortnight, the insect was quite ready to have another suck.

—Charles Darwin, The Voyage of the Beagle, journal entry dated 26 March 1835. It has been suggested that Darwin may have been infected with the Chagas disease

parasite during his travels through Chile (Botto-Mahan and Medel, 2021).

In this book, we define **vector-borne transmission** as pathogen transmission facilitated by biting arthropods (arachnids or insects) that feed non-lethally upon a host (usually the host's blood). Vectors transmit an appalling variety of parasites and pathogens to humans (not to mention the many others they transmit to other organisms). To mention just a few: mosquitoes transmit malaria parasites (*Plasmodium* spp.), dengue virus, Zika virus, West Nile virus, and Chikungunya virus; ticks transmit the bacteria that cause Lyme disease and Rocky Mountain spotted fever, Powassan virus, and Crimean-Congo haemorrhagic fever virus; sandflies (e.g. *Lutzomyia* spp.) transmit trypanosomes (*Leishmania* spp.) that cause leishmaniasis; fleas transmit *Yersinia pestis*, the bacteria that causes plague; tsetse flies (*Glossina* spp.) transmit the trypanosomes that cause sleeping sickness (*Trypanosoma* spp.); reduviid bugs (e.g. *Triatoma infestans* or *Rhodnius prolixus*) transmit the parasite that causes Chagas disease (Figs. 2.3 and 2.4); and biting midges (*Culicoides* spp.) transmit bluetongue virus.

Vector behaviour and biology is obviously an important component of understanding pathogen transmission in these systems. For example, some vectors become infected by the parasites or pathogens that they carry. When the pathogen's life cycle requires infection in the vector host, it is called **biological transmission**. In these cases, the infected vector must stay alive long enough for the parasite or pathogen to develop or multiply before biting a susceptible host—an important constraint on transmission! For other parasites and pathogens, transmission from one host to another can occur simply because the pathogen can persist on the vector's mouthparts—**mechanical transmission**.

Outside of this book, you might see other interpretations of the word 'vector', whose Latin roots mean 'carrier'. For example, some professionals use the term 'vector' to refer to hosts that can play a role in transmission simply by moving the pathogen around, e.g. dogs that transmit rabies virus, badgers that transmit bovine tuberculosis, or deer mice that transmit hantavirus (Wilson et al. 2017). We

**Figure 2.3** Three species of kissing bugs commonly found in Texas, USA: (from left) *Triatoma protracta*, common in the western US; *Triatoma gerstaeckeri*, the most commonly found kissing bug species in Texas; *Triatoma sanguisuga*, the most frequently found species in the eastern US. Scale bar represents 25 mm or approximately 1 inch.
Photo by Gabriel L. Hamer and published in Curtis-Robles et al. (2015); used under the terms of the Creative Commons Attribution License.

would argue that these animals should be regarded as hosts, and that the transmission is technically via direct contact in the case of rabies, environmental transmission via fomites (contaminated fields) in the case of bovine tuberculosis from badgers to cattle, and airborne in the case of aerosolized hantavirus. We are not saying that using the word 'vector' to refer to rabid dogs is wrong (at least we are not saying this *out loud*); but this is a reminder that definitions can vary and so you should be clear about your own interpretations and word use (Box 1.1).

## 2.6 Vertical or congenital transmission

In 2001, in Geneva, Switzerland, a woman delivered a full-term, apparently healthy baby. However, blood microscopy, serology, and PCR screening revealed congenital infection with *Trypanosoma cruzi*, the protozoan that causes Chagas disease. This was quite odd because *Trypanosoma cruzi* is not endemic to Switzerland, but rather the Americas. The parasite is predominantly a vector-borne disease, transmitted by triatomine insects (also

known as kissing bugs, conenose bugs, assassin bugs, vampire bugs, barbeiros, vinchucas, pitos, and chinches). Tropical assassin bugs do not typically lurk in Swiss hospitals, so it is unlikely that an infant could have become infected through the normal transmission route. Instead, the pathogen was transmitted vertically from mother to baby after the mother had become infected previously in Santa Cruz, Bolivia (Jackson et al. 2009).

**Vertical or congenital transmission** typically involves pathogen transmission from mother to offspring, i.e. specifically between hosts of different generations. In contrast, most other transmission mechanisms (e.g. direct contact, airborne, vector-borne) are **horizontal transmission modes**— dispersal methods that allow pathogens to move among hosts regardless of generation.

Vertical transmission happens in a variety of host species, and the exact details vary with host and pathogen biology. In eutherian mammals, vertical transmission can occur across the placenta (e.g. *Trypansosoma cruzi*), through direct contact during the birth process (e.g. *Streptococcus* B bacteria), during breast-feeding (e.g. HIV-1), or potentially from all

**Figure 2.4** Street public health announcement and art from walls in Oaxaca, Mexico that depict kissing bugs (*Triatoma* spp.), *Trypanosoma cruzi*, and health impacts of Chagas disease.

Photos by Kristin Sherwood.

three mechanisms (e.g. HIV). For many arthropods, vertical transmission occurs from mother to eggs—**trans-ovarial transmission**. For example, a protozoan parasite (*Ophryocystis elektroscirrha*) causes mortality and population declines in monarch butterflies (*Danaus plexippus*), and this parasite is primarily vertically transmitted (Majewska et al. 2019); infected female monarchs are covered with dormant parasite spores after they pupate, and these spores rub or fall off on the eggs and surrounding leaves as the females oviposit.

Vertical transmission (or the lack thereof) in vector species can have important implications for human health. For example, adult female black-legged ticks (*Ixodes scapularis*) can transmit the bacterium *Borrelia miyamotoi* to their eggs, and therefore newly hatched larvae possess infections even though they have never taken a bloodmeal from a host (Scoles et al. 2001, Han et al. 2019). In contrast, the closely related Lyme disease spirochete (*Borrelia burgdorferi*) cannot be transmitted vertically in the tick host. This disparity is important: if an animal or human is bitten by a larval black-legged tick (see tick life cycle in Chapter 4), then it cannot be exposed to *Borrelia burgdorferi*, but it can potentially be exposed to *Borrelia miyamotoi*.

---

### Box 2.1 Multiple modes of transmission for the same disease agent

Parasites and pathogens have not always read the textbook on disease transmission (not even if the book is as interesting as this one), and thus they do not necessarily feel restricted to just one transmission mechanism. Many infectious agents use multiple modes of transmission.

For example, Chagas disease is caused by a flagellate protozoan, *Trypanosoma cruzi*, which was discovered by Carlos Chagas in 1908 in Minas Gerais, Brazil.

Worldwide incidence of Chagas disease is currently estimated at 200,000 cases per year. As described in the main text, the trypanosome is typically transmitted by triatomine bugs via **vector-borne transmission** (Pereira et al. 2009). *Trypanosoma cruzi* grows in the bug's digestive tract, and the infectious form is eliminated in their faeces and enters animal or human hosts via cuts, scratches, or mucous membranes. In mammalian hosts, the protozoan normally

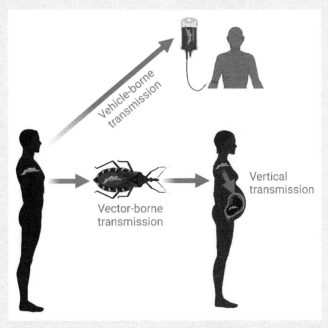

**Figure 2.5** Pathogens can be transmitted via multiple mechanisms. This diagram illustrates vertical transmission and two forms of horizontal transmission (vector-borne transmission and vehicle-borne transmission) of the parasite that causes Chagas disease. Created with BioRender.com.

---

**Box 2.1** *Continued*

circulates in the blood where it is infectious to subsequent assassin bug bloodmeals. Chronic illness can occur when *Trypanosoma cruzi* lodges in muscle and/or heart tissue.

*Trypanosoma cruzi* also has several other transmission routes. These include **vertical transmission** from mother to child (see main text) and **vehicle-borne transmission** during blood transfusions, organ transplants, or consumption of contaminated food (Fig. 2.5). For example, vehicle-borne transmission in a school in metropolitan Caracas, Venezuela, caused 128 cases, including one death, and was likely triggered by common exposure to *Trypanosoma cruzi* in fresh fruit juice prepared at a household with many triatomine bugs in its surroundings. Oral vehicle-borne transmission might also have been historically important in northern Chile and Peru during consumption of infected, not-perfectly-cooked guinea pigs (*Cavia porcellus*). Other clusters of Chagas disease have been blamed on consumption of açai juice, sugarcane juice, and food contaminated by white-eared opossum or mucura (*Didelphis albiventris*) faeces (Pereira et al. 2009).

---

**Table 2.2** Examples of other zoonotic pathogens with multiple modes of transmission.

| Disease agent (disease) | Transmission modes | | | | |
| --- | --- | --- | --- | --- | --- |
| | Direct contact | Vehicle-borne | Airborne | Vector-borne | Vertical |
| *Babesia* spp. (babesiosis) | | ✔ Blood transfusion | | ✔ Ticks | ✔ |
| *Bacillus anthracis* (anthrax) | ✔ Cutaneous | ✔ Needle-sharing | ✔ Inhalational | | |
| *Francisella tularensis* (tularaemia or rabbit fever) | ✔ e.g. skinning infected animals | ✔ Contaminated water | ✔ Respiratory droplets; mowed rabbits | ✔ Ticks or deerflies | |
| Human immunodeficiency virus (HIV) (AIDS) | ✔ Sexual intercourse | ✔ Needle-sharing | | | ✔ |
| Tick-borne encephalitis virus | | ✔ Consumption of milk | | ✔ Ticks | ✔ |
| *Yersinia pestis* (plague) | ✔ Septicaemic plague | ✔ Consumption of raw camel or marmot organs | ✔ Pneumonic plague | ✔ Fleas: bubonic plague | |
| Zika virus | ✔ Sexual intercourse | | | ✔ Mosquitoes | ✔ |

*Sources:* Bin Saeed et al. (2005), Foley and Nieto (2010), Joseph et al. (2012), Kehrmann et al. (2020), Kerlik et al. (2022)

---

## 2.7 Portals of host entry

*Florida, USA, 2020*—A 13-year-old boy went swimming in a lake while camping in northern Florida in late July. Days later, the boy complained of a headache, which worsened throughout the week. During a visit to the emergency room, the boy was incorrectly diagnosed with strep throat (sore throat caused by *Streptococcus* bacteria) and discharged.

He developed increasing symptoms of meningitis (inflammation of the membranes covering the brain and spinal cord), and upon subsequent admittance to a hospital for monitoring, his brain fluid was found to contain an amoeba that should not be there, *Naegleria fowleri*. This amoeba feeds on bacteria in nature, where it is common in warm freshwater (e.g. rivers, lakes, hot springs, pools). Humans often swim in waters containing the amoeba, but rarely, when waters are especially warm, the amoeba invades human hosts by entering the nose—a relatively unusual portal of entry for a parasite.

Once inside a host, *Naegleria fowleri* migrates along the olfactory nerve to the brain; perhaps the worst place to have a parasite or pathogen, because pathogens may cause serious inflammation, benefit from the 'immune privilege' of the brain as they are protected from the peripheral immune system, and due to the related difficulty in treatment. Inside the brain, the amoeba's pathogenicity and the intense immune response mounted by the host causes primary amoebic meningoencephalitis. The resulting damage to nerves and central nervous system tissue nearly always ends in death, though there are recent developments that may be increasing chances of successful treatment (Yoder et al. 2010, Grace et al. 2015). Sadly, only one *Naegleria fowleri* case patient survived from 111 cases in the US from 1962–2008, and the 13-year-old Florida boy died less than two weeks after his swimming vacation.

All parasites and pathogens must enter their host, and there are several common entry portals. Several parasites penetrate skin directly, such as hookworm larvae penetrating the skin of barefoot humans. Vector-borne pathogens get help invading the blood or lymph system when their arthropod vector either inserts its proboscis through the skin (e.g. mosquitoes, ticks, fleas) or cuts the skin (e.g. horse flies). And some vehicle-borne pathogens can also negotiate the skin barrier via needles or thorns, like anthrax or tetanus caused by soil-borne *Bacillus anthracis* and *Clostridium tetani*. Other pathogens invade via the mouth and intestinal route (vehicle-borne via food or liquids), or through inhalation to the lungs and airways (airborne transmission). (Mucosal membranes exist to protect hosts at these entry portals, but ultimately, most pathogens still invade via mucosal membranes; this is why you should often wash your hands and avoid touching your eyes/nose/mouth with contaminated hands!) And, of course, there is the ability for pathogens to cross from the placenta to the foetus through vertical transmission.

A less common but certainly unforgettable entry portal is the anus or cloaca. For example, several species of 'ectoparasites' find homes inside the anuses of thick-skinned mammals, because that is the only location where the parasite can successfully access the host's blood. This includes *Placobdelloides*

*jaegerskioeldi*, a leech that inhabits hippopotamus anuses, and *Cosmiomma hippopotamensis*, a beautiful tick that is endangered with extinction due to the global shortage of rhinoceros anuses (Mihalca et al. 2011, Fig. 2.6). Perhaps more surprisingly (could we be more surprised at this point?), several helminth parasites enter hosts via their exits and then continue further inside, such as many species of entomopathogenic nematodes (roundworms that infect insects) and the larval cercariae of the trematode *Echinostoma trivolvis*, which enter tadpoles through the cloaca on their way to the tadpoles' kidneys. And then there are some disturbingly large marine parasites, like *Sarcotaces arcticus*, a copepod (relatively large crustacean) that makes a home in rockfish anuses by inducing the host's rectum to grow a gall or protective case around the parasite. Photographs of these parasites are all easily found via internet search, but we recommend choosing your search terms carefully.

Having found one [a sea cucumber] by following its smell, a pearlfish will dive into the anus headfirst, 'propelling itself by violent strokes of the tail', according to Eric Parmentier. If the sea cucumber objects and closes down its anus… well, it still has to breathe. Oh yeah, sea cucumbers breathe through their anuses.

—Ed Yong, *National Geographic*, 2016

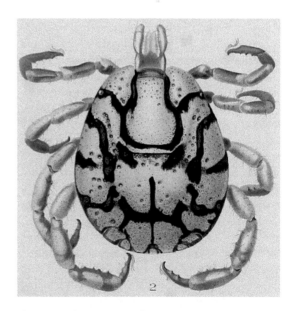

**Figure 2.6** Rhinoceros anus tick, *Cosmiomma hippopotamensis*. By Friedrich Karl Wilhelm Dönitz, https://commons.wikimedia.org/w/index.php?curid=89375233, public domain.

As the examples above illustrate, knowing how infectious agents enter their hosts can help determine how to prevent infections: keep your nose above water or avoid swimming in amoeba-contaminated water during high-risk months, wear masks to prevent airborne pathogens, cook food to kill pathogens, screen blood before use in transfusions, etc. Therefore, determining portals of entry and transmission modes is often an urgent priority for emerging infectious diseases. This can be difficult to accomplish, and sometimes initial best guesses are overturned by later data (e.g. pathogens initially thought to be transmitted by mucous membrane invasion or vehicle-borne transmission after contact with contaminated surfaces turn out to be predominately transmitted by inhalation of airborne particles).

The site of pathogen infection is also important because it can determine the severity of the subsequent disease. For example, anthrax, caused by the bacterium *Bacillus anthracis*, has many possible entry portals to human hosts. The bacterial spores can enter the skin through a cut or scrape (causing cutaneous anthrax); through vehicle-borne transmission by eating raw or undercooked meat from an animal infected with anthrax (causing gastrointestinal anthrax); by inhalation (causing inhalation anthrax), which may occur in workplaces where exposure to contaminated animal products occurs, such as wool mills, slaughterhouses, and tanneries; or by vehicle-borne transmission through needles (causing a faster-progressing cutaneous infection), which occurred in outbreaks of heroin-injecting drug users in northern Europe. Among these entry routes, cutaneous anthrax is typically the most common but least dangerous form of anthrax infection. Most patients with cutaneous anthrax fully recover with treatment, though without treatment, up to 20% of cutaneous anthrax cases can result in death. In contrast, only 60% of treated patients survive gastrointestinal anthrax, and untreated gastrointestinal anthrax kills more than half of cases. And for inhalation anthrax—which begins in the chest's lymph nodes and then disseminates throughout the body, culminating in severe breathing problems and shock—only 55% of treated patients survive, and in the absence of medical treatment, case fatality ratio is 85–90%.

## 2.8 Host exits

Pathogens must also exit the host to infect new naive organisms.

Often, exit from the host is similar to the site of entry. Vector-borne diseases normally require a new bite from the same vector species to exit the host (e.g. malaria and the *Anopheles* mosquito, or *Borrelia burgdorferi* and *Ixodes* ticks). Pathogens that enter the lungs often exit from the lungs (e.g. primary pneumonic plague caused by *Yersinia pestis*). Pathogens that can be spread by intravenous injection (e.g. HIV) can continue chains of transmission from further needle-sharing.

As with entry portals, there may be multiple exit portals. For example, food-borne noroviruses are capable of infecting hosts by entry to the oesophagus (vehicle-borne transmission) and can also exit from the oesophagus during vomiting. In addition, noroviruses can exit from the other end of the intestinal tract in faeces. Both exits can be forceful and spectacular. Subsequent transmission occurs because of airborne exposure to droplets or through the oral-faecal route (also called faecal-oral, or orofaecal) when food or water are contaminated through inadequate sanitation and/or hygiene practices.

We are not aware of hard-and-fast rules to predict how entry and exit mechanisms correspond to each other, and the entry and exit might be quite different. For example, schistosomes—a form of flatworm—enter human hosts through the skin but then, depending on the species of schistosome, migrate to the blood systems of the bowels and exit the host in faeces (e.g. *Schistosoma mansoni*) or migrate to the blood systems of the bladder and exit the host in urine (e.g. *Schistosoma haematobium*).

## 2.9 Infectious, latent, incubation, and symptomatic periods

Now that we understand the ways that hosts become **exposed** to parasites and pathogens (the transmission routes and portals of entry), we can discuss what happens after exposure. Critically, many, if not most, exposures to parasites and pathogens do not cause subsequent **infections**, where the infectious agent is able to survive and multiply in or on the host. Many infectious agents are killed by hosts' immune systems, and other forms of defence against infectious diseases also exist, such as grooming to remove ectoparasites. When host defences eliminate all invading infectious agents, exposure does not lead to infections.

When pathogens do successfully establish in a new host, that infected host is unlikely to be **infectious** immediately; often, the parasite or pathogen must multiply or reproduce until sufficient infectious agents are available for subsequent transmission. For example, for vector-borne parasites and pathogens, there must be enough infectious agents circulating in the blood that some are likely to be sucked up when a vector takes a blood-meal from the host. The time between initial infection and when the pathogen is infectious to new hosts is called the **latent period** (Fig. 2.7). Shorter latent periods allow for faster **generation times** for pathogens, which determines how rapidly an infectious disease can spread in a host population.

After hosts become infectious, they may effectively stay infectious for the rest of their life.

Alternatively, the host could eventually fight off the infection, ending the infectious period (e.g. most seasonal human pathogens, like influenza). There is also an intermediate case, where the parasite or pathogen persists for most or all of the host's life, but it cycles through dormant stages, where it 'hides' from the host immune system, and active stages of replication and infectiousness called recrudescence, e.g. herpes simplex virus. (Dormant stages are also sometimes referred to as 'latent infection', which can be confused with 'latent periods' (see previous paragraph); remember to clearly define your terms!) In all cases, the duration that the host is infectious is called the **infectious** or **contagious period**.

Sometime after infection, infected hosts may develop disease symptoms. The time between infection and the onset of symptoms or clinical illness in the host is called the **incubation period**. Of course, determining incubation periods can be difficult because the precise date of exposure may not

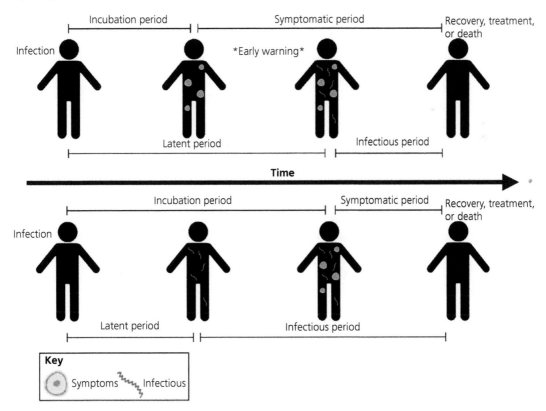

**Figure 2.7** Progression of disease and infectiousness in hosts. Symptoms may occur before or after the pathogen is infectious to new hosts, and the timing has implications for public health: asymptomatic spread can occur if the host is infectious before symptoms arise, whereas treatment and prevention to curtail transmission are easier when symptoms precede infectiousness.
Made by Hanna D. Kiryluk using BioRender.com.

be known. People often remember when they were bitten by a bat or dog, but it may be harder to know when a mosquito bite or contact with rodent faeces occurred. After the onset of symptoms, the duration that a host remains symptomatic is called the **symptomatic period**. When individuals or species do not have noticeable symptoms of disease when they are infected, they are **asymptomatic** (see next section).

For some pathogens, the symptomatic period precedes the infectious period, and for others, the infectious period precedes the symptomatic period (Fig. 2.7); this has important implications for disease control. For example, for smallpox, the symptomatic period comes first; infected people suffer from the rash prior to becoming infectious and shedding the virus. Therefore, people with rashes could be quarantined and treated before they caused subsequent infections—interventions that helped to successfully eradicate smallpox, the last known natural case of which was diagnosed on 26 October 1977, in Merka, Somalia. In contrast, for HIV/AIDS,

the infectious period begins long before the symptomatic period, such that infectious people often do not know that they are shedding the virus. This is one of the reasons why HIV spreads so rapidly and why routine testing in vulnerable populations remains so important for preventing transmission and identifying and treating HIV-positive people before the disease progresses to AIDS.

The periods defined above can be characteristic properties that consistently differ between parasite and pathogen species. For example, people infected by noroviruses are usually symptomatic within 12–48 hours after exposure, and this short incubation time can be used to differentiate norovirus food poisoning from food-borne pathogens with longer incubation periods, such as *Campylobacter* (Fig. 2.8). There is also natural variation in incubation periods and symptomatic periods within parasite and pathogen species; for example, some people with rabies develop symptoms within weeks and others do not develop symptoms for months (Fig. 2.8).

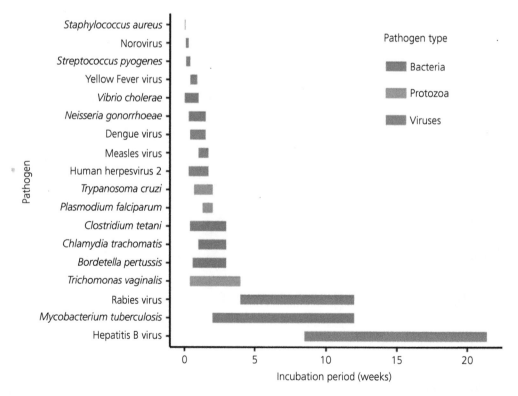

**Figure 2.8** Incubation periods for several pathogens that infect humans. The ranges shown here describe the incubation periods documented for most cases, but there may be outlying cases.
Data from Hopkins et al. (2022).

Variation in incubation period is influenced by factors that influence susceptibility (see below), including the 'dose' of the exposure, transmission route, state of the immune system, etc.

*Yosemite National Park, California, USA, summer 2012*—On 10 June, a 49-year-old woman from the Los Angeles area took a vacation in Yosemite National Park and stayed in a cabin in Yosemite Valley's Curry Village (Fig. 2.9). The cabin was a new format of accommodation—a 'signature cabin'—designed to be more comfortable than older cabins: there were propane heaters and double walls with insulation, so that the cabins were warmer and quieter (Barcott 2012). In late June, the visitor experienced flu-like symptoms: chills, myalgia, fever, headache, dizziness, fatigue. Unlike flu though, the symptoms persisted, causing the woman—'Visitor One'—to seek advice from her doctor.

Myalgia [mīˈalj(ē)ə]—pain in a muscle or group of muscles.

'When presented with Visitor One's symptoms, most physicians would have dismissed it as the flu or, at worse, low-level pneumonia. Her doctor didn't. They talked about what she might have picked up and where' (Barcott 2012). The woman mentioned her visit to Yosemite, prompting the doctor to investigate the possibility of hantavirus infection. Whilst awaiting test results, the doctor started treatment for the virus—better to be safe and proactive rather than sorry. It proved a clever

**Figure 2.9** A hantavirus outbreak occurred in Yosemite National Park (top left), California, USA among visitors to tent cabins (top right) (Núñez et al. 2014). Hantavirus is maintained in deer mice (*Peromyscus maniculatus*) (bottom left) and is a member of the Bunyavirales family (bottom right).
Created using canva.com by Hanna D. Kiryluk.

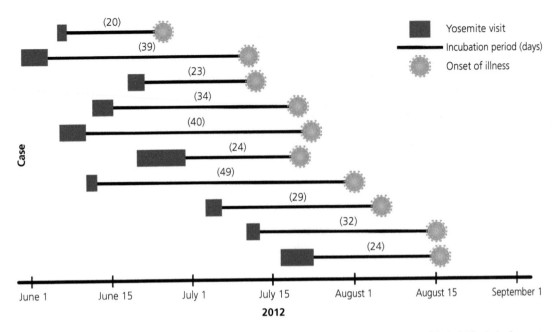

**Figure 2.10** Hantavirus incubation periods illustrated by the time between the dates spent in Yosemite National Park, California (and exposure within the signature cabins) and the date of the onset of illness.
Based on Núñez et al. (2014) and recreated by Hanna D. Kiryluk using canva.com.

response: the woman did indeed have hantavirus, and she recovered.

A second California visitor—a 36-year-old man from near the San Francisco Bay—also stayed in a signature cabin at roughly the same time (June) as Visitor One. He also suffered flu-like symptoms, severe enough to check himself into a hospital on 30 July. The following day he succumbed to hantavirus pulmonary syndrome.

A third person, a 45-year-old man from Pennsylvania, died on 12 August. Little is known about the illness progression in his case, though he did stay in a signature tent cabin in July.

Together, these examples formed a cluster of hantavirus infections. All in all, ten people were infected with hantavirus during a visit to Yosemite National Park in June–July 2012 (Nunez et al. 2014) (Fig. 2.10). The median incubation period was 30.5 days (range 20–49 days). Eight case-patients developed hantavirus pulmonary syndrome; five of the case-patients required intensive care with ventilatory support, and three people died from the disease.

Exposure to hantavirus was from deer mouse nests and tunnels observed in the insulation

material of luxury tent-cabins (Núñez et al. 2014). As Bruce Barcott (2012) phrases it: 'How to attract mice: 1) Offer cozy nests; 2) Provide food'.

## 2.10 Disease

When parasites and pathogens do cause substantial (pathological) harm to their hosts, i.e. disease, they are considered **pathogenic**, whereas parasites that cause limited noticeable harm are considered **non-pathogenic**. As a reminder, parasites and pathogens do not always cause disease in their hosts. For example, the man infected by the tapeworm did not know he was infected until his infection was dramatically ending whilst on the toilet (Section 2.4). Some infectious agents can switch from being non-pathogenic to pathogenic as circumstances change, e.g. the bacterium *Staphylococcus aureus* is normally harmlessly commensal but is also an opportunistic pathogen that can cause life-threatening pneumonia, toxic shock, and sepsis; or group A *Streptococcus* bacteria that often cause sore throats but in rare cases cause necrotizing fasciitis of 'flesh-eating' disease. (To an extent, then, and to be pedantic, many case studies of **disease ecology** are suffering under a misnomer, because the infectious agent doesn't always harm its host!)

Disease by pathogens incurs physiological changes reported as **disease signs and symptoms**. (Medical professionals distinguish between 'symptoms', which can only be reported by the host (e.g. headaches), and 'signs', which can be observed and measured by a medical professional (e.g. elevated blood pressure).) These signs and symptoms can constitute a **clinical diagnosis**; there is no diagnostic test performed for the disease agent.

For example, after being infected by Lyme disease spirochaetes (*Borrelia burgdorferi*) from an infected tick bite, 70–80% of infected humans (Burlina et al. 2020) develop a red expanding rash, called *erythema migrans* (which means 'migrating redness'). This rash, which sometimes resembles a bullseye, is caused by the progression of the disseminating bacteria in the body and the resulting immune reaction. The rash is quite distinctive, so observations of *erythema migrans* can help diagnose Lyme disease. But the absence of a rash cannot be used to conclude that a person is not infected, since the rash does not always occur.

Whereas *erythema migrans* is a distinctive symptom for a single pathogen, many pathogens cause the same 'flu-like symptoms': fever, coughing, runny or stuffy nose, myalgia (muscle ache), and lethargy or fatigue. Since these symptoms are so common, diagnosis using symptoms alone can be problematic as the disease agent could be seasonal influenza, or Lyme disease, or hantavirus, or COVID-19. . . For example, in Arizona in the USA, a man with flu-like symptoms was diagnosed with 'viral syndrome' and told to go home and return if symptoms deteriorated. Unfortunately, the man had plague, caused by a bacterium (*Yersinia pestis*), and was found dead on his couch three days later (Wong et al. 2009); appropriate antibiotics would have suppressed the bacterial infection and possibly saved his life.

Why do so many pathogens cause this same suite of symptoms? Actually, for the most part, the pathogens do not cause the flu-like symptoms—the host does! For example, fever is often a host's physiological response to infection because it activates immune responses and may allow the host's body to kill temperature-sensitive pathogens. But there can be a fine line between a helpful immune response

to infection and an immune response that has gone overboard. For example, bats infected with the fungal pathogen that causes white nose syndrome may survive winter hibernation, during which their immune systems are partially suppressed, only to experience extreme pathology and even death when their immune responses ramp up to fight the fungal infections just weeks after bats emerge from hibernation in the spring (Chapter 6; Meteyer et al. 2012). In this disease system and many others, some of the most severe physiological damage and even death following infections are caused by the host's own immune system.

However, parasites and pathogens are also directly responsible for damaging host tissues or impairing host physiology, as might be best illustrated by how severe disease can be in immunocompromised hosts. For example, most healthy people infected by the fungus *Histoplasma capsulatum* will remain asymptomatic, as macrophages in the lungs quickly engulf the fungal microconidia and transport them to lymph nodes, where they are destroyed. Roughly 20% of exposed persons might suffer temporary acute influenza-like symptoms, 3–21 days after exposure, and a few will develop acute pneumonia that may progress to chronic granulomatous pneumonia lasting months (Diaz 2018). The impact of the infection is very different if the host is immunocompromised. For example, histoplasmosis occurs in 2–25% of HIV-infected patients, and it is the first manifestation of acquired immunodeficiency syndrome (AIDS) in 50–75% of patients in *H. capsulatum* endemic countries. In Latin America, approximately 30% of patients with AIDS who contract histoplasmosis will die of it. Even with antiretroviral therapy, case fatality ratios of AIDS patients with histoplasmosis are about 10% (Diaz 2018). This is just one example of how disease severity can vary among individuals infected by the same pathogen.

Oligosymptomatic [ol′i-gō-simp′tō-mat′ik]—having few or minor symptoms.

Disease severity can also differ across species. For example, in pigs, Ebola virus's impacts are mainly on the respiratory tract; pigs inoculated with Ebola virus by oro-nasal exposure experienced fever and increased respiratory rate (two signs of disease), and pigs recover around nine days after infection. In contrast, primates develop systemic infection associated with immune dysregulation, which results in severe haemorrhagic fever and death (Weingartl et al. 2012). Thousands of western gorillas and chimpanzees have died from Ebola virus infections since the 1990s, and human epidemics have begun after people handle dead primate carcasses (Leroy et al. 2004).

These observations and questions (why do some hosts—whether individuals or species—get sick whilst others remain asymptomatic?) can be viewed using the disease or epidemiological triangle—a conceptual model that represents the interactions between the host, the environment, and the pathogen (Fig. 2.11). Disease only occurs when a susceptible host, an infectious parasite or pathogen, and environmental conditions favourable for disease all occur together in the same place and time.

For example, one framework identifies 11 attributes that influence infection and disease progression (Casadevall and Pirofski 2018): microbiome, immunity, sex, temperature, environment, age, chance, history, inoculum, nutrition, and genetics (cunningly arranged to generate the mnemonic acronym *'misteaching'*). Here we briefly introduce these attributes, which will come up again in later chapters. One should bear in mind that these factors can affect each other.

**Microbiome:** In recent decades, one major scientific advance has been discovering how organisms are critically dependent on their **microbiomes**: all the microorganisms living on or inside of a host, which can vary in space and time and interact with the host and other microorganisms. For example, disruption of the microbiome by antibiotics can result in progression of infectious diseases. In Minnesota, in 2002, a 52-year-old woman received various courses of antibiotics to treat a suspected case of chronic Lyme disease. Five weeks into a prescribed two-to-four-month treatment plan, the patient developed diarrhoea for three days and was diagnosed with *Clostridium difficile* colitis (an infection of the colon). Two days later, the woman was hospitalized while suffering severe abdominal pain, and the following morning she succumbed to cardiac arrest (Holzbauer et al. 2010). In this case and many others, a 'healthy' gut microbiome can prevent invasive microbes from establishing in the body, providing a mechanism for host resistance to infection, whereas a disrupted microbiome may increase host susceptibility to disease.

**Immunity:** A properly functioning immune system is our major defence against infectious diseases. This is most clearly illustrated by strongly immunocompromised hosts, such as pregnant people or people with autoimmune diseases (as illustrated in the above example of HIV/AIDs and histoplasmosis). But even among healthy hosts, there is natural variation in immune function, which can be influenced by factors like stress and diet.

**Sex:** The sex of a host can influence exposure to pathogens via anatomical, behavioural, hormonal, and immune differences. For example, sexual transmission is the major transmission mode for several pathogens, and in humans, several sexually transmitted pathogens are more easily able to invade vaginal tissue than penis tissue due to differences in skin thickness.

**Temperature:** Pathogens have temperature-dependent survival, growth, and reproduction, as do hosts; there are temperatures too cold and too warm for organisms to survive and flourish. For example, the fungus that causes white nose syndrome in bats grows faster on bats roosting in relatively warm sites (Hopkins et al. 2021).

**Environment:** Besides temperature, there are many environmental factors that influence interactions between hosts and parasites. These include things like precipitation, habitat (e.g. stagnant water versus stream), urbanization, and environmental toxins. For example, when amphibians are stressed by pollutants in their aquatic environments, they

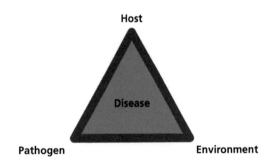

**Figure 2.11** Very basic disease or epidemiological triangle showing the conceptual links between host, pathogen, and environment.

may be more susceptible to parasitic infection and resulting disease (Kiesecker 2002).

**Age:** Host age can affect the probability of exposure to parasites and pathogens and the probabilities of subsequent infection and disease. In Sierra Leone, during the largest ever Ebola virus disease outbreak, the probability that infected patients died from the disease increased with age, where patients younger than 21 years old had a significantly lower case fatality ratio compared to patients >45 years old (57% vs. 94%) (Schieffelin et al. 2014). Disease is often more severe in infancy and old age; at younger ages, immune systems may be immature, whilst at older ages, immune senescence can reduce immune system strength. However, for some infectious diseases, disease is unexpectedly severe for young adults, due to their exuberant inflammatory responses and the damage caused by the immune system. Age and the process of ageing also interact with multiple social, cultural, political, and psychological factors, which also act as important determinants of microbial exposure.

**Chance:** Stochastic processes play a large part in exposure to pathogens, susceptibility to infection, and subsequent disease. For example, a woman happened to be running on a trail in Arizona at the same time as a rabid fox; two species that rarely interact coming together during a chance encounter. The fox bit her and refused to let go, so the jogger ran back to the car, 'slavering animal clamped to her arm', and transported the fox to a medical centre so that she could have the animal tested for rabies (Herrel 2017).

**History:** Prior exposure, prior vaccination, recent travel, and other host experiences can all influence susceptibility to infection and subsequent disease. As an interesting example, infection by the measles virus appears to alter the immune system's memory—immune amnesia—erasing the body's memory of previous infections, and thereby making the body more prone to subsequent infection with other diseases (Mina et al. 2019). Measles infections caused elimination of 11–73% of the antibody repertoire across individuals (Mina et al. 2019) and restricted the ability of the immune response to new pathogens (Petrova et al. 2019). Vaccination thus does more than safeguard children against measles; it also stops other infections from taking advantage

of measles-induced immune damage. This process would also help to explain why rates of childhood deaths from diseases other than measles declined in the US starting in the 1960s, when the measles vaccine was introduced (Mina et al. 2015).

Inoculum [iˈnäk·yə·ləm]—(1) Cells used in an inoculation, such as cells added to start a culture. (2) A biological material (like a virus or toxin or immune serum) that is injected into a human to induce or increase immunity to a particular disease.

**Inoculum/Dose:** The dose of infectious agents that hosts are exposed to varies widely across individuals, populations, and species; for example, did you eat one bite of *Listeria*-contaminated ice cream or did you eat the entire gallon container? When the initial inoculation or dose is small, host immune reactions may be more successful in preventing subsequent infections (e.g. all the cells will be engulfed by macrophages). When infections successfully establish, larger initial doses can cause more severe disease due to direct damage from the parasite/pathogen or indirect damage from the host's immune response.

**Nutrition:** The state of the host's well-being affects physiology, including immune function, which can allow successful immune defence or tolerance to infection that reduces the severity of the disease. The links between malnourishment and infectious disease present a huge problem in low-income countries, where both issues are common and intertwined. For example, malnourished infants are more susceptible to life-threatening diarrheal diseases, and diarrheal diseases can cause or exacerbate malnutrition (e.g. Walker et al. 2013).

**Genetics:** The genetic makeup of the host and the pathogen will influence whether infection and disease will occur. For example, in some bacteria, such as *Escherichia coli* (e.g. food-borne transmission from eating cattle) or *Vibrio* spp. (food-borne transmission from shellfish), serious disease in people only occurs when specific virulence genes are present in the pathogen (Zhao et al. 2018, Sarowska et al. 2019). Similarly, coronaviruses appear ubiquitous in bats around the world, yet only a few have caused human disease; SARS-CoV and SARS-CoV-2 (the cause of COVID-19) appear to have genetic adaptations that allow them to bind to ACE-2 receptors in human respiratory tracts (Liu et al. 2021, Yan et al. 2021). In another example, some people with specific human leukocyte antigen (HLA) types appear resistant to HIV, despite repeated exposure (Goulder and Watkins 2008, Hardie et al. 2008). And the Black Death likely altered the evolution of European people's genome, as genes that increased the chances of survival from *Yersinia pestis* infection in England and Denmark also increase the risk of autoimmune diseases (e.g. Crohn's disease or rheumatoid arthritis) today (Klunk et al. 2022). And

of course, hosts and pathogens often share a co-evolutionary history, where the host species has evolved to become increasingly resistant or tolerant to the pathogen and the pathogen has evolved to become increasingly successful at infecting the host species.

**A combination of factors:** Often, susceptibility to infection and subsequent disease will depend on interactions between several of the components just listed—as described in Box 2.2.

---

### Box 2.2  The disease triangle: lions, canine distemper, ticks, and drought

*Tanzania, 1994*—Six lions (*Panthera leo*) in the Serengeti National Park suffered grand mal seizures (loss of consciousness and violent muscle contractions, also referred to as tonic-clonic seizures), and three other lions exhibited recurrent twitching (myoclonus) in their faces and forelimbs (Roelke-Parker et al. 1996). Other lions were disoriented, ataxic (displaying abnormal, uncoordinated movements), and profoundly depressed, and still others began to die from encephalitis and pneumonia. The culprit was discovered to be canine distemper virus, a single-stranded RNA virus belonging to the genus *Morbillivirus* (which also includes rinderpest and measles viruses).

There were 'only' 39 *documented* lion deaths attributed to canine distemper virus, but the Serengeti ecosystem's lion population declined by a third over the course of the outbreak, from an estimated 3000 to 2000 lions. Other carnivore species in the Serengeti were also affected, including spotted hyaenas (*Crocuta rocuta*), bat-eared foxes (*Otocyon megalotis*), leopards (*Panthera pardus*), and black-backed jackals (*Canis mesomelas*). But the impacts on those species were harder to document because they are less monitored than lions and, consequently, baseline comparison data were lacking.

The source and subsequent spread of canine distemper virus in the lion population presents a complicated puzzle. Canine distemper virus is a generalist pathogen able to infect a broad range of carnivore hosts and during the 1994 Serengeti outbreak, a single canine distemper virus variant was circulating in lions, hyaenas, bat-eared foxes, and domestic dogs, suggesting a common spillover source for all species. Approximately 30,000 domestic dogs live in nearby villages, and canine distemper virus seroprevalence in these dogs rose in the two years prior to the lion

outbreak. A domestic dog spillover event therefore seems probable, but how it occurred is unclear. Direct dog-to-lion contact is not common in the Serengeti ecosystem. Spotted hyaenas may have been a more likely conduit from dogs to lions; they are known to range among human dwellings and travel long distances within the Serengeti National Park, so perhaps they became infected and exposed lions at shared kill sites. Sick jackals were also noticed at the time, and like hyaenas, jackals interact with lions when feeding on carcasses. It is therefore likely that canine distemper virus spread through the Serengeti via multiple host species, and that other wildlife species repeatedly introduced the pathogen into different lion groups, rather than lions spreading the virus pride-to-pride (Roelke-Parker et al. 1996, Craft et al. 2009). Though some of these details are not definitively known, it's clear that the large dog population represents a potential pathogen reservoir source, which could continue to seed outbreaks even as lion populations dwindle.

Seven years later, the virus was once again associated with lion die-offs, this time in the nearby Ngorongoro Crater. Before these two outbreaks, canine distemper virus had not been considered especially pathogenic in either lions or hyaenas. Analyses of stored blood samples from 510 Serengeti lions revealed that there had been seven canine distemper virus outbreaks over the course of 23 years, punctuated by repeated absences of the pathogen, though five of the outbreaks had not actually been recognized as they did not cause noticeable mortality in the kings of beasts, i.e. disease and mortality is not always a consequence of canine distemper virus infection (Munson et al. 2008).

So, why was canine distemper virus suddenly causing mortality in lions and other carnivore populations? Perhaps because the Serengeti lion population was immunologically

**Box 2.2** *Continued*

naive before the virus was observed in 1994 (one seropositive lion was sampled in December 1993; 33 lions sampled between 1990 and 1993 wereseronegative).

Intriguingly though, during the 1994 and 2001 outbreaks, lions were co-infected with *Babesia*, a malaria-like blood parasite that is transmitted by ticks. Lion prides with

**Figure 2.12** Serengeti lion feeding on prey (top) and featuring a necklace of engorged ticks (bottom).
Photos by Peter Hudson.

---

**Box 2.2** *Continued*

higher canine distemper virus prevalence had a greater proportion of animals with high loads of *Babesia*, and what's more, prides with the highest levels of *Babesia* in their blood suffered the highest mortality rates during the canine distemper outbreaks (>67% mortality). Indeed, the only Serengeti pride to remain uninfected by canine distemper virus in 1994 had only moderate *Babesia* levels.

*Babesia* prevalence appeared to be unusually high in 1994 and 2001 due to exceptionally good environmental conditions for ticks. For example, in 1993, the Serengeti experienced its worst drought in four decades, followed by heavy rains. In 2000, similar conditions walloped the Ngorongoro Crater. At the same time, pasture management in the Ngorongoro Crater had also suppressed grassfires, allowing tick numbers to amass. Herbivores were inundated with ticks; indeed, several black rhinoceros (*Diceros bicornis*) in the Ngorongoro Crater died during 2001 due to fulminating babesiosis. Large die-offs of buffalo also occurred, and

lions feeding on tick-infested buffalo carcasses could have been exposed to high numbers of *Babesia*-infected ticks (Fig. 2.12). In fact, two prides that did not prey on buffalo had lower levels of *Babesia* infection, and they did not experience increased mortality in 1994 when exposed to canine distemper virus.

In summary, it seems that environmental conditions (drought, fire suppression) affected tick numbers, and the abundant ticks were able to infect lions with *Babesia*. Co-infections with *Babesia* and canine distemper led to more severe disease outcomes than had been seen in prior canine distemper outbreaks (Munson et al. 2008). Thus, disease was due to a combination of environmental factors, host history (including co-infections), immunity, and chance. This example illustrates how outbreaks and disease can occur because of idiosyncratic interactions between many contributing factors.

---

## 2.11 Disease agent groups

We have managed to get all the way into the latter parts of Chapter 2 without even mentioning the different taxonomic groups of parasites and pathogens. Let's broach that topic now.

Many pathogens of human health importance are **bacteria**, e.g. *Yersinia pestis*, which causes bubonic plague, and the spirochaete *Borrelia burgdorferi* that causes Lyme disease. If available and appropriate, antibiotics can be used to cure hosts of these pathogens and prevent further transmission.

**Viruses** are also important pathogens, separated into RNA viruses and DNA viruses. A large proportion of RNA viruses that infect humans are zoonotic (160/180, Woolhouse et al. 2013); these originate predominantly from other mammals, though our knowledge of the full host range of most viruses is patchy and incomplete (Woolhouse et al. 2013, Streicker and Gilbert 2020).

Other pathogens include **protozoa**, a diverse group of single-celled organisms including *Plasmodium* spp. (the cause of malaria), *Leishmania* spp. (the cause of leishmaniasis), and *Toxoplasmosis gondii* (the cause of toxoplasmosis, Chapter 6). *Plasmodium* spp.

and *Toxoplasma gondii* are examples of **complex life cycle parasites**: they sequentially infect different host species over the course of their life cycle.

**Multi-celled organisms** that cause disease are typically called **parasites**, rather than pathogens. These metazoan parasites include parasitic worms (e.g. tapeworms, nematodes) (Fig. 2.13), parasitic insects (e.g. fleas, kissing bugs, lice), and parasitic arachnids (e.g. ticks).

Many parasites have complex life cycles. For example, the lancet liver fluke, *Dicrocoelium dendriticum*, is a trematode flatworm that spends the adult stage reproducing within ruminant hosts such as sheep and cows (Fig. 2.14). The parasites' eggs pass from the sheep/cow in the faeces, which are later consumed by snails. Inside the snail, the parasite multiplies through several larval stages (miracidia, sporocysts, cercariae). The last stage, the cercariae, migrate to the snail's mantle cavity, where they irritate the skin and cause the snails to produce mucus/slime balls. The snail excretes the cercariae in the slime balls, which are subsequently swallowed by ants. Inside the ants, the cercariae encyst (forming metacercariae), and some end up in the nerve cells where they can manipulate the ant's

**Figure 2.13** The tapeworm *Diphyllobothrium nihonkaiense*, which can be transmitted by eating salmon sashimi (Section 2.4), from a case in Japan. The white scale bar is 10 cm long!
From Ikuno et al. (2018), public domain.

behaviour. As the cool of evening arrives, the ant will climb to the top of a blade of grass, clamp its mandibles, and remain fastened there until dawn. Opportunely for the parasite, grazing sheep or cows accidentally ingest the ant and the fluke is then able to mature into the adult stage and inhabit the ruminant's bile duct. And thus, the cycle begins again!

### Definition of slime ball

'In brief, a slime ball may be described as an aggregation of cercariae in a matrix which is surrounded by a layer of material, gelatinous and hyaline in nature, referred to as slime. Neuhaus and Mattes used the German word Schleimbail to describe what they found to be the transfer agent from the snail intermediate host to the definitive host in the life cycle of *D. dendriticum*. This word may be translated as "slimeball". These writers gave it this name, apparently, because they observed that the outer communal covering of the cercariae consisted of slime. We believe that there is some question as to the origin and composition of the slime ball but consider the term adequate for our purposes.'

—Krull and Mapes (1952)

**Fungi** cause many important wildlife diseases, but they are usually less important as zoonoses (though they can cause fatalities if untreated or if they infect immunocompromised people, such as when *Histoplasma capsulatum* causes mortality in HIV/AIDS patients). Fungal disease is also extremely important for plant hosts and the people that rely on them; for example, the fungus-like oomycete *Phytophthora infestans*, also known as potato late blight, devastated potato crops and caused the infamous Great Famine in Ireland in 1845–1849 (also known as the Irish Potato Famine).

**Prions** are a relatively newly recognized disease agent; they are not genetically evolving organisms or viruses, but misfolded proteins that cause neurodegenerative diseases. The term *prion* stems from *proteinaceous infectious particle*. In the 1990s, mad cow disease, or bovine spongiform encephalitis (BSE), was recognized as a disease agent affecting cattle in the United Kingdom. Believed to have originated from scrapie—a prion disease of sheep—and introduced to the cattle food chain after they were fed the processed animal remains of infected sheep,

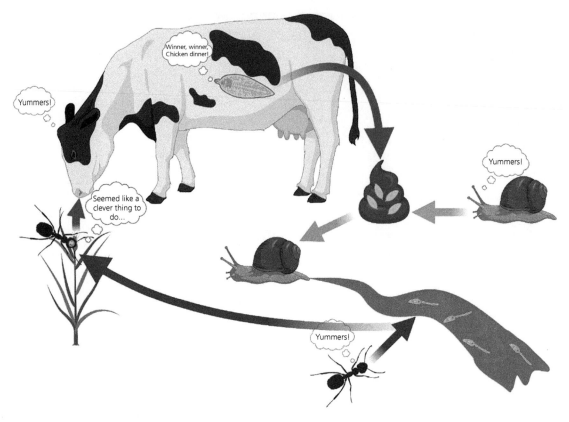

**Figure 2.14** Complex life cycle of *Dicrocoelium dendriticum*.
Created using BioRender.com.

infected cattle embarked upon a horrifying descent into neurologic impairment and deteriorating body condition. In certain countries (e.g. USA, Australia), people who lived in the UK for 3–6 months or more from 1980–1996 are forbidden from donating blood, due to possible exposure to BSE when eating beef. (Full disclosure: two out of three of this book's authors lived in the UK during the BSE outbreak, and as thrifty undergraduate students, they both took full advantage of the precipitous drop in the price of beef to avail themselves of choice beef products.)

Typically, **cancers** are not transmissible, but there are some truly transmissible tumours, the most famous among disease ecologists being Tasmanian devil facial tumour (DFTD) (Loh et al. 2006, McCallum and Jones 2006). There are two strains of DFTD, and the cancerous cells are transmitted among Tasmanian devils when susceptible

Tasmanian devils bite the facial tumours on infected devils (Pearse and Swift 2006, Pye et al. 2016, Patchett et al. 2020); devil populations are so in-bred that their immune cells are not able to recognize the cancerous cells as foreign. Two other transmissible cancers include a leukaemia-like cancer of soft-shell clams (Metzger et al. 2015) and canine transmissible venereal tumours (CTVT) (Rebbeck et al. 2009). There are also several infectious agents that can cause cancer. In humans, *Helicobacter pylori* can cause stomach cancer and human papillomavirus (HPV) can cause cervical cancer (now vaccine-preventable).

## 2.12 Summary

Why is the anatomy of disease important? Because understanding the processes involved in transmission and disease outcomes allows for rapid

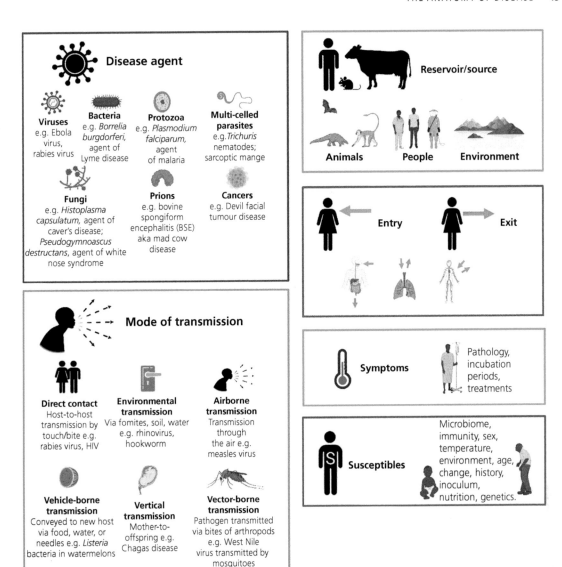

**Figure 2.15** Graphical representation of the anatomy of disease—properties of disease agents that can help in understanding transmission and control interventions. We have tried to come up with a mnemonic to help you remember them all, but not terribly successfully. Perhaps DREEMSS in a Gollumesque fashion, or 'Dress me' in an anaemic, viraemic, squeaky voice? We use the main icons to denote the components in boxes throughout this book.
Created with BioRender.com.

treatment and control responses, even in cases where not everything is well known about the pathogen (Fig. 2.15). For example, what would you do to control pathogen spread if an unidentified pathogen was causing rapid disease in lungs and appeared to be spreading to people in close proximity to infected cases? You could start by

encouraging basic hygiene and use of respiratory masks for patients and people in the vicinity (medical professionals, visitors, nearby patients). These interventions would make sense based on likely exit/entry of the pathogen (airways) and mode of transmission (airborne transmission of respiratory droplets). These control efforts might be especially targeted towards immunocompromised people who would suffer the worst disease outcomes if they became infected. Control strategies might later change following diagnostic investigations of the pathogen, but initial information regarding the 'anatomy of disease' can provide opportunities for critical rapid responses to outbreaks.

## 2.13 Notes on sources

Information on hedgehogs and salmonella: https://www.cdc.gov/salmonella/typhimurium-01-19/index.html (accessed 3/8/2021).

Information on sea cucumber anuses: https://www.nationalgeographic.com/science/article/how-this-fish-survives-in-a-sea-cucumbers-bum (accessed 2/28/2021).

Information on anthrax: https://www.cdc.gov/anthrax/index.html (accessed 17/7/23); https://www.cdc.gov/anthrax/basics/types/inhalation.html, accessed 17/7/23.

## 2.14 References

Barcott, B. (2012). The story behind the hantavirus outbreak at Yosemite. Outside Magazine: https://www.outsideonline.com/1930876/death-yosemite-story-behind-last-summers-hantavirus-outbreak, accessed 17/7/23.

Bin Saeed, A.A., Al-Hamdan, N.A., Fontaine, R.E. (2005). Plague from eating raw camel liver. Emerging Infectious Diseases, 11, 1456–7.

Bloch, A.B., Orenstein, W.A., Ewing, W.M., et al. (1985). Measles outbreak in a pediatric practice: airborne transmission in an office setting. Pediatrics, 75, 676–83.

Boone, S.A., Gerba, C.P. (2007). Significance of fomites in the spread of respiratory and enteric viral disease. Applied & Environmental Microbiology, 73, 1687–96.

Botto-Mahan, C., Medel, R. (2021). Was Chagas disease responsible for Darwin's illness? The overlooked eco-epidemiological context in Chile. Revista Chilena de Historia Natural, 94, 7.

Burlina, P.M., Joshi, N.J., Mathew, P.A., et al. (2020). AI-based detection of erythema migrans and disambiguation against other skin lesions. Computers in Biology and Medicine, 125, 103977.

Casadevall, A., Pirofski, L. (2018). What is a host? Attributes of individual susceptibility. Infection and Immunity, 86, e00636-17.

Centers for Disease Control and Prevention (CDC). (2005). Lymphocytic choriomeningitis virus infection in organ transplant recipients—Massachusetts, Rhode Island, 2005. Morbidity and Mortality Weekly Report (MMWR), 54, 537–9.

Clark, N.J., Soares Magalhães, R.J. (2018). Airborne geographical dispersal of Q fever from livestock holdings to human communities: a systematic review and critical appraisal of evidence. BMC Infectious Diseases, 18, 218.

Craft, M.E., Volz, E., Packer, C., Ancel Meyers, L. (2009). Distinguishing epidemic waves from disease spillover in a wildlife population. Proceedings of the Royal Society, B., 276, 1777–85.

Curtis-Robles, R., Wozniak, E.J., Auckland, L.D., et al. (2015). Combining public health education and disease ecology research: using citizen science to assess Chagas disease entomological risk in Texas. PLoS Neglected Tropical Diseases, 9, e0004235.

Dhondt, A.A., Dhondt, K.V., Hawley, D.M., Jennelle, C.S. (2007). Experimental evidence for transmission of *Mycoplasma gallisepticum* in house finches by fomites. Avian Pathology, 36, 205–08.

Diaz, J.H. (2018). Environmental and wilderness-related risk factors for histoplasmosis: more than bats in caves. Wilderness and Environmental Medicine, 29, 531–40.

Düx, A., Lequime, S., Patrono, L.V., et al. (2020). Measles virus and rinderpest virus divergence dated to the sixth century BCE. Science, 368, 1367–70.

Foley, J.E., Nieto, N.C. (2010). Tularemia. Veterinary Microbiology, 140, 332–8.

Goulder, P., Watkins, D. (2008). Impact of MHC class I diversity on immune control of immunodeficiency virus replication. Nature Reviews Immunology, 8, 619–30.

Grace, E., Asbill, S., Virga, K. (2015). *Naegleria fowleri*: pathogenesis, diagnosis, and treatment options. Antimicrobial Agents and Chemotherapy, 59, 6677–81.

Han, S., Lubelczyk, C., Hickling, G.J., et al. (2019). Vertical transmission rates of *Borrelia miyamotoi* in *Ixodes scapularis* collected from white-tailed deer. Ticks and Tick-borne Diseases, 10, 682–9.

Hardie, R.-A., Knight, E., Bruneau, B., et al. (2008). A common human leucocyte antigen-DP genotype is associated with resistance to HIV-1 infection in Kenyan sex workers. AIDS, 22, 2038–42.

Herrel, K. (2017). Rabid fox vs. jogger. Backpacker.com: https://www.backpacker.com/news-and-events/rabid-fox-vs-jogger, accessed 9/11/2019.

Holzbauer, S.M., Kemperman, M.M., Lynfield, R. (2010). Death due to community-associated *Clostridium difficile* in a woman receiving prolonged antibiotic therapy for suspected Lyme disease. Clinical Infectious Diseases, 51, 369–70.

Hopkins, S.R., Hoyt, J.R., White, J.P., et al. (2021). Continued preference for suboptimal habitat reduces bat survival with white-nose syndrome. Nature Communications, 12, 166.

Hopkins, S.R., Jones, I.J., Buck, J.C., et al. (2022). Environmental persistence of the world's most burdensome infectious and parasitic diseases. Frontiers in Public Health, 10, 892366.

Hotez, P.J., Bethony, J., Bottazzi, M.E., et al. (2005). Hookworm: 'The Great Infection of Mankind'. PLOS Medicine, 2, e67.

Ikuno, H., Akao, S., Yamasaki, H. (2018). Epidemiology of *Diphyllobothrium nihonkaiense* diphyllobothriasis, Japan, 2001–2016. Emerging Infectious Diseases, 24, 1428–34.

Institute of Medicine. (2002). Scientific and Policy Considerations in Developing Smallpox Vaccination Options: A Workshop Report. The National Academies Press. https://doi.org/10.17226/10520

Jackson, Y., Myers C., Diana A., et al. (2009). Congenital transmission of Chagas Disease in Latin American immigrants in Switzerland. Emerging Infectious Diseases, 15, 601–03.

Joseph, J.T., Purtill, K., Wong, S.J., et al. (2012). Vertical transmission of *Babesia microti*, United States. Emerging Infectious Diseases, 18, 1318–21.

Kehrmann, J., Popp, W., Delgermaa, B., et al. (2020). Two fatal cases of plague after consumption of raw marmot organs. Emerging Microbes & Infections, 9, 1878–80.

Kerlik, J., Avdičová, M., Musilová, M., et al. (2022). Breast milk as route of tick-borne encephalitis virus transmission from mother to infant. Emerging Infectious Diseases, 28, 1060–61.

Kiesecker, J.M. (2002). Synergism between trematode infection and pesticide exposure: a link to amphibian limb deformities in nature? Proceedings of the National Academy of Sciences, USA, 99, 9900–04.

Klunk, J., Vilgalys, T.P., Demeure, C.E., et al. (2022). Evolution of immune genes is associated with the Black Death. Nature, 11, 312–19.

Krull, W.H., Mapes, C.R. (1952). Studies on the biology of *Dicrocoelium dendriticum* (Rudolphi, 1819) Looss, 1899 (Trematoda: Dicrocoeliidae), including its relation to the intermediate host, *Cionella lubrica* (Muller). III. Observations on the slimeballs of *Dicrocoelium dendriticum*. Cornell Vet, 42, 253–76.

Kuchta, R., Oros, M., Ferguson, J., Scholz, T. (2017). *Diphyllobothrium nihonkaiense* tapeworm larvae in salmon from North America. Emerging Infectious Diseases, 23, 351–3.

Lei, H., Li, Y., Xiao, S., et al. (2017). Logistic growth of a surface contamination network and its role in disease spread. Scientific Reports, 7, 14826.

Leroy, E.M., Rouquet, P., Formenty, P. et al. (2004). Multiple Ebola virus transmission events and rapid decline of Central African wildlife. Science, 303, 387–90.

Liu, K., Tan, S., Niu, S., et al. (2021). Cross-species recognition of SARS-CoV-2 to bat ACE2. Proceedings of the National Academy of Sciences, USA, 118, e2020216118.

Loh, R., Bergfeld, J., Hayes, D., et al. (2006). The pathology of devil facial tumor disease (DFTD) in Tasmanian Devils (*Sarcophilus harrisii*). Veterinary Pathology, 43, 890–95.

Majewska, A.A., Sims, S., Schneider, A., et al. (2019). Multiple transmission routes sustain high prevalence of a virulent parasite in a butterfly host. Proceedings of the Royal Society, London, B., 286, 20191630.

McCallum, H., Jones, M. (2006). To lose both would look like carelessness: Tasmanian devil facial tumour disease. PLoS Biology, 4, e342.

Meteyer, C.U., Barber, D., Mandl, J.N. (2012). Pathology in euthermic bats with white nose syndrome suggests a natural manifestation of immune reconstitution inflammatory syndrome. Virulence, 3, 583–8.

Metzger, M.J., Reinisch, C., Sherry, J., Goff, S.P. (2015). Horizontal transmission of clonal cancer cells causes leukemia in soft-shell clams. Cell, 161, 255–63.

Mihalca, A.D., Gherman, C.M., Cozma, V. (2011). Coendangered hard-ticks: threatened or threatening? Parasites & Vectors, 4, 71.

Mina, M.J., Kula, T., Leng, Y., et al. (2019). Measles virus infection diminishes preexisting antibodies that offer protection from other pathogens. Science, 366, 599–606.

Mina, M.J., Metcalf, C.J.E., de Swart, R.L., et al. (2015). Long-term measles-induced immunomodulation increases overall childhood infectious disease mortality. Science, 348, 694–9.

Munson, L., Terio, K.A., Kock, R., et al. (2008). Climate extremes promote fatal co-infections during canine distemper epidemics in African lions. PLoS ONE, 3, e2545.

Núñez, J.J., Fritz, C.L., Knust, B., et al. (2014). Hantavirus infections among overnight visitors to Yosemite National Park, California, USA, 2012. Emerging Infectious Diseases, 20, 386–93.

Pancic, F., Carpenter, D.C., Came, P.E. (1980). Role of infectious secretions in the transmission of rhinovirus. Journal of Clinical Microbiology, 12, 567–71.

Patchett, A.L., Coorens, T.H.H., Darby, J., et al. (2020). Two of a kind: transmissible Schwann cell cancers in the endangered Tasmanian devil (*Sarcophilus harrisii*). Cellular and Molecular Life Sciences, 77, 1847–58.

Pearse, A.-M., Swift, K. (2006). Allograft theory: transmission of Devil Facial-Tumour disease. Nature, 439, 549.

Pereira, K.S., Schmidt, F.L., Guaraldo, A.M., et al. (2009). Chagas' disease as a foodborne illness. Journal of Food Protection, 72, 441–6.

Petrova, V.N., Sawatsky, B., Han, A.X., et al. (2019). Incomplete genetic reconstitution of B cell pools contributes to prolonged immunosuppression after measles. Science Immunology, 4, eaay6125.

Prather, K.A., Marr, L.C., Schooley, R.T., et al. (2020). Airborne transmission of SARS-CoV-2. Science, 370, 303–04.

Pue, H.L., Turabelidze, G., Patrick, S., et al. (2009). Human rabies—Missouri, 2008. Morbidity and Mortality Weekly Report (MMWR), 58, 1207–09.

Pye, R.J., Pemberton, D., Tovar, C., et al. (2016). A second transmissible cancer in Tasmanian devils. Proceedings of the National Academy of Sciences, USA, 113, 374–9.

Rebbeck, C.A., Thomas, R., Breen, M., et al. (2009). Origins and evolution of a transmissible cancer. Evolution, 63, 2340–49.

Roelke·Parker, M.E., Munson, L., Packer, C., et al. (1996). A canine distemper virus epidemic in Serengeti lions (*Panthera leo*). Nature, 379, 441–5.

Romo, V. (2018). Man pulls 5 1/2-foot-long tapeworm out of his body, blames sushi habit. The Two Way: https://www.npr.org/sections/thetwo-way/2018/01/19/579130873/man-pulls-5-1-2-foot-long-tapeworm-out-of-his-body-blames-sushi-habit, accessed 28/7/2022.

Rosenthal, P.J. (2019). Be careful what you eat! American Journal of Tropical Medicine and Hygiene, 101, 955–6.

Sarowska, J., Futoma-Koloch, B., Jama-Kmiecik, A., et al. (2019). Virulence factors, prevalence and potential transmission of extraintestinal pathogenic *Escherichia coli* isolated from different sources: recent reports. Gut Pathogens, 11, 10.

Schieffelin, J.S., Shaffer, J.G., Goba, A., et al. (2014). Clinical illness and outcomes in patients with Ebola in Sierra Leone. New England Journal of Medicine, 371, 2092–100.

Scoles, G.A., Papero, M., Beati, L., Fish, D. (2001). A relapsing fever group spirochete transmitted by *Ixodes scapularis* ticks. Vector Borne and Zoonotic Diseases, 1, 21–34.

Sinclair, J.R., Carroll, D.S., Montgomery, J.M., et al. (2007). Two cases of hantavirus pulmonary syndrome in Randolph County, West Virginia: a coincidence of time and place? American Journal of Tropical Medicine and Hygiene, 76, 438–42.

Streicker, D.G., Gilbert, A.T. (2020). Preventing zoonotic emergence from bats requires integrative research. Science, 370, 172–3.

Tang, J.W., Marr, L.C., Li, Y., Dancer, S.J. (2021). Covid-19 has redefined airborne transmission. BMJ, 373, n913.

Vora, N.M., Basavaraju, S.V., Feldman, K.A., et al. (2013). Raccoon rabies virus variant transmission through solid organ transplantation. JAMA, 310, 398–407.

Walker, C.L.F., Rudan, I., Liu, L., et al. (2013). Global burden of childhood pneumonia and diarrhoea. Lancet, 381, 1405–16.

Wang, C.C., Prather, K.A., Sznitman, J., et al. (2021). Airborne transmission of respiratory viruses. Science, 373, eabd9149.

Weingartl, H.M., Embury-Hyatt, C., Nfon, C., et al. (2012). Transmission of Ebola virus from pigs to non-human primates. Scientific Reports, 2, 811.

Wilson, A.J., Morgan, E.R., Booth, M., et al. (2017). What is a vector? Philosophical Transactions of the Royal Society of London, B., 372, 20160085.

Wong, D., Wild, M.A., Walburger, M.A., et al. (2009). Primary pneumonic plague contracted from a mountain lion carcass. Clinical Infectious Diseases, 49, e33–8.

Woodroffe, R., Donnelly, C.A., Ham, C., et al. (2016). Badgers prefer cattle pasture but avoid cattle: implications for bovine tuberculosis control. Ecology Letters, 19, 1201–08.

Woolhouse, M.E.J., Adair, K., Brierley, L. (2013). RNA viruses: a case study of the biology of emerging infectious diseases. Microbiology Spectrum, 1, 1.

Yan, H., Jiao, H., Liu, Q., et al. (2021). ACE2 receptor usage reveals variation in susceptibility to SARS-CoV and SARS-CoV-2 infection among bat species. Nature Ecology & Evolution, 5, 600–08.

Yoder, J.S., Eddy, B.A., Visvesvara, G.S., et al. (2010). The epidemiology of primary amoebic meningoencephalitis in the USA, 1962–2008. Epidemiology and Infection, 138, 968–75.

Zhao, W., Caro, F., Robins, W., Mekalanos, J.J. (2018). Antagonism toward the intestinal microbiota and its effect on *Vibrio cholerae* virulence. Science, 359, 210–13.

# CHAPTER 3

# Descriptive epidemiology of disease outbreaks

## 3.1 Primary and index cases

Disease experienced by a single host (Chapter 2) can be transmitted and progress to cause outbreaks within host populations. After one host is infected, what does subsequent pathogen spread look like?

*Amoy Gardens estate, Hong Kong, 2003*—On 14 and 19 March, a man from mainland China twice visited his brother in a large residential apartment complex. The man was unknowingly infected with the SARS coronavirus, because he was mostly asymptomatic. He complained only of diarrhoea. Following his visits, 324 people subsequently contracted the virus; the outbreak began on 21 March, and the incidence peak occurred from 24 to 26 March. Surprisingly, the infections were spread throughout the large apartment complex, including multiple apartment buildings and on multiple floor levels. The clusters of infection in the other buildings occurred almost simultaneously, which would not be possible if chains of infection propagated by human-to-human contact, delayed by a latent period in each individual host (the time between exposure and infectiousness). So how did so many people become exposed at once?

A detailed analysis of the building plans, the drainage system, hourly meteorological data and fluid dynamics suggested that the infected visitor had created a common source of infection via **aerosolized diarrhoea** (Yu et al. 2004). What?! No! How?! Though sewage drainage pipes might be an obvious culprit, they were not connected between buildings and thus were unlikely to have contributed to the observed outbreak. Instead, malfunctioning drains appeared to be to blame for allowing moist, warm air to flow up into the air shaft from the bathroom used by the initial SARS case. Each time toilets were flushed, huge numbers of aerosols were generated by the hydraulic action in vertical soil stacks. Consequently, a plume of airborne virus rose upwards to the top of the building and then drifted towards the other buildings on a northeasterly breeze, contaminating susceptible people in the downwind buildings—especially in the middle-level floors and especially if apartments had open windows. Furthermore, horizontal airflows between apartments within buildings were driven by wind pressure and exhaust fans in the bathroom or kitchen units.

In this incident, the visitor with SARS was the **primary case**—the person who initially introduces an infectious disease to a population, e.g. a school class, community, or country (Giesecke 2014). Though useful to know, in practice it is often hard to distinguish who or what the primary case was.

In contrast, the **index case** is the first patient identified by health authorities. The index case needn't be the source of an infectious disease outbreak, but rather the first noticed case. There may be times when the index case is indeed also the primary case, but the two terms are not synonymous (Giesecke 2014).

## 3.2 Epidemic curves

One of the most useful ways to describe an outbreak is to use an **epidemic curve** (Box 3.1, Fig. 3.1)—a graphical depiction of the number of new cases by date of illness onset (or positive test result, onset of symptoms). This is simply a histogram with time/date along the x-axis, and the number

*Emerging Zoonotic and Wildlife Pathogens.* Dan Salkeld, Skylar Hopkins, and David Hayman, Oxford University Press.
© Oxford University Press (2023). DOI: 10.1093/oso/9780198825920.003.0003

## Box 3.1  Constructing epidemic curves

To construct an epidemic curve (also referred to as an epi curve), one should first choose a time unit for the *x*-axis that is useful for counting cases. Daily increments *normally* work best because they are easy to interpret. However, hours or half-days may be more useful if the pathogen's incubation period is short, e.g. for some agents of food-borne disease. One rule of thumb is that the time unit on the *x*-axis should be approximately ¼ of the incubation period. For example, the incubation period for noroviruses is about 24–48 hours, so half-day time units would make sense: ¼ of 48 hours. Following this logic, weeks or months may be more appropriate time units for pathogens with long incubation periods. If the incubation period is unknown, one can create several epidemic curves using different time bins and decide which is most useful and informative; there will usually be a threshold where time bins have become too large, such that too many cases are lumped together and patterns are obscured.

After choosing a time unit, one can choose the time interval. Obviously, the interval should include all known cases from the start to the end of the epidemic. Including a pre-epidemic period—approximating two incubation periods—can also be useful. This allows the graph to illustrate baseline or endemic infection rates for pathogens that already existed in a population but are now causing an epidemic. Similarly, including a post-epidemic period can be useful for illustrating when the epidemic ended and infection rates returned to their baseline levels.

Standard graph etiquette applies to epidemic curves. Axes should be clearly labelled using informative and descriptive titles. Labels can indicate time points of interest (e.g. Christmas party and consumption of thin soup; see below). Conventionally, there is no space between the bars on the *x*-axis; whether you want to have spaces or not depends on your willingness to conform and/or your ability to wrestle with your graph-making computer program. You can create epidemic curves in R or Excel, or you can draw rough epidemic curves by hand on the back of an envelope.

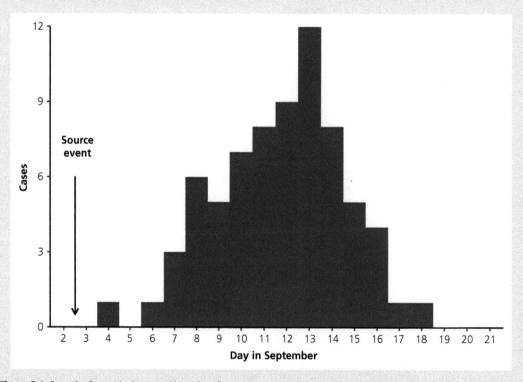

**Figure 3.1** Example of an epidemic curve with number of new cases (*y*-axis) plotted against time (*x*-axis) and using labels to allow interpretation of the outbreak.

of *new* cases on the *y*-axis (e.g. Fig. 3.2). These simple graphs can be beautiful things that offer clues to an outbreak's behaviour and the properties of the disease agent (Torok 2003).

Epidemic curves can provide information on:

- Outbreak magnitude: how many people are infected?
- The trend of the outbreak: are cases increasing or decreasing? Is the outbreak still spreading?
- Disease incubation period: how long from pathogen exposure to onset of symptoms or clinical illness in the host (see Chapter 2)?
- Mechanisms of exposure and spread: was the outbreak caused by person-to-person spread, a common exposure event like aerosolized diarrhoea, or something else? As described below, we can infer mechanisms of exposure and spread based on the shape of the epidemic curve (see next section).
- Outbreak comparisons: were outbreak dynamics different in different populations, e.g. among hospital workers versus members of the public?

## 3.3 Interpreting epidemic curve patterns

The SARS outbreak in the Amoy Gardens estate had a particular shape: an abrupt peak of cases, an apparently brief exposure period and an abrupt decline in incidence after the peak (see Figs. 3.3 and 3.5). This suggested to the investigating epidemiologists that there had been a **common source** of infection, i.e. the plume of aerosolized diarrhoea emanating from one of the buildings. In common source outbreaks, all cases can be linked to a particular place or event, i.e. all cases acquire the infection from the same source. The duration of the event/exposure can vary, so two types of common source outbreaks can be distinguished: common point source and common continuous source outbreaks.

During **common point source** outbreaks, the exposure period is brief, such that most or all cases occur within a single incubation period. *Typically,* these epidemic curves show an abrupt upward slope, with a subsequent swift decline in the number

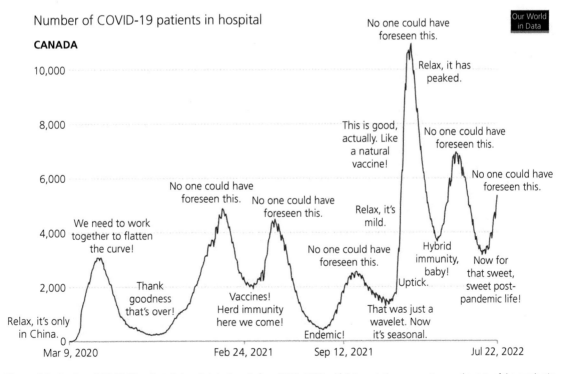

**Figure 3.2** Number of COVID-19 patients in hospitals in Canada from 2020–2022, with interpretative commentary on the state of the pandemic. By Ryan Gregory.

of cases (Fig. 3.3), as occurred in the aerosolized diarrhoea example.

During **common continuous source** outbreaks, the exposure period is longer, such that cases occur over multiple incubation periods. For example, during the famous 1854 Soho cholera epidemic, people were exposed over a long period to the contaminated Broad Street water pump in London, until John Snow had the pump removed. In these long exposure common source outbreaks, the epidemic curve looks less like a mountain peak and more like a gentle plateau (Fig. 3.3).

**Intermittent common source** outbreaks combine aspects of common point sources and common continuous sources, but the disease source is sometimes interrupted. Intermittent exposure often results in epidemic curves with irregular peaks (Fig. 3.3), which reflect the timing and extent of the exposure.

**Propagated outbreaks** are not caused by a common source, but rather by pathogens that are spread from host-to-host, resulting in long-lasting 'waves' of infection. *Typically*, each successive peak is slightly taller than the previous one and each peak is separated by an incubation period (Fig. 3.3).

Note that while there are several distinct types of epidemic curves with *typical* properties, disease outbreaks don't always follow the rules. Outbreak patterns depend on transmission modes and exposure rates, so the exact shape might not match one of the four cookie-cutter examples. And of course,

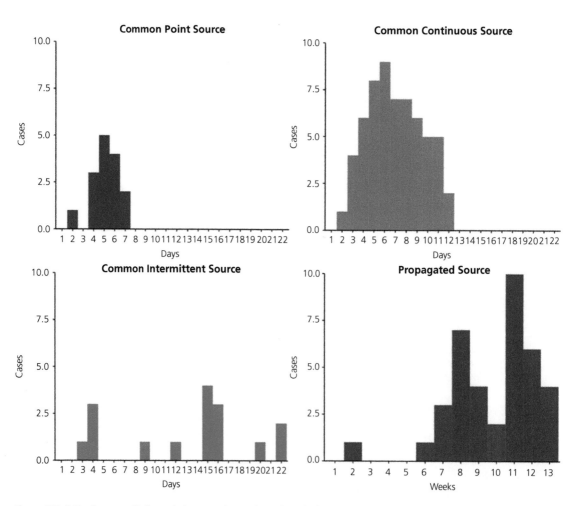

**Figure 3.3** Epidemic curves with four typical patterns that can be used to infer how hosts were exposed to a pathogen.

*missing data* may greatly alter the observed pattern, such as if cases from early in an outbreak were missed because epidemiologists weren't yet conducting surveillance (Chapter 4).

## 3.4 Common source outbreaks

No day ever started well with six dozen raw chickens …
— *Shuggie Bain* by Douglas Stuart

*Campylobacter* is a bacterium that infects animals, including livestock, and humans become infected by accidentally ingesting the bacteria (Fig. 3.4); you may remember that castrating lambs with teeth can expose shepherds to infection (Chapter 1), but vehicle-borne transmission by consuming undercooked meats, particularly poultry, is more common (O'Leary et al. 2009). After exposure, the average incubation period is two to four days, and

it seldom lasts more than a week. Symptoms of *Campylobacter* infection include diarrhoea (sometimes bloody), vomiting, nausea, and fever. In rare cases, people may develop Guillain–Barré syndrome and reactive arthritis.

*Forth Valley, Scotland, 2006*—A *Campylobacter* outbreak infected 47 people during December (see Fig. 3.5). Cases occurred over three weeks, which spans multiple incubation periods (i.e. longer than two to four days); therefore, this outbreak did not seem to have been caused by a point source of exposure. Furthermore, there were no steadily increasing peaks, suggesting that, as is typical for this bacteria, *Campylobacter* wasn't being transmitted between people. By examining the epidemic curve alone, we would guess that the source of infection was either intermittent or continuous; indeed, the infected patients had all dined at the

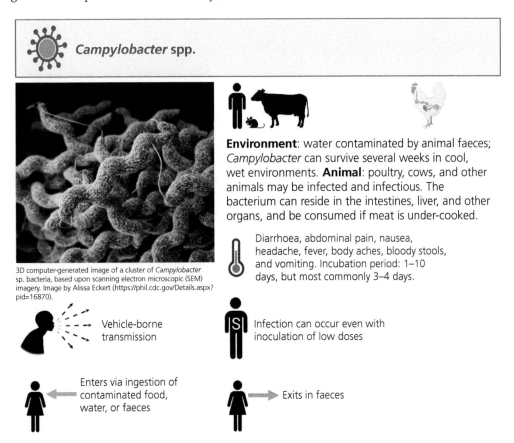

**Figure 3.4** The disease anatomy of *Campylobacter*, a genus of bacteria that causes campylobacteriosis (Gilpin et al. 2020). Created with BioRender.com.

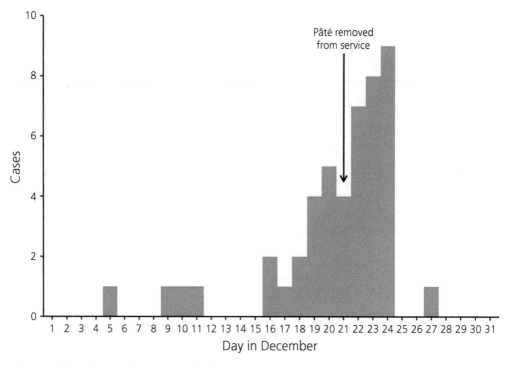

**Figure 3.5** Cases of *Campylobacter* infection associated with Christmas work parties at a Scottish restaurant, based on O'Leary et al. (2009).

same restaurant for various Christmas work parties (O'Leary et al. 2009).

Most diners that succumbed to the illness did so two days after eating at the restaurant, and all of them had eaten the chicken liver pâté. No other food items were associated with illness. When inspecting the restaurant, Environmental Health Officers noticed that the chicken liver pâté had a pinker hue than normal. The more aesthetically pleasing (pinker) product was achieved using a new pâté recipe (introduced on 1 December), where chicken livers were only lightly sautéed and then blended with the other ingredients. (If pan-frying, livers remain bloody until three minutes, remain pink until five minutes, and then turn an unappetizing grey.) This new recipe looked better, but it was also a better vehicle for food-borne transmission—*Campylobacter* is not completely inactivated until cooked for more than five minutes. The chef removed the chicken liver pâté from the menu on 21 December. Subsequent cases occurred because the diners had eaten prior to 21 December, and then the outbreak rapidly ended.

*North Yorkshire, England, 2016*—Christmas and *Campylobacter* joined forces again in a North Yorkshire hotel in 2016. Three Christmas parties held on 17 December partook of a set menu. Once again, chicken liver pâté had the strongest association with diarrhoea, abdominal pain, nausea, headache, fever, body aches, bloody stools, and vomiting (Wensley et al. 2020).

This outbreak had a common point source pattern, which is shaped like a mountain peak (Fig. 3.6). There was a steep initial increase in the number of cases, because everyone was exposed on the same day. The outbreak only lasted as long as the typical *Campylobacter* incubation period of two to four days. Symptoms occurred more quickly for some patients, who might have had higher susceptibility, higher inoculating doses, or higher proclivity for seeking medical attention upon first noticing bloody stool.

As a side note, in both the Forth Valley and North Yorkshire outbreaks, investigators knew that sick people had eaten pâté, but they never found the smoking gun: *Campylobacter*-infected chicken livers. Positive samples might not have been confirmed

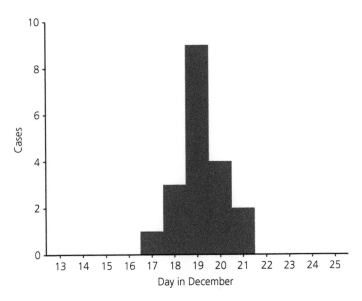

**Figure 3.6** Cases of *Campylobacter* infection associated with Christmas work parties at an English hotel, based on Wensley et al. (2020).

because the infected livers had already been consumed or because there were errors in the testing process. This further illustrates how difficult it can be to identify pathogen sources in abiotic or biotic reservoirs (Chapters 6, 7, 8), even when everyone knows that chicken liver pâté is the most suspicious food on any menu.

*United Kingdom, May 1991*—Choosing chicken liver pâté is not the only way to get exposed to *Campylobacter* in the UK. In a day nursery, 15 children had bouts of diarrhoea (Riordan et al. 1993). *Campylobacter jejuni* was confirmed in stool samples of two of the more seriously ill children. Investigators ruled out the possibility that infections were passed child-to-child, because different age-classes were not exposed to each other. They also noted that the premises were cleaned scrupulously, which left food-borne transmission as the most likely culprit.

Each day the children had a mid-morning snack of milk and fruit, a cooked lunch, and an afternoon snack of milk and sandwiches. Illness was significantly associated with consumption of milk on a single morning. However, the milk was pasteurized and *Campylobacter jejuni* is heat-sensitive, so how were the children exposed to the bacteria?

The most likely explanation was the traditional method of milk-delivery in the UK: glass bottles of milk that were sealed with a foil top used to be delivered to household door-fronts early in the morning, often daily. Cleverly, birds such as magpies (*Pica pica*) and blue tits (*Parus caeruleus*) learned to peck open the foil tops and drink the creamy milk from the top of the bottle. Birds had pecked open the foil lids and contaminated the milk with *Campylobacter* in the 1991 school outbreak; the school cook revealed that milk from bird-pecked bottles had indeed been provided to the children.

## 3.5 Incubation periods and outbreak exposures

Another Christmas time story of gastroenteritis (diarrhoea, cramps, nausea, vomiting, and low-grade fever) was reported from an Italian residential home for the elderly on 22 December (Medici et al. 2009). All the sick individuals had consumed a thin soup at a Christmas party on 20 December.

Pathogen screening of faecal samples from 24 individuals revealed that all were positive for a single strain of norovirus. This included 19 residents and five staff members. Among the staff members, four were asymptomatic and had not suffered gastroenteritis, and three of those four were food-handlers. This information from the pathogen alone could have allowed the investigators to identify the

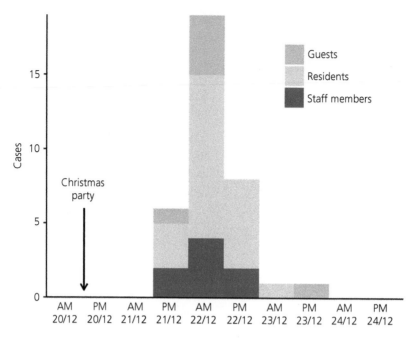

**Figure 3.7** Epidemic curve for norovirus gastroenteritis in a residential care facility in Italy (December 2006), based on data in Medici et al. (2009).

Christmas party as the exposure event, because the party was 24–48 hours prior to the onset of illness—the incubation period for noroviruses.

Similarly, without pathogen screening and sequencing, we might have learned some of these details from the epidemic curve. For example, the epidemic curve exhibited a pattern consistent with a common point source (Fig. 3.7), and notice that the x-axis interval is half-days (a.m. versus p.m. of each day). The very short duration of the outbreak suggests a short incubation period of 24–48 hours. This is an incubation rate characteristic of norovirus and can thus rule out some longer-incubating pathogens that cause gastroenteritis.

## 3.6 Propagated transmission

*Maine, United States, 2020*—On 12 August, Maine's Center for Disease Control and Prevention received PCR test results for two people positive for SARS-CoV-2 (Mahale et al. 2020). Ominously, both cases had been at a wedding party five days previously (7 August). Their symptoms (fever, cough, and sore throat) had commenced the day prior to the

COVID-19 test. (If you swiftly calculated that the incubation period for COVID-19 is about four days, you are correct!) The day after the first two cases were identified, three more cases tested positive for SARS-CoV-2, all of whom had been at the same wedding reception. This potentially shared exposure event prompted an outbreak investigation by Maine's health authorities.

The investigation revealed that some precautions against COVID-19 had been taken. For example, staff at the facility had monitored guests' temperatures as they entered, which all fell within normal temperature ranges. And since the wedding couple and the groom's family travelled to Maine from California, they had tested for SARS-CoV-2 shortly after arriving and all received negative results.

However, many other COVID-19 precautions were eschewed. The wedding reception on 7 August was held indoors, and the number of guests exceeded the state's temporary 50-person limit for indoor gatherings. Furthermore, guests did not follow guidelines related to mask-wearing and social distancing. These decisions were all that was required to create a major outbreak (Fig. 3.8).

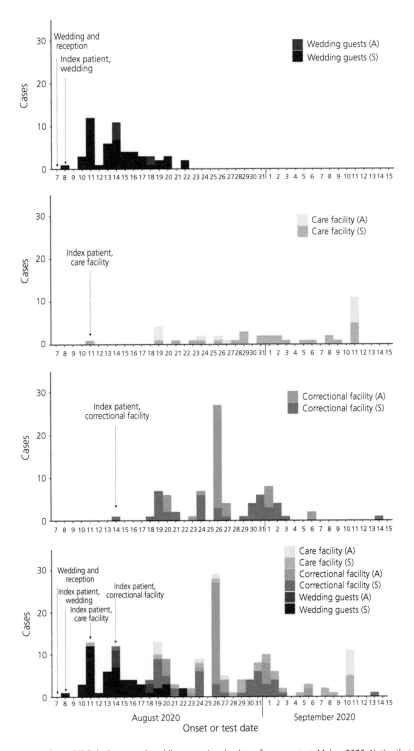

**Figure 3.8** COVID-19 cases (n = 177) linked to a rural wedding reception, by date of onset or test, Maine, 2020. Notice that there were several waves of cases, as is common for propagated outbreaks. However, the waves do not have perfectly identical shapes and timing, because there is always random variation (or 'noise') in case data that we cannot explain.

Based on Mahale et al. (2020).

Among the guests was a Maine resident who was asymptomatically harbouring SARS-CoV-2. Their symptoms (fever, runny nose, cough, and fatigue) manifested on 8 August and they received a positive SARS-CoV-2 test result on 13 August. Of the 55 guests at the wedding, 27 (49.1%) eventually tested positive for SARS-CoV-2. Three other cases were linked to the wedding venue: a staff member, a vendor, and a patron dining at the venue by happenstance. In the local Maine community, 27 further cases were identified (17 secondary and 10 tertiary cases), causing one death and a school shutdown.

Furthermore, despite being sick (cough, myalgia, runny nose, sore throat, and a loss of taste sensation), one wedding guest subsequently turned up to work at their shifts at two separate correctional facilities. On 19 August, four correctional facility staff members received confirmation of COVID-19 diagnoses, and by 1 September, SARS-CoV-2 had been transmitted to inmates (n = 48), colleagues (n = 18) and household contacts of colleagues (n = 16). Apparently, the facility had neither implemented daily symptom screening for staff members nor instigated mask-wearing after the first case was identified.

It gets worse. Another guest at the wedding transmitted the virus to their parent, who was a healthcare worker at a long-term care facility. The healthcare worker became sick a couple of days later, but despite experiencing symptoms (fever, chills, cough, myalgia, runny nose, and headache), continued working for two days before getting tested for SARS-CoV-2, and . . . tested positive. An outbreak emerged at the care home, comprising 38 cases among residents and staff. Tragically, six people in the care home died from their infections.

This outbreak example is a poster child of how quickly and aggressively SARS-CoV-2 and other pathogens can spread between community groups. The viral exposure at the wedding resulted in 177 cases, seven hospitalizations and seven deaths. Though the outbreaks at the three different locations were 100–200 miles from the wedding party, thorough and diligent contact tracing identified common epidemiologic links between the three seemingly unrelated outbreaks. The resulting epidemic curve shows waves of cases, indicative of a propagated outbreak with a roughly four-day incubation period (Fig. 3.8).

### Box 3.2 Measures of disease

**Prevalence** is the proportion of a particular population infected with a disease agent at a specific time. To calculate a prevalence, one divides the numerator—the number of people/animals infected (or testing positive, or showing symptoms), which we can call $a$—by the denominator—the total animals/people tested or exposed (i.e. the population), which we can call $n$. The resulting prevalence can be expressed as a fraction ($a/n$; e.g. 50/100 hosts are infected), a percentage ($a/n*100$; 50% of hosts are infected), or the number of cases per 10,000 or 100,000 individuals ($a$ cases per $n$ hosts; 50,000 cases per 100,000 hosts). For example, if surveillance efforts catch 310 eastern water skinks (*Eulamprus quoyii*) during a field trip ($n$) and 150 of them are positive for haemogregarine blood parasites ($a$), then prevalence of infection is 150/310 = 48.4%. We should also add uncertainty or confidence intervals around these prevalence values, e.g. 95% confidence intervals (95% CI), so that we can interpret how robust our prevalence estimate is. Here, 95% CI tells us that prevalence is 42.7–54.1%, but a smaller sample size would give a larger possible prevalence range, e.g. for the same prevalence, 15/31 has a 95% CI of 30.6–66.6% (see Chapter 8). We'll leave this for now for simplicity, but uncertainty estimates are central to how we interpret data.

Whereas prevalence quantifies current infections, **seroprevalence** is the proportion of a particular population that has been previously exposed to a pathogen and mounted an immune response. Seroprevalence estimates are based on antibodies, which may persist in the body longer than infection and thus represent an accumulation of exposed animals. Imagine a month-long outbreak of Sin Nombre hantavirus in a population of deer mice. Each week, 10% of mice have active hantavirus infections (detectable by PCR) before recovering and seroconverting (developing antibodies) (we're over-simplifying here). If you trapped the mice each week, prevalence of infection would always be 10%, but as the month progressed, seroprevalence would increase from 0% (before seroconversion) to 10%, to 20%, 30%, etc.

**Incidence** is another way of describing infection patterns but conveys how many *new* cases occur in a *population at risk* in a *specified time period*. This differs from prevalence, which is more of a snapshot in time and doesn't necessarily encompass information on whether cases are new or old. For example, the WHO estimates that ~67% of people under the age of 50 are infected with herpes simplex virus 1 (HSV-1). HSV-1 causes lifelong infections, and most people become infected as children. Even though the prevalence is high, the incidence in adults is actually low; if a population of 5000 uninfected adults were surveilled for two years, far fewer than 67% of them would acquire HSV-1. To reflect the time period aspect of the measurement, incidence estimates have units such as 'person-years' (e.g. 43 new cases in a population of 500 over a two-year study would be 43/1000 person-years).

## 3.7 Test validity

To determine whether someone or something is infected with an infectious disease agent, one can use a diagnostic test. Diagnostic tests include direct observations (e.g. using microscopes to observe the agent, including using fluorescent dyes to target the disease agent); culturing samples so that you can observe growth and viability of the disease agent; serology tests for the presence of antibodies (i.e. the immune system's response to prior exposure to a disease agent, Box 3.2); and/or genetic tests for the disease agent (Box 3.3, Fig. 3.9, Box 3.4). Though incredibly useful, diagnostic tests are not perfect things, because biology is messy (Box 3.5) and positive tests are generated by complicated machines and imperfect people wearing white coats and safety goggles.

---

**Box 3.3  PCR**

PCR—polymerase chain reaction—is a process used to create many copies of a specific DNA segment. PCR can also be used to amplify RNA, with the addition of an intermediate step that converts the RNA to DNA. During PCR, cycles of DNA denaturation, annealing and extension are repeated over and over to amplify the DNA segment of interest (Fig. 3.9). After a certain number of these cycles (n), the 'PCR product' contains $2^n$ copies of the DNA segment of interest. These DNA copies can be quantified, visualized (e.g. using gel electrophoresis) and/or used in subsequent analyses.

In epidemiology and disease ecology, PCR is typically used to amplify pathogen DNA or RNA for pathogen detection and/or identification. Pathogens are often so tiny that they cannot be seen (e.g. viruses), in which case detecting their RNA or DNA may be the only way

to confirm that they are present. Additionally, by making many copies of the pathogen's DNA, investigators have more material to work with for subsequent analyses. For example, host faecal or blood samples will be predominantly host material, containing just a tiny bit of pathogen, so targeting and amplifying pathogen DNA fragments can provide larger test samples. This is often an important preparatory step for genetic sequencing, which can be used to identify the pathogen species or the specific strain that caused an infection.

During quantitative PCR (also called qPCR and 'real time' PCR), the quantity of DNA copies is monitored after each amplification cycle. When this process is standardized and validated for a specific pathogen, the quantity of DNA present after a specified cycle number can serve as a diagnostic test. For example, there may be a threshold number of copies that is used to distinguish uninfected from infected samples. Alternatively, qPCR results may be used as a measure of pathogen load, where more copies of the DNA segment indicate high pathogen loads. For example, in Colorado, qPCR was used to discover that just 2% of SARS-CoV-2 individual cases carried 90% of the circulating virus, and that viral loads were no different between symptomatic and asymptomatic people (Yang et al. 2021).

---

A positive result from a diagnostic test can occur for more than one reason:

- the disease agent (or antibodies to the disease agent) truly was present
- the sample was contaminated (e.g. by another sample or by the positive control)
- the disease agent was not present, but a similar, related (or sometimes not even similar or related!) agent caused the result
- the process has created strange, possibly unexplained, results.

Similarly, a negative test result from a diagnostic test can occur for more than one reason:

- the disease agent (or antibodies to the disease agent) was truly not present
- the disease agent was present, but there was not enough infectious agent or antibody to detect its presence
- the diagnostic test failed to work
- the process has created strange, possibly unexplained, results.

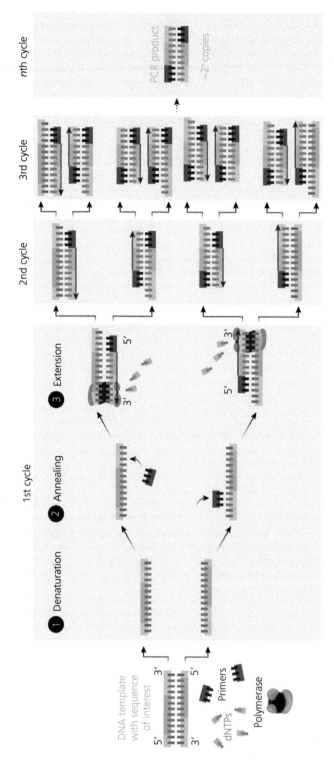

**Figure 3.9** Representation of the PCR process.
Created by Enzoklop, used under Creative Commons Attribution-Share Alike 4.0 International licence.

These technical details might seem a bit dry, but they can be extremely important for understanding and controlling infectious disease dynamics. For example, there are various ways to test if bison (*Bison bison*) have brucellosis (*Brucella abortus*), including serological testing (for live bison) and post-mortem tissue culture. Serological tests can be important tools for sampling living wildlife, but in this case, the serological test is often inaccurate: >50% of bison that test positive for serology have no active infection based on post-mortem tissue culture (Bienen and Tabor 2006). One would not want to base a test-and-slaughter control programme on a test that inaccurately identified bison as being infected! Depending on how diagnostic test results will be used, a more or less accurate test may be acceptable.

Therefore, when developing a diagnostic test or interpreting test results, it is important to investigate **test validity**: how well the test can distinguish between presence and absence of a pathogen or other signs of infection. As outlined in the list above, test validity combines information regarding positive and negative test results. **True positives** are cases where a test correctly identifies the presence of the diagnostic test target, whereas

**false positives** are cases where a positive test result was erroneously caused by contamination, incorrect testing, etc. Likewise for negative test results. **True negatives** are cases where the test correctly identified the absence of the diagnostic test target, and **false negatives** are cases where the test incorrectly reported that the disease agent (or antibodies) was absent, when it was, in fact, present.

This information is usually combined into two metrics of test validity: test sensitivity and test specificity. **Test sensitivity** is the ability of a test to correctly identify cases that have the disease, and is defined as the number of true positives divided by the summed number of true positives and false negatives. **Test specificity** is the ability of the test to correctly identify cases that *do not* have the disease, and is defined as the number of true negatives divided by the summed number of true negatives and false positives. These metrics are illustrated further in Box 3.4.

$$Sensitivity = \frac{True\ positives}{True\ positives + False\ negatives}$$

And:

$$Specificity = \frac{True\ negatives}{True\ negatives + False\ positives}$$

---

### Box 3.4  Interpreting test results

Imagine you have a population of 1000 people (or chipmunks, or willow warblers, or whatever organism you like, even worm snakes). In this population, 100 people are infected with a lurking disease agent; 900 people are not infected (at least not yet).

Imagine you can test the entire population of 1000 individuals and you have a diagnostic test that has a good ability to correctly identify cases that have the disease: let's say 90% sensitivity. The test will correctly identify 90 of the infected individuals (90% of 100) and the remaining cases would be erroneously identified as uninfected (10% of 100).

Similarly, your diagnostic test has a good ability to correctly identify cases that do not have the disease. Let's say 95% specificity: 95% of 900 uninfected cases (= 855) are correctly identified as being uninfected, but 5% of 900 uninfected cases (= 45) are incorrectly labelled as infected.

Overall, the diagnostic test would suggest that 135 individuals are infected and 865 individuals are uninfected.

Therefore, the diagnostic has overestimated the number of infected cases and underestimated the number of uninfected individuals (Table 3.1). This is why confirming test results is a good idea!

In this example, we already knew how many people were infected and the test sensitivity and specificity. That will not usually be the case! Instead, one can calculate the sensitivity

**Table 3.1** Results for diagnostic test examining 1000 people, of which 100 are infected, and 900 are not infected. The test validity is 90% sensitivity and 95% specificity.

|  | Infected | Uninfected | Total |
|---|---|---|---|
| **Positive test result** | 90 TRUE POSITIVES | 45 FALSE POSITIVES | 135 |
| **Negative test result** | 10 FALSE NEGATIVES | 855 TRUE NEGATIVES | 865 |

**Box 3.4** *Continued*

and specificity of a new diagnostic test by creating a table like Table 3.1; to do so, one must be able to compare the new test to another method for identifying infections.

Now that you have insights into the uncertainties associated with diagnostic tests, another way to regard positive and negative test results is to ask: given a positive test result, how likely is it to be a true positive? This is called the **positive predictive value** (PPV), and it is defined as the number of true positives divided by the sum of true and false positives. That is,

*Positive predictive value*

$$= \frac{True\ positives}{True\ positives + False\ positives}$$

And similarly, if you receive a negative result, how likely is it to be a true negative? That is called the **negative predictive**

**value** (NPV) and it is defined as the number of true negatives divided by the sum of true and false negatives.

That is,

*Negative predictive value*

$$= \frac{True\ negatives}{True\ negatives + False\ negatives}$$

In the example above, the positive predictive value would be just 67% (90/90+45); there is a 67% chance that any given positive is a true positive, and thus one-third of positive results would be in error. The negative predictive value would be a more reassuring 98.8% (855/855 + 10).

Crucially, though, these predictive values are sensitive to the prevalence of infection, whereas the test sensitivity and specificity is not.

## 3.8 Test validity, within-host pathogen dynamics, and test type

The same diagnostic test (e.g. a COVID-19 rapid antigen test) might give different results depending on which sample type is used (e.g. nose vs. cheek swab), when the test is completed (e.g. first vs. third day of symptoms) and how much user error is involved. Test validity also varies over time among different types of tests; some diagnostic tests are

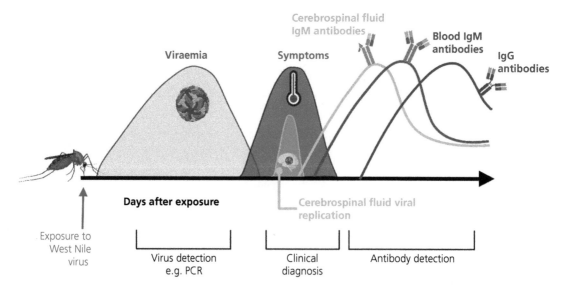

**Figure 3.10** Schematic diagram showing timing of viraemia, symptoms, and antibody response following infection with West Nile virus and encephalitis. Diagnostics to show infection will only succeed if they are carried out at the appropriate stage of infection, e.g. viral detection must occur early, or antibody tests must be carried out when antibodies have had time to be produced.
Made using BioRender.com.

designed to detect early signs of infection (e.g. virus presence) and other tests are designed to detect later signs of current or prior infection (e.g. antibody presence, Fig. 3.10).

Consider the example of a person bitten by a mosquito infectious with West Nile virus. Though the initial inoculum is large enough to establish an infection, it is still relatively small compared to the host's blood volume. Therefore, a blood test that day or during the early stage of infection will likely fail to detect the pathogen, even though it does exist at low abundance (false negative). As the virus multiplies inside the host, the likelihood of detecting the virus with a blood test will increase (effectively increasing a test's sensitivity).

However, as the host's immune system mounts a response, virus abundance will decline, which will once again reduce test sensitivity (Fig. 3.11). In this case, a negative test might correctly deduce that the pathogen is no longer present, *but* not that infection *did* occur! At this point, a different type of test is needed to determine whether a host was infected.

Antibody tests detect the immune system's usually long-lasting response to infection, rather than

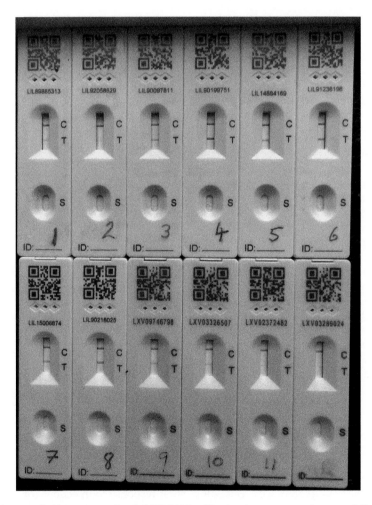

**Figure 3.11** Temporal pattern of antigen concentrations in SARS-CoV-2 rapid antigen home-tests over the course of 12 days from a COVID-19 case. The 'C' line always shows in valid tests, whereas the 'T' line only shows in positive tests. Notice that the T line is not always visible during the course of infection.
Credit: Ian Buchan.

---

**Box 3.5  Koch's postulates**

During an epidemic of unknown cause, one of the first priorities is identifying the causative pathogen (if there is one!). The microbiological gold standard for determining the agent responsible for causing disease used to be demonstrating Koch's postulates. Robert Koch (1843–1910) was a German physician and bacteriologist who outlined three conditions for this process (Fredricks and Relman 1996):

1. The pathogen occurs in every case of the disease in question and under circumstances that can account for the pathological changes and clinical course of the disease.
2. The infectious agent occurs in no other disease as a fortuitous and non-pathogenic organism.
3. After being fully isolated from the body and repeatedly grown in pure culture, the pathogen can induce the disease anew.

If these three conditions were satisfied, Koch stated 'the occurrence of the parasite in the disease can no longer be accidental, but in this case no other relation between it and the disease except that the parasite is the cause of the disease can be considered'.

Koch's approach was a rigorous way to establish cause and effect. His experimental approach could demonstrate that: (1) the infectious agent (Koch studied bacteria) is associated with the infection; (2) infection is indeed the cause of the disease; and (3) the disease agent is transmissible or infectious. For a bacterial disease like tuberculosis (TB), caused by *Mycobacterium tuberculosis*, these criteria can be demonstrated. Indeed, Koch's work on *Mycobacterium tuberculosis* 'slowly changed the romantic perception of consumptive patients as tragically beautiful victims of a wasting disease to dangerously infectious carriers whose cough or sputum transferring as few as ten bacilli could be an ultimately fatal contact' (Tabrah 2011).

When I have fears that I may cease to be
Before my pen has glean'd my teeming brain
      —John Keats (1795–1821), who succumbed to
      bacterial infectionby *Mycobacterium tuberculosis*

Koch's postulates were ground-breaking in the late 19th century, but 'have had a rough ride intellectually, both at their inception and in their application to the many diseases to which they were applied' (Tabrah 2011). Complicating Postulate #3, many parasites and pathogens cannot be grown in culture in the laboratory or if there is no known (or easy) animal model of infection for the disease agent. For example, scientists are still unable to grow the disease agent of leprosy, *Mycobacterium leprae*, in pure culture (Fredricks and Relman 1996). Furthermore, disease (pathological effects) is not always associated with infection, and pathogens are not always able to infect hosts (host immunity). Despite these limitations (even recognized by Koch himself, who was busy enough establishing the bacterial causes of anthrax, tuberculosis, and wound infections (Fredricks and Relman 1996)), one lesson from Koch's postulates remains critical: simply associating a *suspected* pathogen's presence with disease is not sufficient to identify it as *the* pathogen.

> Life has changed since the 1880s when Robert Koch elucidated his guidelines ... for determining whether a microorganism is the cause of a disease. The horse-drawn buggy bumping over dirt roads has been replaced by the computer-assisted automobile speeding along paved highways. It would be absurd to expect modern cars to abide by traffic rules and standards designed for horse-drawn carriages. Yet, many continue to hold Koch's postulates as the unchanging standard for determining causation in medicine, despite a revolution in biotechnology and leaps in medical knowledge. ... The power of Koch's postulates comes not from their rigid application but from the spirit of scientific rigor that they foster. The proof of disease causation rests on the concordance of scientific evidence, and Koch's postulates serve as guidelines for collecting this evidence.
>      —Fredricks and Relman 1996

---

detecting the pathogen itself. As with diagnostic tests for pathogens, the timing of antibody tests will heavily influence detection. If the test is completed too early there will not be enough antibodies present for successful detection. Furthermore, the host's immune response changes over time, where different antibody types rise and fall in abundance.

Immunoglobulin M (IgM) antibodies are often the first response to infections (or antigens) and those are later replaced by different classes of antibodies, such as immunoglobulin G (IgG). Therefore, when antibody tests are used, they must be primed to detect the right kind of antibody at the right time.

## 3.9 Test validity and local pathogen prevalence

As if diagnostic tests weren't complicated enough already, we must now report something that is rather non-intuitive: test interpretation is affected by how common or rare a pathogen is in a population.

Imagine the same population of 1000 individuals (sea otters, Spaniards, wolverines...) from Box 3.4, but this time, disease prevalence is high: 50% prevalence, therefore 500 infected individuals. Using the same diagnostic test as before (with 90% sensitivity and 95% specificity), we can calculate the numbers of true positives (450), false positives (25), true negatives (475) and false negatives (50). We can also calculate the positive predictive value = 94.7% (450/475) and negative predictive value = 90.5% (475/525). You should try these calculations and make sure that you get the same numbers that we did.

Now, recalculate these values imagining that the population of 1000 individuals (mandrills, spring peepers, Kenyans) has several other prevalences of infection: 25% and 1%. Seriously. On an envelope. Or a spreadsheet. Or in R. Work these out and get some practice and familiarity with the process. Some of the numbers are in Table 3.2 so that you can check your answers.

Then look at the positive predictive value for each of these prevalence values. What happens? Oddly, the positive predictive value declines as the pathogen becomes rarer—a positive test result is less reliable if the disease is rare. The converse is true for the negative predictive value; as a pathogen becomes rarer, negative test results become more reliable.

**Table 3.2** Changing positive and negative predictive values as a function of changing prevalence of a pathogen, for a population of 1000 people, using a diagnostic test with 90% sensitivity and 95% specificity.

| Pathogen prevalence | Positive predictive value | Negative predictive value |
| --- | --- | --- |
| 50% | 94.7% | 90.5% |
| 25% | 85.9% | 96.6% |
| 10% | 66.7% | 98.8% |
| 1% | 15.5% | 99.9% |

## 3.10 Test validity and repeat tests

Imagine that you are ill—fevers, malaise, etc.—and you're convinced that you have a case of viral lurgy (fictional). A diagnostic test gives you a negative result, but you're doubtful that the test has worked (you suspect a false negative), so you immediately repeat the test at the same stage of infection. How would your thinking change if the second test was negative? What if the second test was positive?

Repeat tests can result in oddities of interpretation, both for individual patients and population prevalence. For example, if the test has 90% specificity and is performed on 100 truly uninfected people, ten individuals will be given false positive results and 90 people will be correctly identified as uninfected. Some of those 90 individuals may question their uninfected status and quickly take another test during the same stage of infection. For simplicity, let's say that all 90 people are doubtful of their negative results and take a second test: given test sensitivity, 81 of them are correctly deemed negative, but another nine people are incorrectly diagnosed as positive, i.e. another nine false positives. A third test? Another 73 true negatives, a further eight false positives. If this continues, nearly the entire population of uninfected people will have received at least one positive test!

It is not uncommon for patients or medical professionals to halt diagnostic testing once the result matches their expectation. In the example above, they would ignore true negative results and interpret the false positive as the 'proper' result, but they could also ignore a true positive result and interpret the first false negative result as the 'proper' result. To avoid this conundrum, repeatability of results is crucial, i.e. repeat or duplicate tests and ensure that they are providing the same results.

## 3.11 Pooled samples

In a perfect world, there would be infinite time and resources for pathogen diagnostic tests, but as we will see in the next chapter about surveillance, that is not the case! Investigators are often limited by the number of tests that they can run, and they must carefully decide how to design their testing regime around those constraints. This can be especially difficult when a pathogen is known to occur at relatively low prevalence; for example, regardless of test validity, a random sample of 10 individuals is highly unlikely to include a single infected host when the prevalence in a population is 1%, whereas the same random sample would be likely to include infected hosts if the prevalence were 50%. Typically, sample size calculations are made based on our expected infection prevalence, which is fine if you know roughly what you are looking for.

In some cases, it makes sense to pool samples—instead of testing each sample individually, samples are combined together and one test is run on the pooled sample. For example, imagine that you are surveying for a tick-borne pathogen that infects roughly 2% of ticks. You can collect 100 ticks at a site without too much effort, but testing is time consuming and expensive. Should you test each individual tick to determine if the pathogen is present at a site? To save time and money, you could instead pool samples—take 10 ticks, homogenize them in a blender and then do a single diagnostic test on the tick soup. Using this method, you have only to run and pay for 10 pooled tests (instead of 100 individual tests) to determine whether the tick-borne pathogen is present at the site.

The drawback to pooled samples is that you lose some of the accuracy of interpretation. If your pooled sample is positive, anywhere from 1–10 of those ticks might have been infected. So, prevalence of the pathogen could be anything from 1–10% of the population—quite a range, not even including the uncertainty estimates. There are a few ways to deal with this uncertainty. If the exact prevalence is not particularly important to you, you could accept the uncertainty and conservatively report the minimum prevalence of infection (i.e. the pathogen is present with at least 1% prevalence). With many pooled samples, you might be able to better quantify the uncertainty using statistical inferences, such as by determining the probability that a given number of pools will be positive for any given population prevalence. In some cases, identifying infected individuals or accurate prevalences is considered very important, so testing can be done in two parts: pooled samples followed by individual samples. For example, you could construct your pools using half samples (e.g. bisect the

tick and only include half of it in a pooled sample), and then after identifying a positive pool, you run individual tests on the remaining half specimens that contributed to each positive pool (Cross et al. 2018).

## 3.12 Summary

At the beginning of an outbreak, information is at a premium. Quickly interpreting data is crucial to be able to understand infection trends, and epidemic curves are a useful tool to identify potential infection sources and the disease agent. Likewise, accurate identification of the disease agent—by clinical diagnoses, PCR or antibodies—is important for describing outbreak dynamics. Increasingly, and especially if it's a new organism, methods such as genomics can now be used to identify organisms and can be particularly powerful if combined with epidemiological data (see Chapter 8). Either way, interpreting test results and epidemic curves needs to be done with nuance, so that issues associated with missing data or test validity do not obfuscate outbreak patterns.

## 3.13 Notes on sources

Quote from *Shuggie Bain* by Douglas Stuart reproduced with permission of the Licensor through PLSclear. It is an excellent book.

Information on herpes simplex virus from WHO: https://www.who.int/news-room/fact-sheets/detail/herpes-simplex-virus#:~:text=An%20estimated%203.7%20billion%20people,that%20can%20recur%20over%20time (accessed 3/11/2022).

## 3.14 References

Bienen, L., Tabor, G. (2006). Applying an ecosystem approach to brucellosis control: can an old conflict between wildlife and agriculture be successfully managed? Frontiers in Ecology and the Environment, 4, 319–27.

Cross, S.T., Kapuscinski, M.L., Perino, J., et al. (2018). Co-infection patterns in individual *Ixodes scapularis* ticks reveal associations between viral, eukaryotic and bacterial microorganisms. Viruses, 10, 388.

Fredricks, D.N., Relman, D.A. (1996). Sequence-based identification of microbial pathogens: a reconsideration of Koch's Postulates. Clinical Microbiology Reviews, 9, 18–33.

Giesecke, J. (2014). Primary and index cases. Lancet, 384, 2024.

Gilpin, B.J., Walker, T., Paine, S., et al. (2020). A large scale waterborne Campylobacteriosis outbreak, Havelock North, New Zealand. Journal of Infection, 81, 390–95.

Mahale, P., Rothfuss, C., Bly, S., et al. (2020). Multiple COVID-19 outbreaks linked to a wedding reception in rural Maine—August 7–September 14, 2020. Morbidity and Mortality Weekly Report (MMWR), 69, 1686–90.

Medici, M.C., Morelli, A., Arcangeletti, M.C., et al. (2009). An outbreak of norovirus infection in an Italian residential-care facility for the elderly. Clinical Microbiology and Infection, 15, 97–100.

O'Leary, M.C., Harding, O., Fisher, L., Cowden, J. (2009). A continuous common-source outbreak of campylobacteriosis associated with changes to the preparation of chicken liver pâté. Epidemiology and Infection, 137, 383–8.

Riordan, T., Humphrey, T.J., Fowles, A. (1993). A point source outbreak of campylobacter infection related to bird-pecked milk. Epidemiology and Infection, 110, 261–5.

Tabrah, F.L. (2011). Koch's Postulates, carnivorous cows, and tuberculosis today. Hawai'i Medical Journal, 70, 144–8.

Torok, M. (2003). Epidemic curves ahead. Focus on Field Epidemiology, 1(5): https://nciph.sph.unc.edu/focus/vol1/issue5/1-5EpiCurves_issue.pdf, accessed 24/10/2022.

Wensley, A., Padfield, S., Hughes, G.J. (2020). An outbreak of campylobacteriosis at a hotel in England: the ongoing risk due to consumption of chicken liver dishes. Epidemiology and Infection, 148, E32.

Yang, Q., Saldi, T.K., Gonzales, P.K., et al. (2021). Just 2% of SARS-CoV-2–positive individuals carry 90% of the virus circulating in communities. Proceedings of the National Academy of Sciences, USA, 118, e2104547118.

Yu, I.T.S., Li, Y., Wong, T.W., et al. (2004). Evidence of airborne transmission of the Severe Acute Respiratory Syndrome Virus. New England Journal of Medicine, 350, 1731–9.

# CHAPTER 4

# Surveillance

In every walk with Nature one receives far more than he seeks.

—**John Muir**

## 4.1 Surveillance approaches

*Lyme, Connecticut, USA, 1975*—Two residents in a New England town, Polly Murray and Judith Mensch, became concerned about the number of unexplained illnesses occurring in their families and the surrounding neighbourhood (Ostfeld 2011). The symptoms included rashes, fever, and juvenile rheumatoid arthritis—a relatively rare condition in children that involves painful joint swelling. Astutely, Murray and Mensch noticed a geographic cluster of cases along the east bank of the Connecticut River—a location that also harboured a high abundance of black-legged ticks (*Ixodes scapularis*). Perturbed, Murray called the Connecticut State Department, prompting the involvement of Allen Steere, a rheumatologist, and eventually Willy Burgdorfer, a medical entomologist. Together, in the early 1980s, they discovered and described Lyme disease, or Lyme borreliosis, the condition caused by the tick-borne bacterium *Borrelia burgdorferi* (Burgdorfer et al. 1982, Benach et al. 1983).

After the causative agent of Lyme disease was identified, many important questions emerged. How broadly distributed was this newly recognized disease? Was it spreading across the region or country, or did it only occur in its namesake town of Lyme, Connecticut? Why was it so prevalent in one part of town? And what should people do to protect themselves from infection? Answering these questions required **surveillance: the systematic collection and analysis of health data**. Multiple types of surveillance could be adopted at multiple scales to understand the distribution of Lyme disease:

**Vector surveillance:** One surveillance approach involves investigating the distribution of the ticks that transmit *Borrelia burgdorferi* by actively searching for ticks in the environment (Fig. 4.1). Questing ticks lurk on vegetation waiting to attach themselves to passing mammalian hosts, so tick surveys are conducted by dragging or flagging a flannel blanket or sheet over leaf litter or vegetation. Ticks mistake the moving blanket for a potential host and attach themselves to the cloth, where they can be carefully removed, identified, and preserved by researchers.

Tick surveys are an example of **active surveillance**, which requires a bona fide effort to search for a pathogen, vector, or host. Active surveillance is advantageous because it can be used to describe the habitat and geographic location where a vector or pathogen occurs. For example, ticks could be collected in oak/beech woodland, on a particular date, and with known geographical coordinates. There are also drawbacks to active surveillance—it requires a lot of effort (time, money, logistics, motivation, risk of infection) to travel and find samples, so the scope of investigation is always limited. Furthermore, survey efforts must be carefully planned to avoid prejudice towards areas and/or habitats where successful sampling is more likely to be guaranteed. Nonetheless, active surveillance provides critical data on the distribution and abundance of vectors, creating an estimate called **vector risk** or, in this case, **acaralogic (tick) risk**.

**Pathogen surveillance:** Once ticks have been collected, they can be examined—by microscopy or by molecular techniques—to determine how many

**Figure 4.1** Active surveillance for vectors involves dragging flannel 'blankets' on leaf litter or vegetation to collect questing ticks. An adult female *Ixodes scapularis* can be seen on this drag.
Photo by Graham Hickling, University of Tennessee, USA.

of them are infected with pathogens. This can provide an estimate of **disease risk** (or **infection risk**) such as the proportion of ticks that are infected (i.e. **prevalence of infection**). Because ticks have multiple life stages (Figs. 4.2 and 4.3) and both adults and nymphs can be infected with *Borrelia burgdorferi*, you could express this estimate of disease risk as **nymphal infection prevalence (NIP)** or **adult infection prevalence (AIP)**. If you have been diligent and fastidious, you may have used transects of known area/length so that you can quantify the density of ticks, e.g. **density of infected nymphs (DIN)** or **density of infected adults (DIA)**.

**Host surveillance:** Since Lyme disease is zoonotic and is transmitted between ticks and wildlife, surveillance also targets wildlife reservoirs. The first such wildlife investigation quickly found that white-footed mice (*Peromyscus leucopus*) were commonly infested by ticks and infected with *Borrelia burgdorferi* (Bosler et al. 1983). This is another example of active surveillance. In fact, trapping vertebrate hosts and testing them for current or prior pathogen infections is even more time consuming and difficult than sampling ticks, so this form of active surveillance involves greater logistical

challenges. For example, because mammals are relatively long-lived and wide-ranging, it can be difficult to interpret when a wildlife host became infected, especially if the tests are for antibodies.

To get more precise measures of risk (e.g. **incidence**), researchers can use **sentinel testing**, where they quantify the rate that known uninfected animals become infected. Much like the 'canary in the coal mine', sentinel hosts can be used to quantify risks in current conditions. To do this, sentinel hosts must be able to be easily monitored and to become infected, but they do not necessarily need to act as a reservoir host or be a wildlife species. For example, in California, domestic chickens have been used as sentinel hosts to help monitor West Nile virus activity. The chickens do not suffer West Nile virus-induced disease but do seroconvert (develop antibodies) after exposure. Because the chickens can be maintained in coops, they are easy to monitor regularly using fortnightly blood sampling. These sentinel surveys can be used to calculate the rates at which susceptible hosts become infected and to describe variation in the force of infection (see Chapter 5) across space and time.

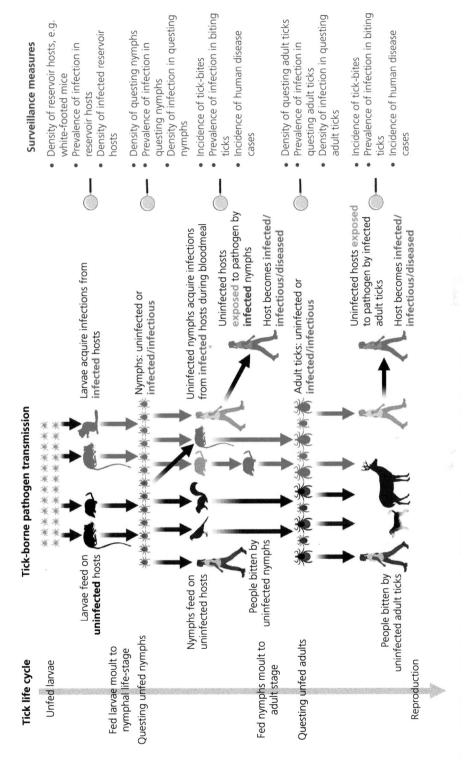

**Figure 4.2** Life cycle of *Ixodes* ticks, transmission of tick-borne pathogens such as *Borrelia burgdorferi*, and relevant surveillance methods.
Created using BioRender.com.

**Figure 4.3** Life stages of the black-legged tick (*Ixodes scapularis*). Questing larvae (note that they only have six legs at this stage) (Top right), questing nymph (Left), and adult female engorged after a bloodmeal (Bottom right).
Photos by Graham Hickling, University of Tennessee, USA.

**Human and veterinary surveillance:** Lyme disease surveillance also focuses on infection patterns in human or pet populations—the populations that we are usually most concerned with protecting. When physicians and/or diagnostic labs observe positive results for Lyme disease, the data on human cases are reported to county health agencies. County health agencies then assimilate the data and send that information to state health agencies, which pass the information on to national health organizations. When test results are collected via voluntary submission or government mandates, the effort constitutes **passive surveillance**.

Passive surveillance is advantageous because it is relatively cheap and geographically broad, but there can also be drawbacks. For example, data are often submitted using the patient's address (county or zipcode) or the medical facility where they were tested as the only geographical information. Such information may not be representative of the place where the person was exposed. Remember Dr Goldberg's tick nostril? It was discovered in Wisconsin,

USA, but exposure occurred in Kibale, Uganda (Chapter 1). In this era of globalization, people can travel far and wide, complicating analyses of passive surveillance data.

Another drawback of passive surveillance is that there is often limited incentive or human-power to curate and submit the data, so cases may be under-reported. Recently in the US, about 30,000–40,000 Lyme disease cases in humans are reported annually (Rosenberg et al. 2018)—30,000–40,000 events where a physician or lab reported a confirmed case to the local health authorities, and that information was then shared with the appropriate federal agency (the Centers for Disease Control and Prevention, CDC). However, *actual* numbers of Lyme disease cases are estimated to be 10 times higher (roughly 300,000–450,000) (Nelson et al. 2015, Kugeler et al. 2021), and most cases simply aren't being reported properly. Under-reporting can be especially problematic for diseases that are difficult to diagnose based on symptoms alone (like Lyme disease and malaria), because pathogen testing is expensive and time consuming. These barriers are extra problematic for diseases that disproportionally affect marginalized groups or countries with limited health infrastructure. Well-designed surveillance must address all these issues.

**Citizen science:** Many of the surveillance approaches described above can be aided by citizen science. Citizen science, also referred to by some as community science, involves the participation of non-professionally trained scientists in scientific activities (Kullenberg and Kasperowski, 2016). There are many benefits to citizen science, but one of the biggest is the sheer number of people willing to contribute to the scientific process; having

---

### Box 4.1 Surveillance biases

Let's switch systems and look at Colorado tick fever virus in Rocky Mountain wood ticks (*Dermacentor andersoni*). Perhaps unsurprisingly, Rocky Mountain wood ticks can be found in montane habitats (Fig. 4.4). A survey examining tick abundance in northern Colorado, USA, exploited the natural altitudinal variation of the Cache la Poudre Canyon (roughly ranging from 1700 to 2500 m) and actively sampled (dragging a white flannel cloth) for ticks on three occasions from 20 April to 1 June, 2006 (Eisen et al. 2007). Ticks were most abundant at mid-range elevations (2200–2400 m) and were largely absent below 2100 m and above 2500 m. At four sites that were below 2100 m (Gray Rock, Hewlett Gulch, Narrows, and Mountain Park), only a single tick was collected across the six-week investigation. The entomologists (a quick tangent: strictly speaking, entomology is concerned with insects, and ticks are arachnids, but tick researchers also refer to themselves as entomologists) responsible for the study speculated that perhaps conditions at lower elevations are simply too dry and too hot to allow tick populations to persist (Eisen et al. 2007).

A few years later, though, an attendee at a 2014 conference on the Ecology and Evolution of Infectious Diseases went on a hike at Lory State Park, at low elevation. (The park's highest point, Arthur's Rock, is 2061 m). On her flight home to Atlanta, the attendee experienced a terrible fever. Once home, she discovered and removed a feeding tick (never formally identified to species) and recovered, before relapsing into fever several days later. Her clinical symptoms—biphasic fever following tick-bite—were emblematic of a case of Colorado tick fever. And yet her sole foray into the Colorado outdoors had been the hike at Lory State Park. This incident suggested that tick surveillance in the Poudre Canyon, though an excellent effort, had not fully described local Rocky Mountain wood tick ecology.

Let's examine some potential sources of bias from the Poudre Canyon study. That surveillance effort occurred in a relatively tight time window—six weeks in a single year—and may have missed tick activity at some altitudes; phenological studies often find that life cycle events, like when flowers bloom, happen earlier at lower elevations. Furthermore, the year of sampling could have been unusual—particularly dry/hot summer or especially wet spring or a longer snow season. Single-year field studies often suffer from these 'what if' questions because they are more of a limited snapshot than a long-term picture. Additionally, despite intensive field work, the Poudre Canyon study could have had design limitations. For example, although the Rocky Mountain wood tick was suggested to peter out at 2500 m (Eisen et al. 2007), only one site looked at ticks at this elevation. Issues like these are a constant struggle for active surveillance.

In more recent years, surveillance efforts have covered a larger temporal window (earlier and later in the year) and a larger elevational range. As a result, Rocky Mountain wood ticks are known to be found across a larger elevational gradient. The ticks have frequently been found higher than 2500 m in the Cache la Poudre Canyon, where the current standing record highpoint is 2893 m. It is possible that climate change and warmer conditions in high elevations have allowed the ticks to spread further up the canyon since the original survey in 2006. But ticks have also been found at lower elevations than in the original survey, including the city of Fort Collins at 1585 m. Sometimes they occur in large numbers at 1585 m, but some years they appear hard to find, perhaps reflecting the importance of annual environmental variability or fluctuations in host abundance.

This example highlights some important considerations for designing surveillance programmes. Most importantly, studies must account for the fact that the ecological distributions of vectors, hosts, and pathogens may fluctuate across space and time. Study design (e.g. how many sites at high elevation) can limit or bias interpretations, so much time and consideration should be spent on careful designs before beginning a surveillance project. And once surveillance has begun, prolonged sampling may be required to properly describe species distributions.

---

**Box 4.1** *Continued*

---

**Figure 4.4** Active surveillance for Rocky Mountain wood ticks taking place in the Rocky Mountains, Colorado, USA.
Photo by Rae Ellen Bichell/KUNC.

thousands of people looking for ticks is better than a team of four researchers! Citizen science approaches can involve either active surveillance (e.g. intrepid citizens deliberately looking for ticks in the field) or passive surveillance (e.g. citizens reporting their observations to scientists using phone applications).

For tick surveillance, citizen science can take many forms. For example, people can use smartphone applications to log tick exposure (Fernandez et al. 2019) or send their ticks in the mail to be examined by a professional scientist in a laboratory running a submit-a-tick programme (Nieto et al. 2018, Porter et al. 2021a). Ticks and tick-borne

diseases also constitute a veterinary health issue, and veterinarians can also catalogue data on ticks found on pets or livestock (Duncan et al. 2021; Pieracci et al. 2019; Saleh et al. 2019). There is some concern that citizen science data can bias tick distribution maps, such as when people report where they were when they found a tick on their body instead of where they were when they were exposed to the tick (Nieto et al. 2018, Eisen and Eisen 2021). However, with enough participation, citizen science data can be very similar to data collected by field researchers (Lyons et al. 2021, Porter et al. 2021a,b, Tran et al. 2021) and has been used in tick modelling efforts (Soucy et al. 2018). Furthermore, there is something to be said about how citizen science data can counter unrecognized biases of professional scientists!

Even better, citizen science that documents where people or their pets encounter ticks can improve our understanding of actual tick exposures; the data show where people *were* exposed. In contrast, active surveillance for ticks can only predict where people and pets *might be* exposed, without incorporating human or pet behaviour and whether ticks are truly encountering and biting human and pet hosts. Similarly, surveillance of human disease cases does not usually lend itself to easily interpreting exposure risk, because case data are

the culmination of exposure, infection, disease and successful diagnosis. Citizen science that measures actual exposures can fill in data gaps.

Combining multiple surveillance methods can address the shortcomings of any one technique and build a more cohesive picture of pathogen dynamics. For example, in the Rocky Mountains, the aptly named Rocky Mountain wood tick is the predominant human-biting tick species (we discussed this tick earlier in the context of bias and surveillance in Box 4.1). At a coarse scale, CDC maps suggest that the Rocky Mountain wood tick is observed across the state of Colorado (Fig. 4.5). However,

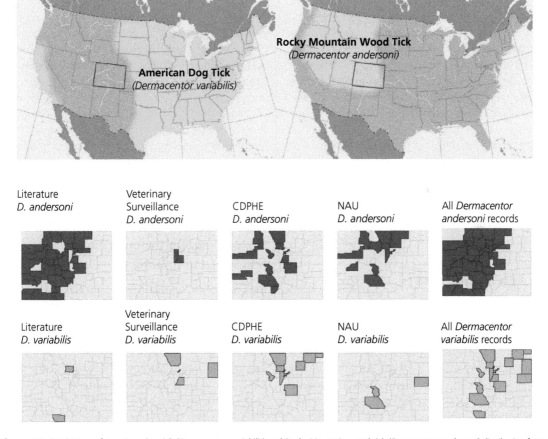

**Figure 4.5** (Top) Maps of American dog tick (*Dermacentor variabilis*) and Rocky Mountain wood tick (*Dermacentor andersoni*) distribution from US Centers for Disease Control and Prevention (CDC), with the Colorado state boundary highlighted in red (https://www.cdc.gov/ticks/geographic_distribution.html, accessed 5/4/2022). (Bottom) Colorado counties with presence of ticks based on different data sources (scientific literature, veterinary surveillance (Duncan et al. 2021), Colorado Department of Public Health and the Environment, Northern Arizona University citizen science, and combined records).

Data are from Freeman and Salkeld (2022).

investigations of different hosts, using different surveillance techniques, suggest a more nuanced picture (Freeman and Salkeld 2022). Active surveillance reports Rocky Mountain wood ticks in 33/64 Colorado counties (James et al. 2006), and citizen science data from pets and people adds another five counties to the tally (Freeman and Salkeld 2022). Most counties with Rocky Mountain wood tick records are west of the continental divide and are in higher elevation areas of the state—so the tick isn't truly distributed everywhere throughout the state.

Furthermore, the American dog tick, *Dermacentor variabilis*, occurs in the eastern US (as well as in a band of the western US), and was considered absent from the Rocky Mountains by the CDC. Published records show scant records of *Dermacentor variabilis* in Colorado from just two counties (Lehane et al. 2020). However, publications can be biased because they reflect scientists' motivations to publish data. Citizen science suggests the presence of American dog ticks in 16 scattered counties in the middle and eastern, lower altitude, parts of Colorado (Freeman and Salkeld 2022). Of course, people may have travelled with dogs from elsewhere, and these tick records could be reporting travel associations. But citizen science submissions of American dog ticks may also be revealing that dog ticks are encroaching and establishing themselves in new territory. Combining active and passive surveillance offers rapid insights into the dynamic patterns of pathogen distribution changes.

## 4.2 Aggregating data

*Colorado, 2015*—During a study of plague dynamics, researchers collected blood samples from several mammal species to understand how interspecific interactions might affect *Yersinia pestis* dynamics in black-tailed prairie dogs (*Cynomys ludovicianus*) (Stapp et al. 2008). In northern grasshopper mice (*Onychomys leucogaster*), 11/253 blood samples contained plague antibodies; that's 4%—not a hugely impressive number. Therefore, the investigators' initial reaction was to dismiss any hypothesized role of grasshopper mice in local plague ecology. However, those 253 grasshopper mice had been sampled from areas with *and* without current plague outbreaks. Looking exclusively at areas with plague

outbreaks, 10/49 of grasshopper mice were seropositive. That's 20%—a higher prevalence that invited a very different appraisal of whether grasshopper mice were important components of plague spread. Indeed, when the data were parsed even further to consider single prairie dog colonies where plague activity was current, *Yersinia pestis* seroprevalence in grasshopper mice was even higher, e.g. 29% (5/17), 33% (3/9), and 60% (3/5). (You may criticize some of these sample sizes, but you only have a right to do so after working on wind-blasted short grass steppe, dragging heavy sleds carrying hundreds of traps through cactus, getting exposed to a deadly zoonosis, rattlesnakes, and sunburn on your fair, fair Celtic skin; see Chapter 14.)

This example illustrates the importance of interpreting data at the appropriate scales, and the potential risks of **aggregating** or grouping data. Though aggregating blood samples across the entire region seemed logical at first, one wouldn't expect grasshopper mice to be seropositive if they were trapped in areas where no plague activity was occurring, and thus trying to infer evidence from broadly aggregated summary statistics was misleading. In this example, aggregating to the colony level was probably the more informative option, but it is always difficult to choose the spatial and temporal scales for aggregating data in ecology and epidemiology (see Box 4.2).

## 4.3 Aggregating data: ecologic fallacy and Simpson's paradox

Data aggregation is further complicated when sampling is organized by human jurisdictional boundaries, e.g. counties, regions, states, and countries. This kind of data aggregation is done to assemble information, to protect individual privacy, and to compare information from different sources at similar scales. But there can be issues with this type of correlational analysis.

For example, there is strong interest in determining where and when and why Lyme disease will affect people in the United States. As described above, human cases of Lyme disease must be reported to health agencies by doctors or laboratories that diagnose the disease, so county- and state-level data are available for Lyme disease cases.

---

**Box 4.2  Scale of surveillance and analysis**

Rabies virus is a persistent problem for people, domestic animals, and wildlife. In the Serengeti district of Tanzania, rabies persists at a very low prevalence; it never surpasses 0.15% overall, yet it is still responsible for tragic loss of human and animal life. For example, in the study described here, 44 local people died from rabies from 2002 to 2016. 'Its persistent circulation at such low prevalence in largely unvaccinated populations is an enduring enigma' (Mancy et al. 2022). This highlights the importance of disease ecology research and control interventions.

In some ways, the Serengeti rabies system is highly amenable to research because it is possible to monitor disease transmission at the individual level: bites from rabid animals can be documented and clinical signs of disease are readily identifiable in individual animals (infected animals typically die within a week of disease onset). Highly resolved genetic data from the viral strains also help researchers trace

transmission networks. In other ways, disease surveillance is quite difficult, such as tracking the activities of the many free-ranging dogs (predominantly owned but unleashed) living in the region.

In a population of roughly 50,000 dogs, Mancy et al. (2022) traced 3612 rabies infections in animals (3081 cases in dogs, 75 in cats, 145 in wildlife, and 311 in livestock). From these cases, they documented 6684 potential transmission events to other animals and 1462 people bitten by rabid animals. Simply astonishing! Using these detailed descriptions of intricate transmission chains, the researchers found that 22 transmission chains accounted for >70% of rabies cases and circulated for more than a year (including two for more than 4 years).

Individual-based models using spatially explicit data were developed to understand and describe rabies dynamics across a range of scales (Fig. 4.6). At the smallest scale,

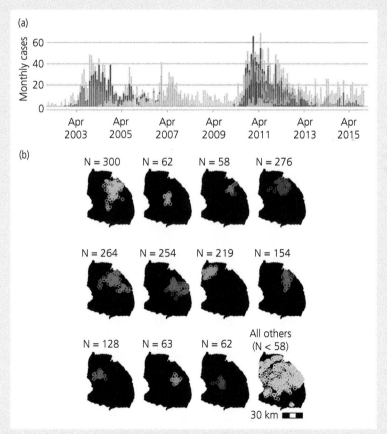

**Figure 4.6**  (A) Monthly reported rabies cases grouped by transmission chain—the 11 chains with most cases are categorized by colour (all smaller chains, <58 cases, are shown in grey). (B) Spatial distribution of transmission chains.
From Mancy et al. 2022.

---

**Box 4.2** *Continued*

they simulated populations within 0.25 km² grid cells, and at the largest scale, they simulated a single well-mixed population across the whole Serengeti district. Only models that used a 1 km² scale could reliably reproduce observed disease dynamics. At this 1 km² scale, susceptible dog numbers were rapidly depleted due to death from rabies and exposed dogs incubating new infections. Individual dog behaviour emerged as a key factor driving transmission: some infected dogs travelled long distances and introduced new rabies lineages into naïve neighbouring communities, facilitating metapopulation dynamics and dogged rabies persistence, whereas others simply bit local animals. A few rabid dogs bit many other dogs (4% of rabid dogs bit >10 other dogs each, and four dogs bit >50 dogs) and some ran long distances before biting other animals (nine rabid dogs ran >10 km).

Individual variation in the likelihood of causing new cases results in transmission dynamics featuring rare but explosive outbreaks combined with more frequent pathogen fade-outs, and overall, longer maintenance of infection. Individual-level differences, at least in this rabies system,

appear more important than environmental or population-level differences such as host density; counterfactual simulations in which dogs behaved more consistently failed to replicate the real-world outbreak patterns.

More typical modelling approaches might aggregate the data, and in doing so, fail to observe and appreciate the behavioural and population processes driving transmission dynamics in spatially structured host populations.

And such insights also affect perspectives on control interventions (see Chapter 14). Dogs can be highly mobile. Culling fails to restrict potential movement of infected dogs and would have to radically reduce dog populations to have any impact on rabies circulation. However, fully comprehensive dog vaccination removes susceptible populations and so a travelling rabid dog fails to instigate new clusters of rabies infections. Model simulations indicate that, even in areas where vaccination coverage was between 10 and 40%, dog vaccinations prevented more than 4000 animal cases, 2000 human rabies exposures, and 50 deaths in the Serengeti district.

---

People have often used these data to test hypotheses about why Lyme disease incidence varies geographically. For example, it is hypothesized that Lyme disease risk declines with mammal biodiversity, and one can estimate mammal biodiversity by counting how many species have geographic ranges that overlap each state (Turney et al. 2014). We could also hypothesize that obesity might impact susceptibility to Lyme disease, and that human behaviour can be reflected by food and diet (Shortridge 2005). Correspondingly, we can collect state-level data on the number of fried chicken restaurants (a proxy measure for human culture) and state-level data on obesity rates, which might reflect comorbidities and health status.

Indeed, state-level Lyme disease incidence decreases as mammalian biodiversity increases, though the relationship is not strong ($R^2 = 0.12$, so variation in mammal biodiversity explains only 12% of variation in Lyme disease incidence) (Fig. 4.7). Lyme disease incidence is slightly better explained by the number of fried chicken restaurants ($R^2 = 0.17$), and there is a strong correlation

between Lyme disease incidence and state-level obesity rates ($R^2 = 0.47$) (Salkeld and Antolin 2020). Of course, correlation does not mean causation, and the mechanisms underlying these relationships aren't well understood. Is there a dietary component in fried chicken that is protective against *Borrelia* infection? Does obesity indicate low rates of physical activity, reflecting less time spent on outdoor activities such as hiking and gardening and thus fewer exposures to ticks in outdoor habitats (Porter et al. 2019)? Perhaps time spent in fried chicken restaurants is time *not* spent in tick habitat! Even though we can't exactly describe the mechanisms, it is easy to create rationales that explain the observations that Lyme disease incidence is lower in states with higher mammal biodiversity, more fried chicken restaurants, and higher obesity rates.

However, a wiser interpretation of these analyses is that the correlations are all spurious because the underlying data are simply unfit for testing the hypotheses. Each Lyme disease infection, like any zoonotic spillover event, occurs at the scale

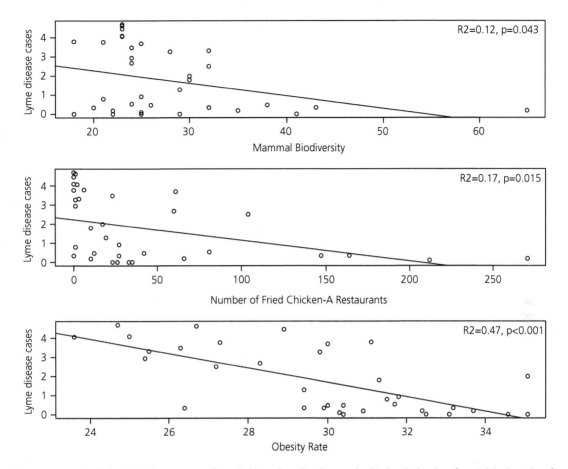

**Figure 4.7** State-level relationships between Lyme disease incidence (ln + 1) and mammalian biodiversity (number of species) (top), number of fried chicken restaurants (chain A) (middle), and obesity rates (bottom).
From Salkeld and Antolin (2020).

of the individual, yet the analyses rely on state-level aggregated data. From these data, one has no idea whether people with Lyme disease are obese, eat fried chicken, or recreate in highly biodiverse habitats. It is entirely possible that in the states with the lowest obesity rates, every person with Lyme disease is obese, and in the states with the lowest obesity rates, none of the people with Lyme disease are obese. If that were the case, our analysis would be preposterous, but there is no way to confirm or deny this possibility with state-level data. Furthermore, our analysis ignored other factors simply because the data were not accessible or didn't exist, such as how much time people actually spend outside during tick season; it included questionable 'proxy' variables, such as the number

of fried chicken restaurants as a proxy for diet; and it did not consider many possible complexities (feedbacks, non-linear relationships, etc.).

Our example of Lyme disease incidence and fried chicken restaurants is plainly ridiculous (see also Nadelman and Wormser 2005, Messerli 2012), but it is not uncommon for analyses to aggregate data inappropriately (Gordis 2009). The attraction of creating correlations between existing data sets arises because it can be easily and cheaply done, across large scales, and the resultant regressions may confirm existing theory or hypotheses; also, it would seem reasonable that large-scale patterns should reflect mechanisms occurring at individual levels.

Above, we noted an example of an '**ecological fallacy**', where inferences about the nature

**Figure 4.8** 'Proxy Variable' by xkcd, http://xkcd.com/2652

of individuals are deduced from attributes of the group to which those individuals belong (Gordis 2009, Webb and Bain 2011; Pollet et al. 2015). It is also possible for aggregated data to produce statistically significant patterns that are diametrically opposed to those occurring at the individual level: a phenomenon called **Simpson's paradox**, where correlations are reversed during data aggregation because the causal driver has not been measured or incorporated into analyses (Bickel et al. 1975, Pollet et al. 2014, 2015). When results have been spuriously created by ecological fallacies and Simpson's paradox, the statistically significant relationships will dissolve when looking at smaller scales.

In summary, correlations based on aggregated data can be easy to do at large scales and can be useful for developing hypotheses, and these approaches are increasingly tempting in our data-rich world. But to demonstrate causality, individual-level data are required, ideally in combination with experimental approaches that investigate the underlying mechanisms (Fig. 4.8).

## 4.4 Surveillance and zoonotic outbreaks

*Équateur Province, Democratic Republic of the Congo (DRC), 2018*—On 8 May 2018, in the market town of Bikoro, an outbreak of Ebola virus started (Seifert et al. 2022). In June, an investigative field team was deployed to determine whether Ebola virus (EBOV) was simultaneously circulating in local bat populations in the Cuvette and Likouala departments in the neighbouring Republic of Congo. PCR analyses of bat samples did not find Ebola virus genomic material (i.e. no active infections), but serological data did suggest recent Ebola virus exposures in 10/144 bats from the Cuvette department and 4/28 bats in the Likouala department. This could imply that bats were the reservoir hosts responsible for the outbreak (Chapter 8). Alternatively, bats throughout West and Central Africa have now been detected to be seropositive at similar percentages and so might have been exposed but not *infectious* when the outbreak started (or possibly were never infectious, as the viral-immune response dynamics in bats are still poorly understood). Unfortunately, many such studies share the same issues, and disease ecologists have been unable to definitively identify the primary Ebola virus reservoir, whether it is one or more species (Olival and Hayman 2014). Moreover, reservoirs may not be directly responsible for transmission to a target host (here people), as other intermediate hosts can be infected (see Section 1.6).

In some ways, attempting to look for an outbreak source *after* the spillover has already occurred is to try and shut the stable door after the (diseased) horse has bolted. However, most ecological investigations do occur *after* spillover events and outbreaks (Leroy et al. 2005, Pourrut et al. 2005), because widespread surveillance for all pathogens is still a thing of the future. Surveillance of wildlife disease is especially rare, though in the wake of COVID-19 and the West Africa Ebola virus outbreak, there is more interest than ever in disease surveillance in wildlife populations, and in doing this cleverly. SMART surveillance—Strategic Monitoring, Assays, Response and Treatment—has three important characteristics. First, to integrate surveillance of humans *and* animals (whether wildlife or livestock), which can help improve animal health, identify reservoir hosts, and serve as an early warning for outbreaks in humans. Second, instead of focusing on one or a few known pathogens, SMART surveillance can use metagenomic testing (an untargeted approach to sequencing) to look for many pathogens at once. This not only increases information gained per surveillance effort, but it also provides a way to surveil for entirely novel pathogens. And third, SMART surveillance should involve consistent and repeated monitoring to establish baseline force of infection in areas of high spillover risk (hot spots). Together, these programme components could improve our general understanding of zoonosis and enable a more anticipatory approach to outbreak prevention (Gardy and Loman 2018).

Of course, there are many barriers to SMART surveillance. Many countries lack in-country disease diagnostic capacity, and delays in sample analysis (e.g. time required to physically ship samples) can hamper outbreak response times. Investing in local lab capacity in concert with improved surveillance would allow quicker outbreak response and treatment (Keusch et al. 2022). There are also significant logistical challenges associated with surveilling hard-to-capture wildlife in hard-to-investigate field environments (Seifert et al. 2022). And of course, infectious diseases do not obey jurisdictional or species borders, forcing government agencies (e.g. health, environment, agriculture, trade, etc.), health professionals (e.g. doctors, veterinarians) and researchers (e.g. entomologists, anthropologists) to work together for successful surveillance (Karesh et al. 2012).

This constitutes an enormous amount of work but given that disease outbreaks can cause staggering economic disruptions and societal upheaval, integrating disease surveillance and control for humans, livestock and wildlife together might be more cost-effective than human surveillance and control programmes alone (Karesh et al. 2012, Torgerson and Macpherson 2011). Reducing the drivers that lead to these outbreaks in the first place would probably be even more smart and cost-effective (Dobson et al. 2020, Bernstein et al. 2022, Chapter 15) (Fig. 4.9).

## 4.5 Outbreak surveillance

The motivation to investigate outbreaks can vary. One might hope to determine whether the outbreak is ongoing, design an intervention (remove the pâté!) to prevent further cases, or advance understanding of epidemiology and disease ecology more broadly. Outbreaks are often over or petering out before they are properly recognized, especially if there is a short incubation period and the outbreak was caused by a common point source. Yet outbreak investigation may still be worthwhile because it could provide knowledge that can be used to prevent or control future outbreaks. Regardless of why the investigation is being conducted or the particular pathogen involved, outbreak investigations often follow similar lines of inquiry (Reingold 1998, Fig. 4.10):

**Recognition that an outbreak is occurring:** There is often no need for a background or training in epidemiology to recognize that an outbreak is occurring. Public health agencies may be primed to recognize outbreaks because they have background data and expertise. But clinicians or nurses or diagnosticians may be the first people who notice an unusual cluster of cases that signify a larger

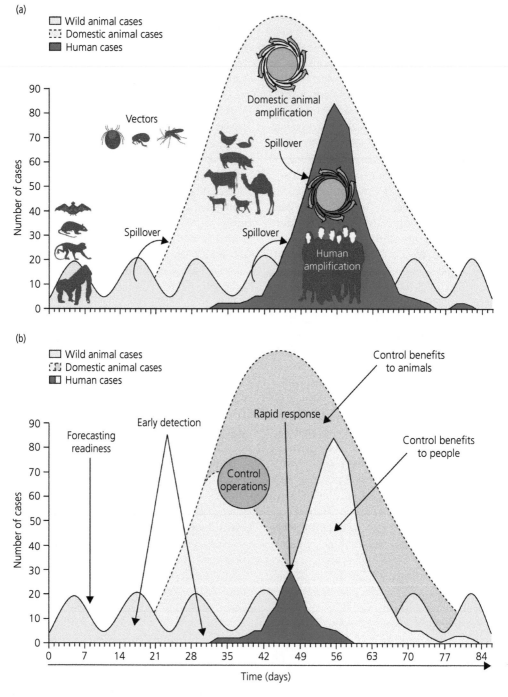

**Figure 4.9** (A) Spillover of infection to humans (red) can occur directly from wild animal populations (pink) or from livestock populations (light green) that convey and amplify the pathogen from wildlife populations to human populations. (B) SMART surveillance that enables early detection and control efforts can prevent and reduce disease incidence in people (light blue) and animals (dark green).
From Karesh et al. (2012).

**An outbreak investigation framework**

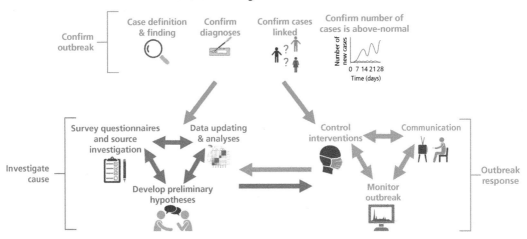

**Figure 4.10** A general framework for confirming outbreaks and the corresponding efforts to investigate the outbreak cause and respond to the outbreak.
Made using BioRender.com.

outbreak. For example, physicians first recognized clusters of atypical pneumonia during the emergence of both the original SARS and SARS-CoV-2. Members of the public might also be the first to suspect a problem and alert medical professionals, as in the discovery of Lyme disease by residents of Lyme, Connecticut. This includes media sources, who may be the first people to catch wind of unusual disease cases and catalyse official investigations. When small outbreaks are widespread but decentralized, data sharing by diagnostic labs and health agencies or analyses of internet searches may trigger recognition that an outbreak is occurring.

To classify as an outbreak, investigators must show that the incidence of cases is above normal; confirm case diagnoses (i.e. this person with supposed Lyme disease is truly infected); and establish epidemiologic links between cases (i.e. common exposures or risk factors). For example, gastrointestinal illness or flu-like symptoms or encephalitis can be caused by many different disease agents, so diagnostic tests are needed to confirm cases have a common cause. Or a new cluster of cases may seem like an outbreak, but it could instead be the result of a new technician in a diagnostic lab who is learning protocols and interpreting test results slightly differently, highlighting the need for standardization.

**Case finding:** Then, investigators need to identify as-yet-unknown cases, both to treat those cases and to prevent further pathogen spread. This might involve public health announcements or **contact tracing**—the process of locating and notifying contacts that they have been exposed to a disease. As a part of contact finding, one must **develop a case definition**—define who (or what) is a case, and almost as importantly, define who (or what) is *not* a case. A case definition can use criteria based on disease symptoms, disease agents and/or strains, likely sources of exposure, etc. At the outset of the outbreak the case definition may be vague—there may be uncertainty regarding the criteria, especially if this is a new pathogen, or it may be beneficial to be less specific in the early case definition to maximize chances of finding and identifying all potential cases. The definition can evolve and narrow during the outbreak as information accumulates, making it easier to rule out cases with similar symptoms but different causes.

Depending on the tests available, case definitions can be categorized by the strength of evidence behind them (suspected case, probable case, confirmed case):

**Suspected case**—a clinically compatible case with epidemiologic linkage without presumptive or confirmatory laboratory results. That is, the signs (e.g. fever) or symptoms (e.g. fatigue) can be described and match the signs and symptoms for the case definition *and* there was a relevant exposure event (e.g. ate chicken liver pâté at the Christmas party), but there was no laboratory testing.

**Probable case**—a clinically compatible case with presumptive laboratory results. 'Presumptive' test results often have potential confounding interpretations. For example, antibody tests might indicate recent exposure or exposure a long time ago, or the test may have cross-reactivity with other similar pathogens (e.g. infected with a bacterium in the genus *Borrelia*, but no confirmation of the species).

**Confirmed case**—a clinically compatible case with confirmatory laboratory results. 'Confirmatory' lab results have a high level of certainty, which usually arises because the disease agent can be cultured, sequenced, or observed during pathological analysis, such as by taking a tissue sample and seeing distinguishing features of the pathogen under a microscope. A confirmatory test might also reflect investigators' abilities to describe host immune responses over timelines germane to the infectious organism (e.g. high levels of antibody that reflect recent infection, or measures of antibodies before and after exposure).

As an outbreak progresses, symptoms and patterns of exposure may be deemed sufficient for describing case numbers and patterns. When this happens, efforts to perform diagnostic tests may decline, and suspected cases may become the primary case definition used by investigators.

**Cause finding** is the attempt to understand what caused the outbreak. Was it a contaminated water source? Undercooked liver pâté? Birds stealing early morning milk? Investigators need to establish preliminary hypotheses quickly, and then rapidly test them. Can existing data be used to reject or support hypotheses? Does an epidemic curve provide clues? How does that epidemic curve appear once new cases are identified and data accumulate?

Importantly, investigators need to maintain an open mind so that they don't become too blinkered as to the root cause of an outbreak. In wise words: 'a modicum of skepticism should be retained because the obvious answer is not invariably correct' (Reingold 1998). Using prior knowledge but investigating

local context of behaviours and customs (especially in circumstances different from your own background) are all ways to finesse your hypothesizing. Open-ended interviews or survey-questionnaires of infected cases can be used to identify all possibly relevant exposures (e.g. a list of all foods consumed, or places visited, or recent unusual behaviours) during a given period. Of course, these efforts will yield a massive quantity of data, and human brains are prone to see patterns even when none exist. Therefore, statistical analyses are a critical tool for evaluating whether there are important relationships among various exposure sources and cases (see Hill's epidemiologic criteria for causal association, Chapter 8). Ideally, **analyses** will be performed rapidly, updating the picture with incoming data so that working hypotheses can be continuously rejected or refined.

Of course, data will not always be perfect. Errors may have occurred during data entry; a case-patient's recall of events may be muddled or they may have simply been unaware of potential exposures, e.g. people exposed to nymphal tick-bites may remain oblivious to their vector encounter; or investigators may have asked the wrong questions, failed to ask the right ones, or made judgemental statements that caused interviewees to clam up or answer falsely. Many of these errors are caused by insufficient understanding of human behaviour and culture, emphasizing that collaborations between social scientists and medical professionals or scientists are critical for 21st century disease control.

In an ideal scenario, the epidemiologic investigation would guide investigators to the outbreak's source, where they would then be able to obtain specimens and demonstrate the 'smoking gun', e.g. a bag of spinach remaining in the fridge and contaminated with the same *Escherichia coli* strain observed in a national outbreak. However, this is not always possible as samples or specimens might no longer be available, or conditions have changed so that the force of infection is now reduced, e.g. the remaining pâté does not test positive for *Campylobacter* or the original source of the SARS-CoV-2 spillover cannot be pinpointed. Though it is rare that evidence and available samples can reliably demonstrate the outbreak's narrative, those rare outbreak investigations that can achieve this are important opportunities to learn more about general epidemiology.

**Control interventions** can occur rapidly, even while an outbreak investigation is just beginning and information is limited. For example, it may be better to presume that antibiotics will prevent full-blown infections and administer them whilst waiting for diagnostics to culture bacteria for 48 hours. Of course, there will always be a trade-off between employing interventions (e.g. food product recalls, cancelling large events where people will gather, deploying vaccines) and the costs associated (evolution of antibiotic resistance, negative side effects of some medical treatments, economic woes, inconvenience, and disappointment) and this can be a difficult line to walk. One especially rapid intervention can be to **communicate findings** throughout the course of an outbreak investigation, sharing updates with the public, the media, colleagues, etc. This can empower people to avoid exposures and reduce further cases (e.g. by getting vaccinated or wearing masks). The downsides are that media attention can promote inaccurate information and inconsistent messaging can create public confusion. We will revisit control methods in Chapter 14, including for wild and domestic animal species.

Surveillance should not end just because a source has been identified or an intervention has been applied; by definition, surveillance is systematic collection and analysis of health data. Continued surveillance allows investigators to evaluate efficacy and be sure that nothing is unexpectedly changing. Have all sources of exposure truly been eliminated? Is the number of new cases declining? Sometimes surveillance must continue for decades, as seen in examples of ongoing eradication efforts for polio virus and Guinea worm.

## 4.6 An outbreak case study

*Arizona, 2007*—On the night of 29 October 2007, a previously healthy, 37-year-old National Park Service biologist became acutely ill, suffering fever, chills, nausea, myalgias, and cough with blood-tinged sputum. He attended a clinic the following morning, where he was examined (high fever, no noticeably swollen lymph nodes, clear-sounding lungs, low oxygen levels) and a rapid influenza

test was negative. The biologist was diagnosed with 'viral syndrome' and given instructions to return for further examination if symptoms worsened. He turned up for work on the following morning, but then went home and failed to show up to work again. Sadly, on 2 November, the wildlife biologist was found deceased at his home in Grand Canyon National Park (Wong et al. 2009).

If you were investigating this case, what would you do to follow up? What type of case definition are you working with? Is it appropriate to do contact tracing? What methods could you use to acquire more data, given how little you currently know?

Investigators interviewed friends and colleagues to investigate the circumstances preceding the biologist's death and were able to access the biologist's camera, cell phone and computer. Interviews revealed that the biologist's work duties included trapping and collaring mountain lions (*Puma concolor*, also known as puma or cougar) and removing rodents from buildings. He was regarded as 'a highly skilled and widely admired researcher, who was a passionate champion for wild cats and their habitats worldwide' (Wong et al. 2009). The biologist's exposures with wildlife were not documented in the patient history from his visit to the clinic—presumably no one had asked, despite his wildlife-oriented occupation.

Perhaps the biologist had been exposed to hantavirus because of his work with rodents? A postmortem examination discovered large amounts of haemorrhagic and frothy fluid in both lungs (right lung, 1.4 kg; left lung, 1.05 kg), which supports the hypothesis that hantavirus was the responsible disease agent. Would this new information change your case definition or the direction of your outbreak investigation?

In subsequent analyses, lung, liver, and spleen tissues were determined PCR-positive for *Yersinia pestis*, and culture of lung and liver samples yielded viable bacteria. This is a **confirmed case** of plague, a disease for which there is mandatory reporting in the United States. Human cases of plague are rare in the US, and its appearance always merits an outbreak investigation because it is deadly, contagious and a recognized bioterror agent. So how was the biologist exposed to plague? Were the rodents still to blame?

Interviews with colleagues revealed that the biologist had recently retrieved the carcass of a radio-collared mountain lion and conducted a post-mortem examination. This mountain lion had been radio-tracked by the biologist for 6 months until a 'mortality signal' (no movement after 6 hours) was transmitted from the mountain lion's collar on 25 October at 3:13 a.m. within Grand Canyon National Park. Photographs taken by the biologist showed that he located the mountain lion carcass on 26 October at ~2:00 p.m. The carcass was in excellent condition with no signs of scavenging, but blood was present in the animal's nostrils and in the soil under its mouth and nose (Fig. 4.11). The biologist proceeded to carry the carcass ~1 km to his vehicle and then home to his garage to perform a necropsy to explain the cause of death. He worked alone and did not use personal protective equipment. The biologist concluded that the animal had been attacked by another mountain lion and had died from thoracic haemorrhage, because the lion's thoracic cavity was filled with blood.

However, when further examined by the outbreak investigators, samples from the mountain lion's liver and submandibular lymph node also tested positive for *Yersinia pestis*. Isolates from the mountain lion and the biologist were indistinguishable. The biologist had developed fever and

haemoptysis 3 days after he was exposed to the mountain lion and died 6 days after the presumed exposure. Exposure seemed to have occurred through inhaled aerosols, either generated while carrying the mountain lion or during the necropsy in the garage when using power tools to dissect the carcass, because the man's lungs appeared to be the principal site of infection. This was a case of 'primary pneumonic plague' (Wong et al. 2009).

Three main forms of plague occur in humans—bubonic, septicaemic and pneumonic. Primary pneumonic plague, acquired by direct inhalation of *Yersinia pestis*, is the rarest form, with only 13 cases in the US from 1925–2006. Those cases had likely been caused by face-to-face contact with infected pets (five cats and one dog as the suspected source) or exposure in laboratory settings (Wong et al. (2009) report three such cases and cite 'Centers for Disease Control and Prevention (CDC), unpublished data'; we, the book authors, regard that as unfairly tantalizing and short on details!). Pneumonic plague is characterized by high mortality and, frighteningly, has the potential for further person-to-person spread.

Persons in close contact (within 2 metres) to the biologist after he had developed symptoms were identified through contact tracing and offered chemoprophylaxis (preventative antibiotics to

(a)

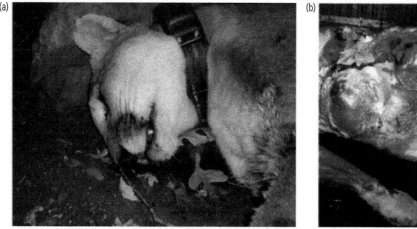

(b)

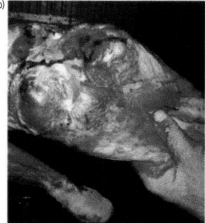

**Figure 4.11** (A) Picture by the National Park biologist of the mountain lion carcass in situ in the field with blood in the animal's nostrils and the soil beneath the head. Taken on 26 October 2007; 1:56 p.m. (B) The skinned head and forelimb of the mountain lion during necropsy and the biologist's unprotected left hand (26 October 2007; 7:40 p.m.).
From Wong et al. (2007).

combat potential infection): 49 contacts opted for precautionary medicines, including 11 emergency responders, 10 work colleagues, 10 clinic patients, eight clinic personnel, and three residential/social contacts. No one developed signs of plague.

This unfortunate case emphasizes the importance of education regarding zoonotic diseases and appropriate use of personal protective equipment (Wong et al. 2009). This is especially important for people who come into close contact with wildlife, such as biologists, hunters, and outdoor enthusiasts (remember the Yosemite National Park campers who acquired hantavirus infections, or cavers who acquired histoplasma infections?). At the most basic level, people contacting wildlife should wear gloves, avoid contaminating their clothing and thus spreading potential pathogens (e.g. by wearing lab coats or decontaminating field gear), and wash their hands frequently. Protections should be even more stringent when working with known reservoir hosts and performing high risk activities like necropsies, which is why laboratory safety is so carefully scrutinized for work with human pathogens, zoonotic or otherwise.

This case also reminds us that a single pathogen may have multiple modes of transmission (Chapter 2), including ones that we have not recognized yet. About three weeks after the mountain lion's death, investigators used global positioning satellite coordinates and the biologist's photographs to identify the exact location of the blood-contaminated soil that could be seen under the animal's mouth and nose. The soil could not be directly tested with *Yersinia pestis* culture diagnostics (too many contaminants in the soil that would inhibit plague bacteria growth), so investigators suspended the soil in sterile physiologic saline and injected it subcutaneously into laboratory mice. Within 12 hours of inoculation, one mouse became moribund, and liver and spleen samples were cultured to confirm presence of *Yersinia pestis* and show that it was the same strain that infected the mountain lion and biologist. This showed that *Yersinia pestis* can survive for at least 24 days in soil under natural conditions. The mechanism of persistence is unknown, but the site's microhabitat was sheltered from UV light and temperatures were cool in late October, perhaps protecting the bacteria (Eisen et al. 2008). Further laboratory experiments found that when laboratory mice with skin damage burrowed through soil contaminated with highly bacteraemic blood, it was possible (but rare) for the mice to acquire infections through soil-borne transmission (Boegler et al. 2012). Plague bacteria may also be able to survive and even replicate inside some soil amoebas (Markman et al. 2018). Though very rare, environmental persistence of *Yersinia pestis* might be sufficient to kick-start new chains of transmission after some period of latency in the soil. This facet of *Yersinia pestis* would not have been discovered or recognized without a thorough outbreak investigation.

## 4.7 Summary

Surveillance—the systematic collection and analysis of health data—appears to be a simple concept, yet the difficulties in obtaining, assimilating, and analysing surveillance data are legion. There are always manifold ways to portray health issues— each with strengths and weaknesses (e.g. passive vs. active surveillance). But simultaneously garnering data at multiple scales, multiple sites, and across the transmission mechanisms—in hosts, vectors, multiple pathogens—provides a more holistic understanding of emerging infectious diseases. Certainly, without strategic, forward-thinking surveillance we are condemned to reactive disease control in the event of novel pathogen spillovers.

Armed with this knowledge of how to describe and investigate outbreak data, in Chapter 5 we will describe how mathematical models can help *predict* and understand disease dynamics based on what we know about pathogens, hosts and their environments.

## 4.8 References

Benach, J.L., Bosler, E.M., Hanrahan, J.P., et al. (1983). Spirochetes isolated from the blood of two patients with Lyme disease. New England Journal of Medicine, 308, 740–42.

Bernstein, A.S., Ando, A.W., Loch-Temzelides, T., et al. (2022). The costs and benefits of primary prevention of zoonotic pandemics. Science Advances, 8, eabl4183.

Bickel, P.J., Hammel, E.A., O'Connell, J.W. (1975). Sex bias in graduate admissions: data from Berkeley. Science, 187, 398–404.

Boegler, K.A., Graham, C.B., Montenieri, J.A., et al. (2012). Evaluation of the infectiousness to mice of soil contaminated with *Yersinia pestis*-infected blood. Vector-Borne and Zoonotic Diseases, 12, 948–52.

Bosler, E.M., Coleman, J.L., Benach, J.L., et al. (1983). Natural distribution of the *Ixodes dammini* spirochete. Science, 220, 321–2.

Burgdorfer, W.A., Barbour, A.G., Hayes, S.F., et al. (1982). Lyme disease – a tick-borne spirochetosis. Science, 216, 1317–19.

Dobson, A.P., Pimm, S.L., Hannah, L., et al. (2020). Ecology and economics for pandemic prevention. Science, 369, 379–81.

Duncan, K.T., Saleh, M.N., Sundstrom, K.D., Little, S.E. (2021). *Dermacentor variabilis* is the predominant *Dermacentor* spp. (Acari: Ixodidae) feeding on dogs and cats throughout the United States. Journal of Medical Entomology, 58, 1241–7.

Eisen, L., Eisen, R.J. (2021). Benefits and drawbacks of citizen science to complement traditional data gathering approaches for medically important hard ticks (Acari: Ixodidae) in the United States. Journal of Medical Entomology, 58, 1–9.

Eisen, L., Meyer, A.M., Eisen, R.J. (2007). Climate-based model predicting acarological risk of encountering the human-biting adult life stage of *Dermacentor andersoni* (Acari: Ixodidae) in a key habitat type in Colorado. Journal of Medical Entomology, 44, 694–704.

Eisen, R.J., Petersen, J.M., Higgins, C.L., et al. (2008). Persistence of *Yersinia pestis* in soil under natural conditions. Emerging Infectious Diseases, 14, 941–3.

Fernandez, M.P., Bron, G.M., Kache, P.A., et al. (2019). Usability and feasibility of a smartphone app to assess human behavioral factors associated with tick exposure (The Tick App): quantitative and qualitative study. JMIR Mhealth Uhealth, 7, e14769.

Freeman, E.A., Salkeld, D.J. (2022). Surveillance of Rocky Mountain wood ticks (*Dermacentor andersoni*) and American dog ticks (*Dermacentor variabilis*) ticks in Colorado. Ticks and Tick-borne Diseases, 13, 102036.

Gardy, J., Loman, N. (2018). Towards a genomics-informed, real-time, global pathogen surveillance system. Nature Review Genetics, 19, 9–20.

Gordis, L. (2009). Epidemiology. 4th edn. Saunders Elsevier.

James, A.M., Freier, J.E., Keirans, J.E., et al. (2006). Distribution, seasonality, and hosts of the Rocky Mountain wood tick in the United States. Journal of Medical Entomology, 43, 17–24.

Karesh, W.B., Dobson, A., Lloyd-Smith, J.O., et al. (2012). Ecology of zoonoses: natural and unnatural histories. The Lancet, 380, 1936–45.

Keusch, G.T., Amuasi, J.H., Anderson, D.E., et al. (2022). Pandemic origins and a One Health approach to preparedness and prevention: solutions based on SARS-CoV-2 and other RNA viruses. Proceedings of the National Academy of Sciences, USA, 119, e2202871119.

Kugeler, K.J., Schwartz, A.M., Delorey, M.J., et al. (2021). Estimating the frequency of Lyme disease diagnoses, United States, 2010–2018. Emerging Infectious Diseases, 27, 616–19.

Kullenberg, C., Kasperowski, D. (2016). What is citizen science? – A scientometric meta-analysis. PLoS One, 11, e0147152.

Lehane, A., Parise, C., Evans, C., et al. (2020). Reported county-level distribution of the American dog tick (Acari: Ixodidae) in the contiguous United States. Journal of Medical Entomology, 57, 131–55.

Leroy, E.M., Kumulungui, B., Pourrut, X., et al. (2005). Fruit bats as reservoirs of Ebola virus. Nature, 438, 575–6.

Lyons, L.A., Brand, M.E., Gronemeyer, P., et al. (2021). Comparing contributions of passive and active tick collection methods to determine establishment of ticks of public health concern within Illinois. Journal of Medical Entomology, 58, 1849–64.

Mancy, R., Rajeev, M., Lugelo, A., et al. (2022). Rabies shows how scale of transmission can enable acute infections to persist at low prevalence. Science, 376, 6592.

Markman, D.W., Antolin, M.F., Bowen, R.A., et al. (2018). *Yersinia pestis* survival and replication in potential ameba reservoir. Emerging Infectious Diseases, 24, 294–302.

Messerli, F.H. (2012). Chocolate consumption, cognitive function, and Nobel laureates. New England Journal of Medicine, 367, 1562–4.

Nadelman, R.B., Wormser, G. (2005). Poly-ticks: blue state versus red state for Lyme disease. Lancet, 356, 280.

Nelson, C.A., Saha, S., Kugeler, K.J., et al. (2015). Incidence of clinician-diagnosed Lyme disease, United States, 2005–2010. Emerging Infectious Diseases, 21, 1625–31.

Nieto, N.C., Porter, W.T., Wachara, J.C., et al. (2018). Using citizen science to describe the prevalence and distribution of tick bite and exposure to tick-borne diseases in the United States. PLoS One, 13, e0199644.

Olival, K., Hayman, D. (2014) Filoviruses in bats: current knowledge and future directions. Viruses, 6, 1759–88.

Ostfeld, R.S. (2011). Lyme Disease: The Ecology of a Complex System. Oxford University Press.

Pieracci, E.G., De La Rosa, J.D.P., Rubio, D.L. (2019). Seroprevalence of spotted fever group rickettsiae in canines along the United States-Mexico border. Zoonoses and Public Health, 66, 918–26.

Pollet, T.V., Stulp, G., Henzi, S.P., Barrett, L. (2015). Taking the aggravation out of data aggregation: a conceptual guide to dealing with statistical issues related to the pooling of individual-level observational data. American Journal of Primatology, 77, 727–40.

Pollet, T.V., Tybur, J.M., Frankenhuis, W.E., Rickard, I.J. (2014). What can cross-cultural correlations teach us about human nature? Human Nature, 25, 410–29.

Porter, W.T., Barrand, Z.A., Wachara, J., et al. (2021b). Predicting the current and future distribution of the western black-legged tick, *Ixodes pacificus*, across the Western US using citizen science collections. PLoS One, 16, e0244754.

Porter, W.T., Motyka, P.J., Wachara, J., et al. (2019). Citizen science informs human-tick exposure in the Northeastern United States. International Journal of Health Geographics, 18, 1–14.

Porter, W.T., Wachara, J., Barrand, Z.A., et al. (2021a). Citizen science provides an efficient method for broad-scale tick-borne pathogen surveillance of *Ixodes pacificus* and *Ixodes scapularis* across the United States. Msphere, 6, e00682–21.

Pourrut, X., Kumulungui, B., Wittmann, T., et al. (2005). The natural history of Ebola virus in Africa. Microbes and Infection, 7, 1005–14.

Reingold, A.L. (1998). Outbreak investigations—a perspective. Emerging Infectious Diseases, 4, 21–27.

Rosenberg, R., Lindsey, N.P., Fischer, M., et al. (2018). Vital Signs: Trends in reported vectorborne disease cases—United States and Territories, 2004–2016. Morbidity and Mortality Weekly Report (MMWR), 67, 496–501.

Saleh, M.N., Sundstrom, K.D., Duncan, K.T., et al. (2019). Show us your ticks: a survey of ticks infesting dogs and cats across the USA. Parasites & Vectors, 12, 595.

Salkeld, D.J., Antolin, M.F. (2020) Ecological fallacy and aggregated data: a case study of fried chicken restaurants, obesity and Lyme disease. EcoHealth, 17, 4–12.

Seifert, S.N., Fischer, R.J., Kuisma, E., et al. (2022). Zaire ebolavirus surveillance near the Bikoro region of the Democratic Republic of the Congo during the 2018 outbreak reveals presence of seropositive bats. PLoS Neglected Tropical Diseases, 16, e0010504.

Shortridge, B.G. (2005). Apple stack cake for dessert: Appalachian regional foods. Journal of Geography, 104, 65–73.

Soucy, J.-P.R., Slatculescu, A.M., Nyiraneza, C., et al. (2018). High-resolution ecological niche modeling of *Ixodes scapularis* ticks based on passive surveillance data at the northern frontier of Lyme disease emergence in North America. Vector Borne and Zoonotic Diseases, 18, 235–42.

Stapp, P., Salkeld, D.J., Eisen, R.J., et al. (2008). Exposure of small rodents to plague during epizootics in black-tailed prairie dogs. Journal of Wildlife Diseases, 44, 724–30.

Torgerson, P.R., Macpherson, C.N.L. (2011). The socioeconomic burden of parasitic zoonoses: global trends. Veterinary Parasitology, 182, 79–95.

Tran, T., Porter, W.T., Salkeld, D.J., et al. (2021). Estimating disease vector population size from citizen science data. Journal of the Royal Society Interface, 18, 20210610.

Turney, S., Gonzalez, A., Millien, V. (2014). The negative relationship between mammal host diversity and Lyme disease incidence strengthens through time. Ecology, 95, 3244–50.

Webb, P., Bain, C. (2011). Essential Epidemiology: An Introduction for Students and Health Professionals. 2nd ed. Cambridge University Press.

Wong, D., Wild, M.A., Walburger, M.A., et al. (2009). Primary pneumonic plague contracted from a mountain lion carcass. Clinical Infectious Diseases, 49, e33–8.

## CHAPTER 5

# Making simple predictions using models

## 5.1 Introduction

*Faridpur District, Bangladesh*—In February and March 2004, several people in western Bangladesh suffered from fever and altered mental status, which can be symptoms of Nipah virus infection. Nipah virus occasionally infects people, but spillover from the virus's hosts, bats, is rare enough that people have no pre-existing immunity to the virus and there is no Nipah virus vaccine. Therefore, the virus was now invading a completely susceptible human population in the Faridpur District. In the following weeks, a total of 36 people developed similar symptoms, 23 people had laboratory confirmed Nipah infections, and most patients died from their infections (Gurley et al. 2007). Contact tracing interviews revealed that most infected people had visited a single infected religious leader (Patient F; Fig. 5.1), suggesting that the virus was spreading from person to person. Given established person-to-person transmission, there seemed to be a risk that a large epidemic and even more human suffering and mortality would follow. However, after April 2004, the outbreak subsided just as quickly as it had begun. How could public health officials have predicted that this outbreak would fizzle out rather than turn into an epidemic?

In this case and many others, public health practitioners, governments, wildlife managers, and the public have pressing questions about how disease dynamics will unfold in the future. Will a pathogen spread to every country, turning into a pandemic? How many wildebeest will become infected? Will a pathogen fade out and disappear from an endangered carnivore population on its own? How can we slow the rate of spread during an epidemic to avoid overwhelming hospitals and other health infrastructure while we develop a vaccine? Unfortunately, infectious disease experts cannot know exactly what the future holds; they don't have crystal balls or superpowers. Instead, they must use available information to make informed guesses (also called 'predictions').

One common method for making predictions is to use mathematics to describe how parts of the disease system are related, creating a simplification of the real world called a **mathematical model**. Models allow researchers to test ideas and control strategies '*in silico*'—a fancy pseudo-Latin term for 'in silicon', or on a computer—before trying them in real systems, where the consequences of bad decisions could be catastrophic. In this chapter, we will overview the simplest epidemiological models, including the assumptions that we make when we use them and the ways that they can be used to improve our understanding and prediction of disease dynamics.

## 5.2 Mathematical models are simplifications of disease systems

A good model does not attempt to reproduce every detail of the biological system; the system itself suffices for that purpose as the most detailed model of itself.
—Simon Levin, 1989 (Levin 1992)

Building *useful* simplifications of complex systems is difficult because one must make many decisions about what is important, often when information

*Emerging Zoonotic and Wildlife Pathogens.* Dan Salkeld, Skylar Hopkins, and David Hayman, Oxford University Press.
© Oxford University Press (2023). DOI: 10.1093/oso/9780198825920.003.0005

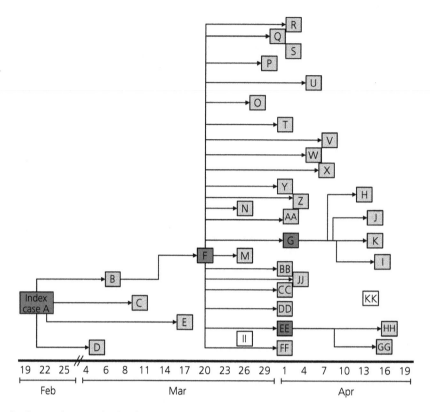

**Figure 5.1** Timeline for a Nipah virus outbreak with person-to-person transmission in Bangladesh in 2004 (figured modified from Gurley et al. 2007). The boxes show infected individuals, and the lines show inferences about who infected whom based on contact tracing interviews. The colours indicate the number of people that each infectious person infected, where the yellow squares indicate an infected person who infected a single other person and the red squares indicate a person who infected more than one other person. It remains unclear who infected individuals II and KK.
Created by Hanna D. Kiryluk in BioRender.com.

is limited. For example, think of a common disease system (e.g. COVID-19, chickenpox): is host death caused by the pathogen so rare that we can assume it never happens, or must we include disease-induced mortality in our model for accuracy? We might need to try the model both ways to see how different the predictions will be before we can decide. Trying it both ways doesn't sound too bad, but if you need to make 20 decisions, trying all possible combinations quickly gets out of hand!

All mathematical models are simplifications of reality that contain simplifying assumptions. Sometimes assumptions made in a model do not reflect reality, and this might be OK if the model assumptions do not greatly affect model predictions. Other times, the assumptions make the difference between accurate and extremely misleading predictions. So, if you take nothing else away from this chapter, let it be this: **when you build a mathematical model, you should be explicit about the assumptions it contains, and you should be very wary of models built by other people if they do not make their inherent assumptions clear.**

Enough disclaimers and warnings! Let's move on to the nuts and bolts. To build or evaluate useful models, we must first decide what 'useful' even means, because the purpose of the model will determine how simple the model should be (Barlow 1996). There are many possible goals, and thus many types of models (Box 5.1).

One common modelling goal is to predict **quantitative** outcomes for a particular disease system,

**Table 5.1** Examples of qualitative and quantitative predictions.

| Qualitative predictions | Quantitative predictions |
|---|---|
| The norovirus outbreak will end without a public health intervention. | The norovirus outbreak will last 4 months without public health intervention. |
| More people will die from COVID-19 in cities than in rural areas. | More than 3 million people will die from the COVID-19 pandemic in cities versus 2 million in rural areas by the end of 2021. |
| Smallpox transmission can be reduced by widespread vaccination. | ~80% of the human population needs to be vaccinated to cause smallpox to be locally extirpated and globally eradicated. |

like the magnitude or duration of an epidemic (Table 5.1). To accurately make quantitative predictions, models usually need to be relatively complicated, containing many system-specific details about the hosts and pathogens that are being modelled. Another common modelling goal is to predict **qualitative** behaviour about disease systems in general, such as whether a pathogen will generally become more or less prevalent over time, without accurately knowing the magnitude or timing associated with the outcome.

---

### Box 5.1 Model goals and types

In this book, we will cover several types of disease models, but we can't cover them all. So we wanted to give you a taste of the variety that exists, in this box and Box 5.2. In this chapter, we mostly focus on **mathematical models**, which describe the theoretical relationships among parts of a disease system. These are usually **dynamic models**, which explain or predict how interrelated variables change over time. When mathematical models are married to probability distributions and data, they become **statistical models**, which can be used to make inferences about samples from populations. For example, if you sample 100 individuals in a population and calculate the prevalence of *Toxoplasma gondii*, you can make inferences about the prevalence in the sampled population.

As another example, simple linear regression is a statistical model that combines a mathematical model—the mathematical equation for a line ($y = mx + b$)—with the normal probability distribution and data to estimate the slope ($m$) and intercept ($b$) of the line (the relationship between the x and y variables). For example, how much does snake mass increase with snake length? If

we just want to 'describe' this relationship (i.e. estimate how big the slope and intercept are or how strongly the mass and length variables are correlated), we can use simple linear regression as a purely **descriptive model**. Alternatively, we could use the observed relationship to predict future outcomes or observations (e.g. assuming that this relationship would be relevant in future scenarios, what will the mass of a random snake be if its length is 20 cm), in which case it would be a **predictive model**. We would use different statistical methods for describing versus predicting.

Models also occur on a continuum from highly mechanistic to highly phenomenological. **Mechanistic models** (also called 'process models') are built using parameters that have a specific biological interpretation. For example, epidemiological models often contain an infectious host recovery rate, which is the inverse of the duration of the infectious period and can be observationally or experimentally measured with real infectious hosts. In contrast, **phenomenological models** (also called 'pattern models') describe relationships with parameters that might not be measurable; for example, in many regression analyses, the slope estimate can tell us something about the strength of the relationship between two variables, but the slope does not have a specific biological meaning. In fact, the two variables might not even be causally related (because correlation $\neq$ causation).

Finally, note that for this chapter, we look at **population-level models**, which describe changes in populations over time using parameters that represent population averages (e.g. the average contact rate). Alternatively, we could use **individual-level models**, which describe the behaviour of individual hosts and can be aggregated to understand dynamics in populations. Individual-level models (also called agent-based models) become more useful as variation among individuals or environments becomes more important.

## 5.3 Basic compartmental models

One (or more) infected person is introduced into a community of individuals, more or less susceptible to the disease in question. The disease spreads from the affected to the unaffected by contact infection. Each infected person runs through the course of his sickness, and finally is removed from the number of those who are sick, by recovery or by death.

—Kermack and McKendrick (1927)

Pick your favourite animal species—pond snails, rough green snakes, tri-coloured bats, humans—and take a minute to imagine all the complexity in one population. The individuals in that population might vary in age or life stage, sex, body condition, movement speed, foraging behaviour, breeding status, or other factors, and all those characteristics could vary across space and time. Some models consider variation among individuals, which we'll come back to in Chapter 14. But for this chapter, we will ignore all that complexity, because the first simplifying assumption in many epidemiological models is that all individuals are identical, *except* that they vary in their infection status.

In particular, many epidemiological models divide the otherwise homogeneous host population into three groups, or 'compartments', by infection status: currently uninfected and susceptible (**S**), currently infected and infectious (**I**), or immune/resistant to infection due to previous infection or vaccination (**R**; sometimes called 'removed' or dead). **SIR compartmental models** are often illustrated using 'stock and flow' diagrams (Fig. 5.2) that show how the three groups of individuals are related. Individuals can flow from one state to another at a rate represented by the size of the connecting pipe. For example, in Fig. 5.2, the larger pipe size for the transmission rate (β) represents a faster transmission rate, where susceptible individuals are more rapidly converted to infected individuals. These diagrams are one way to visualize the relationships in real systems and the maths that we use to describe them.

Despite being drawn in separate compartments, simple epidemiological models usually assume that all hosts mix randomly and homogeneously, with interaction rates between hosts being proportional to the concentrations of each host type (i.e. the **law of mass action**, Hamer 1906, Box 5.2). In other words, each susceptible host is equally likely to make an infectious contact with an infectious host and become infected. That's like assuming that everyone at the New Year's Eve event was equally likely to consume the appetizer with the

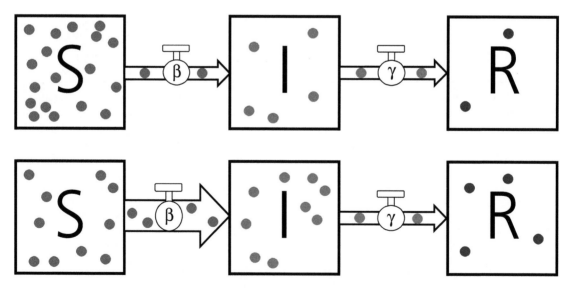

**Figure 5.2** Two SIR compartmental models, where the squares are 'compartments' containing Susceptible, Infected and Infectious, and Recovered and Resistant individuals. Individuals are represented by circles, and the size of the pipes between compartments indicates the rates that individuals are converted from Susceptible to Infected (β) and from Infected to Resistant (γ). The bottom diagram has a higher transmission rate (β), leading to more individuals in the Infected class.

contaminated raw oysters or shake hands with infected Uncle Harold. Just like random mingling and food consumption at a party is unlikely, the law of mass action assumption probably is not realistic, but it might be realistic *enough* to be useful for some modelling applications.

**Box 5.2  Are hosts like molecules?**

You might be noticing that some of the assumptions in mathematical models of disease seem odd or have unusual names, like the 'law of mass action' and 'homogeneous mixing'. These ideas actually originated in physics and chemistry. Models describing the movement and interactions of molecules were already developed and easy to work with. Ecologists borrowed them to build simplified models that could improve understanding of dynamics between predators and prey, mosquitoes and human malaria, and many other ecological systems. In fact, all consumer–resource models can be built from a

single unifying mathematical model depending on which assumptions are made, as shown by Lafferty et al. (2015), in a paper entitled, 'A general consumer–resource population model'. The caveat of adapting models for molecules to be relevant to animals (or other organisms) is that animals are more complicated and difficult to predict, and thus sometimes those complexities greatly affect disease dynamics in ways that simple models cannot predict (e.g. the badger example in Chapter 14).

Though SIR models are the best-known compartmental models, one can use more or fewer than three compartments to realistically model some disease systems. For example, for diseases where susceptible individuals become infected and infectious and then remain infectious their entire lives, such as with people with untreated HIV, there is no Resistant/Recovered class. This simplified model is called an **SI model** (Fig. 5.3). If infectious

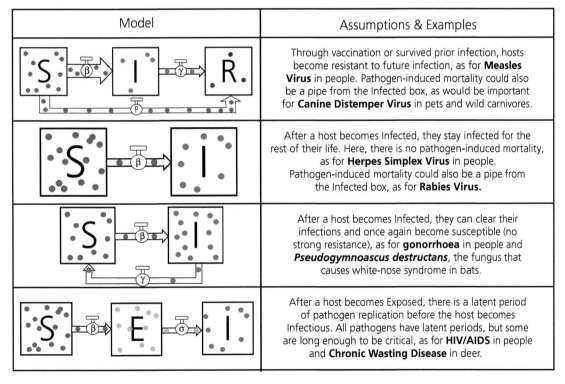

**Figure 5.3** Examples of compartmental models with different complexity from the classical SIR model. Note that depending on the goals of the model, some of the example diseases might best be modelled using a different structure; for example, rabies is used as an SI example here, but it has a long **latent period** and is often represented by an SEI model instead. The parameters controlling the flow of hosts from one compartment to another are the transmission rate (Susceptible to Exposed/Infected; $\beta$), the recovery rate (Infected to Resistant; $\gamma$), the rate that hosts become infectious or the inverse of the latent period (Exposed to Infectious; $\sigma$), and vaccination rate (Susceptible to Resistant; P).

**Table 5.2** Three equations for a simple SIR model that assumes that there are no births, deaths, immigration, or emigration from the constant host population. The inflows into compartments are purple and the outflows are red, and each type of gain/loss is shown in words and the corresponding mathematical symbols.

| Rate of Change | = + | demographic additions | + | transmission gains | − | transmission losses | − | demographic losses | Final Equation |
|---|---|---|---|---|---|---|---|---|---|
| change in Susceptible hosts over time $\dfrac{dS}{dt}$ | = + | no births or immigration $+ 0 + 0$ | + | no transmission or recovery gains $+ 0$ | − | Susceptible to Infectious $- \beta S \dfrac{I}{N}$ | − | no deaths or emigration $- 0 - 0$ | $\dfrac{dS}{dt} = - \beta S \dfrac{I}{N}$ |
| change in Infected hosts over time $\dfrac{dI}{dt}$ | = + | no births or immigration $+ 0 + 0$ | + | Susceptible to Infectious $+ \beta S \dfrac{I}{N}$ | − | Infectious to Resistance $- \gamma I$ | − | no deaths or emigration $- 0 - 0$ | $\dfrac{dI}{dt} = \beta S \dfrac{I}{N} - \gamma I$ |
| change in Resistant hosts over time $\dfrac{dR}{dt}$ | = + | no births or immigration $+ 0 + 0$ | + | Infectious to Resistance $+ \gamma I$ | − | no losses $- 0$ | − | no deaths or emigration $- 0 - 0$ | $\dfrac{dR}{dt} = \gamma I$ |

individuals can recover but do not have immunity to future infection, as with gonorrhoea in humans, we draw an extra flow pipe from I back to S to create an **SIS model** (Fig. 5.3). And for pathogens where **Exposed** susceptible hosts require a substantial incubation or latent period before becoming infectious, we can create an **SEIR model** (Fig. 5.3). In short, compartmental models can contain as many compartments and flows as necessary to adequately describe a disease system, including compartments for free-living pathogen stages, vectors, or other host species.

After we are satisfied with the compartmental model that we have designed, we can use mathematics (usually differential equations) to write our model. Each compartment will be a **state variable** in the model—a variable that describes the current state of the system and can change over time. Each inflow and outflow pipe will be represented by a **parameter** (e.g. $\beta$). Parameters are usually considered constant (i.e. they do not change over time), but they can be dependent on other conditions (e.g. the transmission rate might increase with density, Section 5.4). Parameters and state variables can be found on the right sides of the equations in Table 5.2. Together, they describe the rates of change for each state variable as individuals move into and out of the compartment.

For a basic SIR model, there are three state variables, so the model requires three equations describing the rates of change for each state variable (Table 5.2). These rates of change, shown on the left side of the equations, are written as dX/dt, and can be read as 'change in state variable X over change in time'. These are written out for each equation in Table 5.2.

These equations might not look like much, but they are powerful tools for understanding and predicting infectious disease dynamics. For example, if one picks reasonable values for the transmission rate ($\beta$) and the recovery rate ($\gamma$) and starting values for each state variable, one can use numerical simulation (Box 5.5) to predict how many Susceptible, Infected and Recovered individuals there will be in a population over time (Fig. 5.4). This time series is a prediction for what could happen if no interventions are applied to control transmission. One could similarly plot the time series for the system with an intervention to see how things would change (Fig. 5.4).

However, before we try to make important predictions with SIR models, we need to know more about the assumptions that they make and when it is reasonable to make those assumptions. For example, in the basic SIR model that we have explored so far, we assumed for simplicity that the

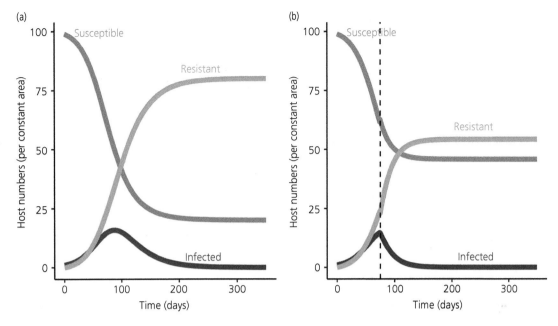

**Figure 5.4** On the left, there is a time series for an SIR model, created by numerically simulating one scenario where β = 0.001 (there's a 0.1% chance of successful transmission given a contact between infectious and susceptible hosts); γ = 0.05 (it generally takes 20 days to recover, so the recovery probability is 1/20); S(0) = 99 susceptible hosts (at time zero); I(0) = 1 infectious host (at time zero); R(0) = 0 recovered hosts (at time zero); and the area occupied by the population is constant, such that numbers and densities are equivalent. On the right, we see a simulation of the same scenario, except that starting at t = 75 days, 5% of the Infected population is treated each day (and moved to the Resistant population).

host population size (N = S + I + R) is constant over time (no births, deaths, or immigration/emigration). But as we'll see in the next section, simplifying assumptions about numbers, densities and areas can have large impacts on model predictions.

## 5.4 How does host density affect pathogen transmission?

*Tanzania, 2000s*—Natal multimammate mice (*Mastomys natalensis*) are native to sub-Saharan Africa and are perhaps best known for two traits. First, they are reservoirs for several human pathogens, such as Lassa virus, leptospirosis, and *Yersinia pestis*, the causative agent of bubonic plague. Second, the female mice have 18 teats, which allows them to raise large litters. Taken together, these traits distinguish multimammate mice as an abundant reservoir species, with population sizes that can vary dramatically across time and from place to place due to abiotic or biotic conditions (Borremans et al. 2017a). One could imagine that places with many multimammate mice are places where humans are more likely to contract zoonotic

diseases, whereas places with few multimammate mice have relatively low zoonotic disease risk. But how, exactly, does the abundance or density of multimammate mice affect disease transmission (Box 5.3, Fig. 5.5)?

In most cases, we expect that pathogen transmission increases as host density increases, because increased host density usually increases **disease-relevant contacts** among hosts (whatever kinds of contact can lead to transmission); this relationship is the **transmission function**, a mathematical assumption about the relationship between transmission and host density. Relationships between host density and transmission are probably already familiar to you, because these relationships were the targets of many interventions during the COVID-19 pandemic, such as restricting large gatherings and enforcing social distancing.

If we assume that disease-relevant contact rates among hosts and thus pathogen transmission rates increase *linearly* with host density, we call the resulting transmission function the **density-dependent transmission function** (Figs. 5.6, 5.7). When contact rates do not change at all with host density, the modelling assumption is called the **density-independent or frequency-dependent transmission function** (Fig. 5.7). Frequency-dependent transmission is often assumed to apply to vector-borne diseases and sexually transmitted diseases (Anderson and May 1991), especially if the sexually transmitted disease is chronic and has a relatively low transmission rate, like HIV. When things do not have frequency-dependent transmission, modellers usually assume that they have density-dependent transmission.

However, it is probably unlikely that contact rates increase linearly to infinity, as would be expected if hosts and parasites bounce around randomly in their environments like molecules (Box 5.2). For example, for multimammate mice, contact rates increase with mouse density at intermediate densities, but at high densities (>160 mice ha$^{-1}$), contact rates plateau, perhaps because mice are spending more time looking for scarce resources and less time

mating (Borremans et al. 2017b) (Fig. 5.6). (Contact rates are also relatively constant at the lowest multimammate mouse densities, but they never go all the way to zero, probably because mice increase efforts to find mates at low densities.) In situations like this, where contacts increase with density but not in a linear fashion, we can use one of many possible **non-linear transmission functions** (Antonovics et al. 1995, McCallum et al. 2001).

It's possible that most, if not all, relationships between contact rates and host density are non-linear, but if one were to zoom in on a small range of host densities, a linear relationship might be a good approximation (Hopkins et al. 2020, Lafferty et al. 2015). For example, maybe a frequency-dependent transmission function is relevant over a small range of very high host densities: the probability of contacting a sick person might be roughly equivalent when one is in New York City (2021 population density = 28,210 people mi$^{-2}$) versus Los Angeles

---

**Box 5.3  Host numbers or densities?**

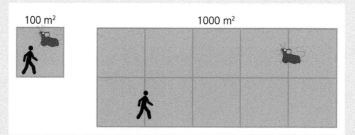

100 m$^2$     1000 m$^2$

**Figure 5.5** Two fields with the same numbers of humans and mice, but with different areas and thus different host densities.

Have you noticed that we keep using host density in our models, rather than host abundance or numbers? Why do you think that might be? It might be easiest to imagine this distinction from a parasite's point of view, so imagine that you're a flea that was just born on one multimammate mouse, and you want to go out and explore the world by hopping onto a human. Would you prefer that your current multimammate mouse host is foraging in a 100 m$^2$ field containing one human host (0.01 human m$^{-2}$) or a 1000 m$^2$ field containing one human (0.001 human m$^{-2}$)?

In both cases, there is one possible human host, but the area and thus the human host density varies between the

two examples. When the search area is lower and thus the host density is higher, the multimammate mouse should be more likely to encounter the human, even though both fields have the same number of multimammate mice and humans.

As this example illustrates, the S, I, and R in SIR models should usually represent host densities, except in two circumstances: (1) the host density or sampling area are constant across space and time (such that numbers and densities would be equivalent); or (2) area (A) is a variable that is explicitly included in the model equations (e.g. N/A is the number of organisms per area, creating a density) (Begon et al. 2002, Hopkins et al. 2020).

(2021 population density = 8,485 people mi$^{-2}$), but it could be much lower in Juneau, Alaska (2021 population density = 12 people mi$^{-2}$). In cases like this, the choice to use a non-linear function instead of a linear function would depend on the range of host densities relevant to the study and the goals of the model.

But how much does the contact rate function really matter in SIR models? Simulations with SIR models based on data from multimammate mice infected with Morogoro virus, an arenavirus related to Lassa virus, found that models with different transmission functions made very different predictions about whether Morogoro virus could invade susceptible mouse populations or drive mouse populations to extinction (Borremans et al. 2017a). These and other simulation studies suggest that choosing the correct function might be crucial for making accurate predictions about epidemics (Hopkins et al. 2020). And though it might be tempting to assume that the more complex and non-linear functions are always better, those functions also

have more parameters and are thus more data-hungry in statistical analyses—a tricky trade-off!

In conclusion, transmission functions exist in all SIR models, and the modeller must choose which assumptions to make about how contact rates change with host density. At this point, you are likely wondering where, exactly, the transmission function goes in an SIR model. The transmission rate, ß, is the product of the contact rate function (g(N)) and the probability of transmission success given a contact between an infectious and a susceptible individual (v). Box 5.4 digs into the maths further.

## 5.5 Using simple models to make predictions

Now that we have covered some basics for making assumptions and building SIR models, let's move on to using those models to understand and predict disease dynamics.

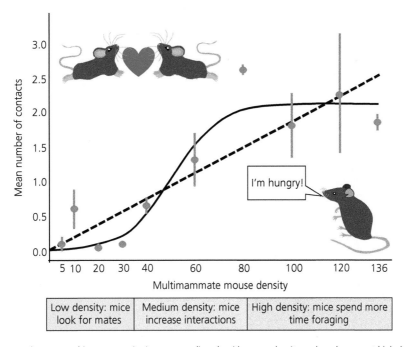

**Figure 5.6** Contact rates between multimammate mice increase non-linearly with mouse density, perhaps because at high densities, mice spend more time foraging and avoiding contacts with other mice. A linear, density-dependent contact rate function (the dashed line) does not fit the observed experimental data as well as the non-linear contact rate function.
Figure modified from Borremans et al. (2017b).

**Box 5.4  Force of infection**

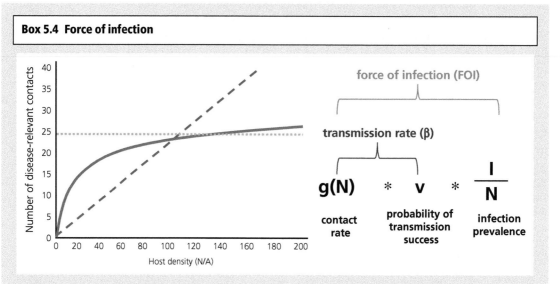

**Figure 5.7** On the left, there are some examples of contact rate functions, g(N), which can be described with mathematical equations and 'plugged in' to the force of infection equation on the right. The two lines are the density-dependent contact rate function (blue) and the frequency-dependent contact rate function (green), and the orange curve shows just one possible non-linear contact rate function.

**Table 5.3** Derivations of SIR model equations when transmission functions are assumed to be density-dependent versus frequency-dependent.

| Steps for plugging the contact rate function into the SIR model | Density-dependent transmission | Frequency-dependent transmission |
|---|---|---|
| Define the contact rate function, g(N) | A line with slope $x$ going through the intercept (0,0), because there are no contacts when density is zero $$g(N) = 0 + x\frac{N}{A}$$ | A horizontal line with a slope of zero, where the intercept x is the density-independent contact rate $$g(N) = x + 0\frac{N}{A}$$ |
| Simplify | $$g(N) = x\frac{N}{A}$$ | $$g(N) = x$$ |
| Plug the contact rate function into the FOI equation: $FOI = g(N) * v * I/N$ | $$FOI = x\frac{N}{A}v\frac{I}{N}$$ | $$FOI = x\,v\,\frac{I}{N}$$ |
| If possible, simplify | $$FOI = \frac{x}{A}vI$$ The Ns cancelled out | $$FOI = x\,v\,\frac{I}{N}$$ |
| Replace the contact rate (now just x) multiplied by the probability of transmission success (v) with the transmission rate, $\beta$ | $$FOI = \frac{\beta}{A}I$$ | $$FOI = \beta\,\frac{I}{N}$$ |
| Plug the FOI term into the SIR model (it gets multiplied by S, because it's the rate that susceptible hosts become infected). We left out the R equation, which is unaffected. | $$\frac{ds}{dt} = -\frac{\beta}{A}IS$$ $$\frac{dI}{dt} = \frac{\beta}{A}IS - \gamma I$$ If area is constant, A = 1 | $$\frac{dI}{dt} = \beta\frac{I}{N}S$$ $$\frac{dI}{dt} = \beta\frac{I}{N} - \gamma I$$ |

---

**Box 5.4** *Continued*

---

What do the contact rate function and transmission function look like in SIR models? They are part of a broader term called the **force of infection (FOI)**, which is defined as the rate that susceptible hosts become infected. The FOI is the product of three parameters or functions in SIR epidemiological models: the contact rate function, g(N), which determines how often a susceptible individual contacts other individuals in the population; the prevalence of infection, I/N, which determines how likely each contact is to be with an infected individual; and the probability of transmission success, v, which determines how likely it is that a contact between a susceptible and an infected individual will lead to successful transmission. Note that because contact rates are a function

of density, they are represented by a function (g(N))—rather than a constant parameter—in the FOI equation (Fig. 5.7).

So where is the transmission rate, $\beta$? We derive that by multiplying the contact rate function and the probability of transmission success. This means that when disease-relevant contact rates are a function of density (whenever we do not have frequency-dependent transmission), the transmission rate ($\beta$) is also a function of density. We could thus write our FOI term as $\beta(N)*I/N$. With that in mind, take a look at Table 5.3 to see how our assumptions are incorporated in SIR models when we assume that contact rates are density dependent or frequency dependent.

---

**Box 5.5  The mathematics behind making predictions**

---

This chapter covers various qualitative and quantitative predictions about disease dynamics that one can make using mathematical models. But how, exactly, does one make predictions with mathematical models? There are generally two methods: analytical solutions and numerical solutions.

**Analytical solutions** are formulas that give exact solutions or predictions, like the equation that describes the exact relationships between the model parameters in the simple SIR model and the basic reproductive rate ($R_0 = \beta/\gamma$). Analytical solutions are derived from a model's differential equations using calculus or matrix algebra; for example, the $R_0$ formula is derived from the dominant eigenvalue of the next generation matrix (Diekmann et al. 2010). If that means nothing to you, don't worry! You can learn more about analytical solutions in an advanced calculus course or textbook, or you can read the rest of this textbook without thinking about them again. All you need to know for now is that analytical solutions derived from simple models have greatly advanced our knowledge of disease dynamics, for example by producing formulas for $R_0$ and vaccination thresholds. However, most ecological and epidemiological models are too complicated to calculate their analytical solutions.

Instead, most predictions are made with **numerical solutions**, which approximate how S, I and R state variables (or other subpopulations of interest) will vary by adding up changes over many tiny steps through time.

There are many methods for numerically simulating ordinary differential equations (e.g. Taylor Series Method, Runge Kutta Method), but you do not need to know any of them to read the rest of this book. When simulating **deterministic models**, which do not incorporate random chance or **stochasticity**, the same model parameters (e.g. $\beta = 0.001$) and starting conditions (e.g. S = 100, I = 1, R = 0) will lead to the same pre-determined, approximated outcome in every simulation. In **stochastic models**, which include random chance, quantitative and qualitative predictions can change between simulations that use the same parameters and starting conditions (Section 5.11). In either case, predictions based on numerical solutions are usually shown as graphs of relationships, such as the number of infected individuals versus time.

## 5.6 The basic reproductive number, $R_0$

Somewhere between a 'stuttering chain' of infections that fizzles out and a full-blown epidemic (see Chapter 1), there exists an **invasion threshold** that separates the two scenarios. This threshold can be defined using values of **$R_0$** (pronounced 'R nought'), the average number of expected cases produced by an infected individual when introduced into a completely susceptible population (Diekmann et al. 1990). Infectious individuals must infect an average of at least one susceptible individual ($R_0 \geq 1$) for a pathogen to continuously spread through a new population. When $R_0 < 1$, the pathogen may

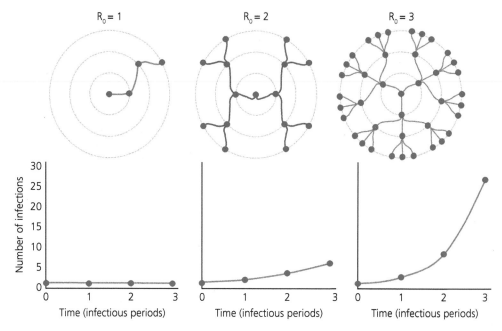

**Figure 5.8** Examples of average transmission chains from a single infected individual (central dot) to other individuals in a completely susceptible population. The concentric circles represent infectious periods and thus show how many new infections each infected individual creates. When $R_0 = 1$, the pathogen maintains a stable level in the population, and when $R_0 > 1$, the pathogen increasingly spreads through the susceptible population in an epidemic. After starting with a single infected individual and passing through three pathogen 'generations', there are 1, 8, or 27 infected individuals when $R_0 = 1$, 2, or 3, respectively.
Figure inspired by a similar figure published in The Guardian (Evershed and Ball 2021).

cause small outbreaks or stuttering chains, but in the long-term, the pathogen will not be able to invade the population and will die out. When $R_0 = 1$, the pathogen will remain stable in the host population, managing to replace itself but not causing sudden outbreaks (Fig. 5.8).

We have covered how $R_0$ can help us make predictions, but where does $R_0$ come from? Historically, the concept was borrowed from ecology, where it is called the **basic reproductive number** and is defined as the average number of female offspring produced by a female in a population (Heesterbeek 2002). Though $R_0$ is often erroneously called the 'basic reproductive *rate*', $R_0$ is a dimensionless number that does not have a time component in the units; it is the number of new infections per infectious period, rather than the number of new infections per day or week (like km per trip vs. km per hour). For example, smallpox and tuberculosis have similar $R_0$s (Fig. 5.9), but their epidemic dynamics are different due to their different infectious periods: smallpox infects an average of approximately six new people while the primary host is infectious for a week or so, whereas for tuberculosis, the infectious period can be several weeks or even years. While 'R' stands for 'reproductive number', the zero or 'nought' subscript reminds us that the basic reproductive number is the average number of new infections produced per infected individual when the first infected individual **enters a completely susceptible population**. Below we will consider what happens in populations that are not completely susceptible.

Mathematically, $R_0$ is derived from the differential equations that are used to formalize SIR compartmental models. When these models are relatively simple, one can assume that all but one individual in the population is susceptible, do some calculus, and come up with a relatively simple equation for $R_0$ (an analytical solution; Box 5.5). For example, for the set of equations in Table 5.2, $R_0 = \beta/\gamma$. Intuitively, increases in the transmission rate

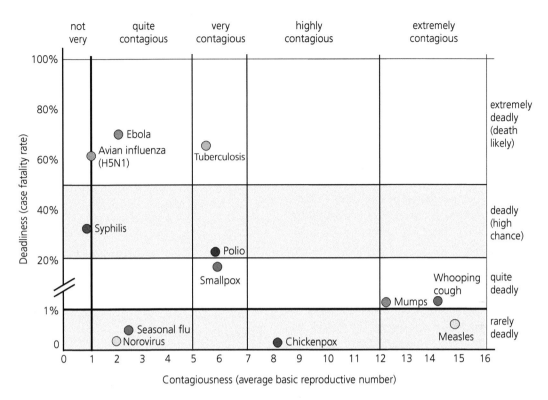

**Figure 5.9** Pathogens' basic reproductive rates range from those near 1 and not very contagious, like one estimate for H1N1, to extremely contagious childhood diseases, like measles. So far, we have only discussed infectious individuals recovering and either becoming immune or susceptible again, but infectious individuals may also die from the disease (i.e. pathogen-induced mortality), and this can be added to SIR models. This graph shows that different pathogens also have different **case fatality rates** (proportion of people who die from infection). This figure was informed by one made by David McCandless (McCandless 2014). For each pathogen, we have chosen point estimates for $R_0$ values and case fatality ratios for illustrative purposes, without showing the natural variation in either of these measures that exists across strains or across populations.

will increase $R_0$; the faster a pathogen spreads from one host to the next, the more likely it is to cause an epidemic in a susceptible population. In contrast, increases in the recovery rate, which is the denominator of the $R_0$ equation, will decrease $R_0$; the faster infected hosts recover or die from their infections (i.e. the smaller the duration of infectiousness), the less likely a pathogen is to successfully spread from one host to another, and thus the less likely that pathogen is to cause an epidemic.

Since $R_0$ is affected by the transmission rate, recovery rate, and other details that are specific to each disease system, $R_0$ varies among disease systems. Some pathogens, like the extremely transmissible 'childhood diseases' such as measles or mumps, tend to have very high basic reproductive numbers (Fig. 5.9). Other pathogens, like *Treponema pallidum*, the bacteria that causes syphilis, tend to

have basic reproductive numbers close to 1. (But note that syphilis may be spreading more rapidly in recent history in some communities.)

$R_0$ can also vary substantially among host populations for the same pathogen (Dietz 1993). For example, $R_0$ estimates for measles are usually around 12–18, but in different study areas and time periods, estimated $R_0$ for measles outbreaks has varied from 3.7 to 203.3 (Guerra et al. 2017)! This seems extraordinary! But remember that local conditions can affect transmission rates and other parameters that factor into the $R_0$ equation. For example, for pathogens with density-dependent transmission, the same pathogen invading a population with 100 susceptible hosts per $km^2$ will have a much lower $R_0$ than it would have when invading a population with 1000 hosts per $km^2$, all else being equal. This explains why epidemics often occur so explosively in major urban areas (e.g. Ebola virus in West Africa in 2013–2014).

## 5.7 Deterministic vaccination thresholds

For many pathogens that we want to make predictions about or control, host populations will already have some pre-existing immunity; some individuals may have been infected and recovered already or some individuals may have been vaccinated, creating a population that is a mix of fully or partially immune individuals and susceptible individuals. Under these circumstances, $R_0$ is no longer a particularly useful concept, because $R_0$ only describes what happens in completely susceptible populations.

When pre-existing immunity exists in a population, one can calculate a different threshold based on the **effective reproductive number** ($R_e$ or $R_t$), which is defined as the average number of new infections produced per infected individual at a specific time (t). The effective reproductive number is related to the basic reproductive number: $R_e = R_0*S/N$, where S/N is the proportion of the population that is susceptible to infection. If 100% of the population is susceptible, S/N = 1, then $R_e = R_0$. In contrast, whenever some individuals in the population have pre-existing immunity, $R_e < R_0$. This means that

when some individuals have pre-existing immunity, the pathogen spreads more slowly in the population than it would if the population was completely susceptible, because some of the potential transmission events between infected hosts and other hosts will involve immune individuals who can't become infected and continue transmission chains (Fig. 5.10).

Just like for $R_0$, 1 is the key threshold for $R_e$ where disease dynamics change from one state to another. When $R_e > 1$, the number of infected hosts in the population will increase, and when $R_e < 1$, the number of infected hosts in the population will decrease towards zero (Table 5.4).

Now that you know that decreasing the proportion of the population that is susceptible (S/N) can decrease $R_e$, you might be wondering how much one must reduce S/N to eliminate a pathogen from a host population. We can derive that! First, we start with the $R_e$ equation:

$$R_e = R_0 * \frac{S}{N}$$

Next, we define the proportion of the population that is immune or vaccinated as P/N, where P/N = 1 − S/N; when 0% of the population is susceptible, P/N = 1, and when 100% of the population is susceptible, P/N = 0. Similarly, S/N is equal to 1 − P/N, which we can plug into our equation:

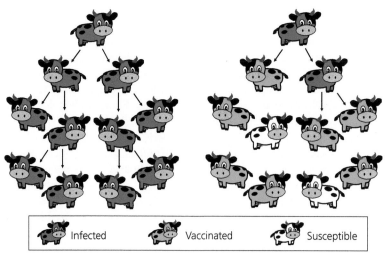

Infected      Vaccinated      Susceptible

**Figure 5.10** On the left, when no cows are vaccinated, no susceptible cows are indirectly protected from transmission by immune individuals. In contrast, on the right, the population has herd immunity: some susceptible cows do not become infected because they are indirectly protected by vaccinated cows. Note that even though herd immunity exists, some susceptible cows still become infected.

**Table 5.4** Some calculations for $R_e$ in a population where $R_0 = 3$ and various proportions of the population are immune/vaccinated (P/N). Here we assume that there is complete protection from infection in immune individuals.

| Population conditions | $R_e$ calculation | Long-term outcome |
|---|---|---|
| $R_0 = 3$, $\frac{S}{N} = 1$, $\frac{P}{N} = 0$ | $R_e = 3 * 1 = 3$ | Bigger epidemic |
| $R_0 = 3$, $\frac{S}{N} = 0.5$, $\frac{P}{N} = 0.5$ | $R_e = 3 * 0.5 = 1.5$ | Smaller epidemic |
| $R_0 = 3$, $\frac{S}{N} = 0.33$, $\frac{P}{N} = 0.67$ | $R_e = 3 * 0.33 \approx 1$ | Endemic |
| $R_0 = 3$, $\frac{S}{N} = 0.2$, $\frac{P}{N} = 0.8$ | $R_e = 3 * 0.2 = 0.6$ | Pathogen elimination |

$$R_e = R_0 * \left(1 - \frac{P}{N}\right)$$

Since the threshold between an epidemic and pathogen elimination is $R_e = 1$, we can replace $R_e$ with 1 in the equation to find the smallest $P/N$ that can shift dynamics towards pathogen elimination:

$$1 = R_0 * \left(1 - \frac{P}{N}\right)$$

And now we can rearrange the equation to solve for $P/N$, which is also called the **vaccination threshold**:

$$\frac{P}{N} = 1 - \frac{1}{R_0}$$

Based on this derivation, vaccination programmes that aim to eliminate a pathogen from a population should shoot for $P/N$ greater than or equal to $1 - 1/R_0$. This simple model therefore predicts that pathogens with large $R_0$ values require a large proportion of the population to be successfully vaccinated to eliminate a pathogen (Fig. 5.11). And because $R_0$ can vary among populations for the same pathogen, the vaccination threshold required for pathogen elimination may also vary among populations.

Cow jokes? Herd them all. Now I'm imoooone to them.
—A dad

The proportion of the population (P/N) that needs to be vaccinated to reduce $R_e$ below 1 is also called the **herd immunity threshold**; you've probably *herd* of it! The term first came from veterinary medicine, where it described how fewer susceptible individuals became infected over time as transmission chains were broken by individuals who had already recovered from their infections and developed immunity (Jones and Helmreich 2020). Therefore, 'herd immunity' is a process where transmission *still exists* but susceptible individuals are less likely to become infected than they would be if no hosts in the population were immune. To say it another way, even when herd immunity exists in a population, transmission will still occur until the pathogen has been eliminated.

A runaway train doesn't stop the instant the track begins to slope uphill, and a rapidly spreading virus doesn't stop right when herd immunity is attained.
—Bergstrom and Dean (2020)

The simple equation for the herd immunity or vaccination threshold can be very useful, but one must not forget the many assumptions inherent in this simple equation. The equation assumes that the pathogen is **contagious**—spread from person to person, rather than from an environmental reservoir (e.g. tetanus)—and that the host population mixes completely and randomly. It also assumes that hosts are vaccinated randomly, that immunity does not wane over time after vaccination, and that the immunization is 100% effective at preventing infection. If any of these assumptions are violated, one could try to account for them by building more complexity into the mathematical model.

Even when there are reliable herd immunity threshold estimates, reaching those thresholds can be difficult (Chapter 14). In human populations, reaching vaccination targets is complicated by logistics of vaccine transport and storage, vaccine misinformation and hesitancy, and limited healthcare capacity. Vaccination is even more difficult in wildlife populations (Barnett and Civitello 2020). And even when the herd immunity threshold is close to being reached, the herd immunity itself can act as an evolutionary pressure on pathogens, leading to new pathogen variants that escape host immunity (Rodpothong and Auewarakul 2012). All of this is to say that if you are looking for a career where you can always work towards solving complex problems, infectious disease control may be the career for you!

## 5.8 Deterministic invasion thresholds

*Greater Yellowstone Ecosystem, US*—After near extirpation in the early 1900s, the bison (*Bison bison*) population in the Greater Yellowstone Ecosystem has
increased to several thousand individuals spread
across a few distinct herds. This great conservation
success has not come without costs, however; *Brucella abortus*, a bacterial pathogen, was first detected
in bison in the region in the early 1900s and has since
increased in prevalence. The risk of bison (or elk)
spreading *Brucella* to cattle herds might be reduced
by vaccination, but in the 1990s, the *Brucella* vaccine
had low efficacy in bison, and it seemed unlikely
that the herd immunity threshold could be reached
(Dobson and Meagher 1996). This left disease ecologists wondering if there might be another way to
reduce pathogen transmission, perhaps to the point
where $R_e$ was below 1 and the pathogen might be
eliminated; could bison herds be reduced to a density below which *Brucella* transmission was too rare
for the pathogen to invade or persist in the herds?

This theoretical bison herd size or density represents a new threshold: the **invasion threshold
density ($N_t$)**, or the population density at which $R_0$
or $R_e$ are equal to 1. (To find this threshold number, one can solve the $R_0$ or $R_e$ equation for N,
the host density, when $R_0$ or $R_e$ are equal to 1.)
Theoretically, when host densities are below the
invasion threshold density ($N < N_t$), pathogens cannot successfully invade a population. Similarly, if

population densities are reduced from above to
below the invasion threshold density, an established pathogen should decline to elimination.

You might be able to imagine how exciting the
invasion threshold density was when it was first
derived from mathematical models; for example,
Ross (1905) showed that mosquito populations only
need to be reduced below a threshold density to
eliminate malaria, whereas people had previously
thought that mosquitoes would need to be completely eradicated to eliminate malaria (Smith et al.
2012).

However, the invasion threshold was not useful
for *Brucella* control in bison. There wasn't much of
a relationship between bison herd size and *Brucella*
seroprevalence. Based on the data available, a simple model predicted that $R_e$ would be less than 1
when the herd size was ~200 bison (Dobson and
Meagher 1996). Yet there was no data from herds
with <200 individuals, so an invasion threshold
could not be successfully documented (Fig. 5.12).
Furthermore, since herd sizes <200 individuals are
very small indeed, it seemed unlikely that efforts
to cull bison herds to below the estimated invasion threshold would be publicly acceptable. All of
this is to say that although it was worth considering with a model, host culling was not a viable
method for disease control in this system (also see
Chapter 14).

In fact, invasion threshold densities have never
been solidly documented in a wildlife disease

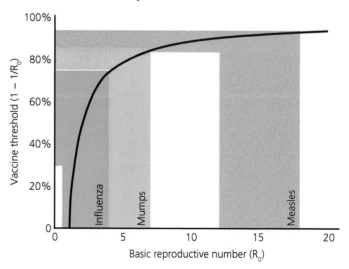

**Figure 5.11** Assuming that vaccines are 100%
effective (an assumption that can be changed with a
slightly more complex model), the vaccination
threshold ($P/N = 1 - 1/R_0$) increases steeply with $R_0$
until the thresholds are in the 80–90% coverage
range. The black line shows the relationship, and the
shaded regions show some estimated ranges for $R_0$
and vaccination thresholds for measles, mumps and
influenza.

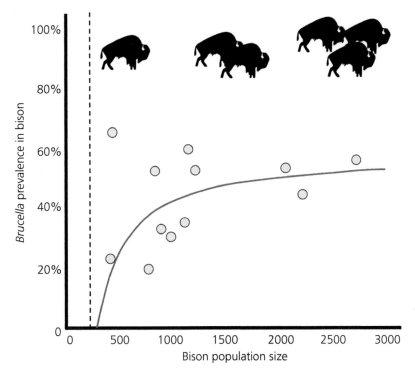

**Figure 5.12** The seroprevalence of *Brucella abortus* in bison in Yellowstone National Park from the 1940s to the 1990s varied with bison population size. The line shows the fitted relationship from an SIR model; based on that SIR model and the corresponding basic reproductive rate equation, a herd size of ~200 bison was predicted to be the threshold density below which *B. abortus* could not invade a population (dashed vertical line).
This figure was modified from Dobson and Meagher (1996).

system that does not involve vector-borne transmission. In contrast to our simple models, real systems are complex and difficult to observe. For example, imagine how difficult it is to document all cases where a pathogen was introduced into a wildlife population and caused a small outbreak before fading out. And in the bison model and many others, things are perhaps over-simplified (e.g. excluding important environmental transmission routes or spatial variation). Given these complexities and challenges, invasion threshold densities are probably less like exact numbers and more like a range of densities that can be difficult to estimate due to data uncertainty (Lloyd-Smith et al. 2004).

## 5.9 When do pathogens drive host species to extinction?

After a pathogen successfully invades a host population, causing an epidemic, four qualitative outcomes could follow, depending on whether the host and pathogen persist or not (Fig. 5.13). The pathogen could cause the host population to decline until it is extirpated, at which point the pathogen would also be extirpated, unless it has alternative host species. These outcomes might not be as common as you expect: only 6% of all species listed as threatened or extinct by the IUCN Red List has infectious diseases listed as one of the threats pushing them towards extinction, and most of those are amphibians threatened by a few pathogens (Smith et al. 2006). If the host species is not extirpated, the pathogen could still be lost from the population sometime after causing the initial epidemic, either through natural processes or due to disease control efforts. Alternatively, the pathogen and host could co-exist in a long-term **dynamic equilibrium**, where the numbers of each may fluctuate or cycle over time, but neither drops to zero (Fig. 5.13). SIR models can predict which of these four outcomes

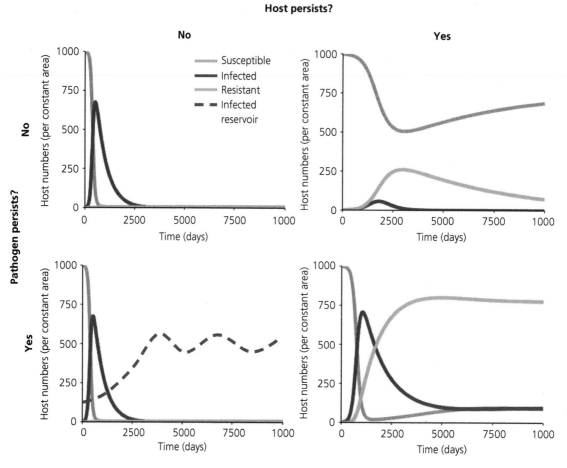

**Figure 5.13** There are four qualitative long-term outcomes when a pathogen invades a completely susceptible host population where individuals develop long-term immunity if they survive their infections: (1: top panels) the pathogen can be lost from the population; (2 and 3: left panels) the host population can be extirpated with or without the pathogen; or (4: bottom right panel) both the host and the pathogen can co-exist. Note that in both left panels, the focal host is extirpated, but in the bottom left, there is a reservoir species that can maintain the pathogen.

will occur, and as we will show below, predictions will vary depending on whether the models incorporate random chance (aka stochasticity).

## 5.10 Predicting long-term dynamics while ignoring random chance

As a host population declines towards extinction, we expect density-dependent pathogen transmission to decline and to perhaps stop altogether after the invasion threshold density is crossed. Therefore, pathogens may often be lost from small host populations before the host population is extirpated, giving the host population a chance to rebound.

But according to deterministic SIR models that ignore random chance, there are two scenarios where pathogens will not be lost from dwindling host populations and can therefore extirpate the host population: (1) the pathogen has a persistent reservoir, or (2) the pathogen has frequency dependent transmission (De Castro and Bolker 2005). (The second case emphasizes how important it is to determine which transmission function is most appropriate for a given disease system.)

Let's dig into the reservoir scenario further. If pathogens can persist in abiotic or biotic reservoirs outside of the host species, then the rate that host individuals are exposed to the pathogen can remain high even as the host population of interest dwindles (De Castro and Bolker 2005). For example, white-tailed deer (*Odocoileus virginianus*) are reservoir hosts that are relatively unaffected by the nematode known as deer meningeal worm (*Parelaphostrongylus tenuis*), which has a complex life cycle involving gastropod hosts. When moose (*Alces alces*) accidentally consume infected gastropods, they become infected as dead-end hosts—they do not spread meningeal worm, but they can still become infected and often die from their infections. Moose populations have been declining as deer have spread northward in North America due to climate change (Lankester 2010, 2018), and there is no reason to expect that moose infection rates will decline alongside moose populations, because thriving deer populations will continue to infect moose (Chapter 10). As we discuss in Chapter 6, many of the most devastating wildlife pathogens have persistent environmental reservoirs and/or resistant reservoir species.

## 5.11 Incorporating random chance when predicting long-term dynamics

Earlier in this chapter, we noted some instances where an outcome could be different from our deterministic prediction due to random chance. For example, when a pathogen has just started to invade a host population and there are few infected individuals, a budding epidemic could fizzle out even if the pathogen has $R_0 > 1$ (Fig. 5.14). Similarly, after a pathogen has successfully invaded a host population, it could disappear from the population as the number of susceptible individuals declines, even if $R_e > 1$. These outcomes cannot be predicted from deterministic epidemiological models, but they can be predicted from stochastic epidemiological models that incorporate random variation.

Before we dive into the models, let's overview the kinds of random variation that are relevant. **Environmental stochasticity** is random variation in environmental conditions, like temperature or precipitation, which can influence hosts and pathogens. For example, especially wet years or sudden heat waves could provide unusually good conditions for pathogen transmission or unusually low host tolerance to infection, resulting in host mass mortality

events (Chapter 10). **Demographic stochasticity** is random variation in births, deaths, immigration or emigration in a population; this type of variation can be especially important in populations that are already small (e.g. a nearly extirpated bird population where every fledgling happens to be female one year, limiting male mate options). Environmental and demographic stochasticity have always existed, but humans have often exacerbated these processes, such as by increasing the frequency of extreme weather events. And though the preceding examples focus on especially extreme or unusual events, we note that any random variation around a mean can be important, depending on the ecology of a system.

To model stochasticity, we usually switch from using **continuous time models** (like the deterministic models we have considered so far) to **discrete models**, which assume that events and individuals in the real world are distinct (e.g. gain one infected host on Tuesday and another infected host on Thursday). Discrete events either happen or do not happen, and whether one event happens impacts subsequent events (i.e. the events are dependent). We can assign probabilities to these events, which allows us to add components of random chance to their occurrence in our models.

To model chains of discrete events, we use a similar mathematical framework to the simple differential equations in Table 5.2, but we use partial differential equations or Markov processes to keep track of each event and/or each individual and the random probabilities associated with them. Note that many people do not work with the partial differential equations directly, but rather use numerical simulations and individual-based models to simulate model behaviour. Unlike simulations from deterministic models, which are always identical for a given set of model parameters (e.g. $\beta = 0.001$) and starting conditions (e.g. S = 1000 individuals $km^{-2}$), simulations from stochastic models vary each time we run them (Fig. 5.14).

As you might imagine, these more complicated stochastic models are more computationally intensive to run and the variability among simulations can make predictions more difficult to interpret. But these complications can be worth it, because they may make predictions more realistic, especially

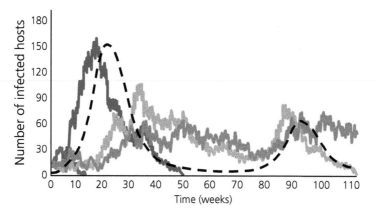

**Figure 5.14** The dotted line simulates a continuous, deterministic SIR model, where the pathogen is never predicted to fade out from the host population (i.e. zero infected hosts), even if the number of infected individuals becomes very low. The solid lines show a few different simulations of a discrete, stochastic SIR model with the same parameters and starting conditions. Note that the stochastic simulations are all quantitatively different from the deterministic model due to random chance. Also, some of the stochastic simulations are qualitatively different from the deterministic model, because transmission stutters after an initial chain of infections and the pathogen fails to invade (red line) or the pathogen fades out after the initial epidemic peak (blue line).

when populations are small or rare events are otherwise important.

In fact, we cannot predict if or when an **epidemic** will **fade out** unless we use stochastic models, because epidemic fade-out is an inherently random process. Fade-out occurs in the period after an epidemic when the remaining population of susceptible individuals becomes so small relative to resistant individuals that by random chance, the infected individuals do not contact susceptible individuals and pass on their infections. For example, in Fig. 5.14, the pathogen sometimes persists after the epidemic (with a second epidemic occurring later, when susceptible individuals are replenished; grey lines), but other times, the pathogen fades out (red and blue lines).

There is no exact formula for calculating how big a population needs to be for a pathogen to persist in the population for a given time without fading out, but the abstract idea of a population that is 'big enough' for persistence is called the **critical community size** (Fig. 5.15). For example, measles epidemics were observed to fade out in cities with less than 250,000–300,000 people, so this was generally considered the critical community size for measles (Bartlett 1957, Black 1966). Unlike the many observed measles epidemics, epidemic fade-outs have rarely, if ever, been observed in wildlife

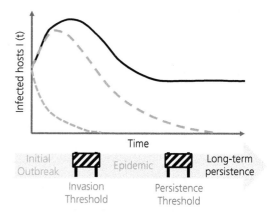

**Figure 5.15** When a pathogen is first introduced into a new host population, it may or may not cause an epidemic, depending on $R_0$ (or $R_e$, if there is pre-existing immunity) and random chance. If $R_0$ is less than 1, the outbreak will fizzle out (blue line). If $R_0$ is greater than 1 (and thus for pathogens with density-dependent transmission, the host density is larger than the invasion threshold density), an epidemic can occur (green line). After a pathogen successfully invades, it might be lost from the host population due to random chance in population sizes below the critical community size (also known as the persistence threshold). If the population is big enough and $R_e$ remains greater than 1, the pathogen may persist in the host population for the long-term (black line).

disease systems (Lloyd-Smith et al. 2005), because we rarely have long-term disease dynamic data from wildlife. Even so, stochastic epidemiological models inform us that pathogens are probably often lost from host populations after epidemics, and especially so when host populations are relatively small.

## 5.12 When models are wrong

All models are wrong, but some are useful.
—Box (1976)

[that] can be rephrased as, 'All models are wrong, and many are not useful'.
—Jones (2021)

---

**Box 5.6  Counting sheep**

So far, this chapter has highlighted how useful models *can* be. However, the world is full of useless and even harmful epidemiological models. Stop us (or stop reading this box) if you've heard this joke already:

While traveling through the United Kingdom, an American researcher becomes enamoured by the beautiful, small-scale goat farms surrounded by ancient stone walls and hedgerows. The researcher decides to start a new career as a goat farmer and stops at one farmhouse to seek advice.

Speaking to the farmer, the researcher decides to try their luck, and raises a bet with the farmer. If the researcher can calculate the exact number of goats on the farm, would the farmer provide two of the flock so that the researcher can start a herd of their own?

The farmer agrees. The researcher painstakingly observes the goat herd, makes notes and calculations, and announces confidently: 178 goats.

'Amazing!' exclaims the farmer. 'How did you do it?'

'I counted the number of legs and divided by four,' explains the researcher.

The farmer counters: 'If I can name your profession, can I keep the animals?'

The researcher, feeling a little like a character in *Rumpelstiltskin*, agrees.

Without skipping a beat, the farmer proclaims: 'Mathematical modeller'.

'Unbelievable! How did you work that out?' demands the researcher, stupefied.

The farmer answers, 'You arrived here unasked for and used overly complicated techniques to provide me with information I already knew. Also, those are sheep, not goats.'

To all the mathematical modellers reading that joke for the first time and now wondering how often you've counted goat legs to quantify sheep, please feel free to have a lie down and return to the book when you are ready. But don't bother taking a nap. For the rest of your life, the thought of counting sheep will keep you up all night, rather than helping you fall asleep (Hopkins and Hayman, pers. comm.).

---

Epidemiological models are used to make important decisions, and thus when the models do not reflect reality, catastrophe can follow. This can be especially problematic in an era when anyone can use readily available tools to build a model and post a simulation to a preprint archive or social media, whereas relatively few people have the expertise to evaluate which simplifying assumptions should be made in epidemiological models. Just as one would trust a trained, licensed dentist to safely remove a broken tooth more than a neighbour wielding rusty pliers who 'watched a YouTube video about this, don't worry', important societal and ecological decisions should be based on models constructed and evaluated by experts; that is, people trained in research on infectious diseases and the species that serve as hosts (such as public health experts). But even models created by trained professionals can be wrong when assumptions do not match reality (Box 5.6).

When an Ebola outbreak was declared in the Democratic Republic of Congo this spring, there were all kinds of predictions about how the epidemic would play out.
—Whitehead (2018)

*Mbandaka, Democratic Republic of Congo*—Early in 2018, an Ebola virus outbreak originated in a relatively rural region of the Democratic Republic of Congo (DRC). The world held its breath, hoping that the outbreak would not develop like the West Africa Ebola virus outbreak in 2014, which infected at least 28,500 people. In May 2018, a case was observed in the large city of Mbandaka, sparking fears that the 2018 epidemic might explode.

In early June, a mathematical epidemiologist used the case data from the outbreak so far to build a model (Whitehead 2018). The model predicted that the rapid public health response to the outbreak would cause the outbreak to fizzle out by the end of June 2018, with a total of ~55 people (plus or minus

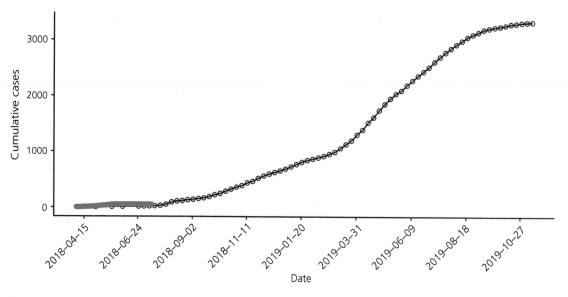

**Figure 5.16** The cumulative cases confirmed and suspected during the 2018–2020 Ebola epidemic in the Democratic Republic of Congo (data from Aruna et al. 2019). The red line shows the plateau in total cases that was predicted early in the epidemic (Whitehead 2018)—a prediction that did not match the epidemic that followed.

some uncertainty) becoming infected before the epidemic ended. The modeller reported that he was 'quite confident' in his quantitative prediction, and other epidemiologists agreed that the epidemic was likely to fizzle out soon (Whitehead 2018). But the epidemic did not end in June 2018. It continued for an additional 18 months and infected thousands of people, becoming the second largest Ebola epidemic in history (Fig. 5.16).

Predictions about infectious disease are very challenging since humans are the primary drivers of infection, and we are so unpredictable. . . Ebola in particular is very random; one 'unlikely' event can change the course of an outbreak [and] spike cases.

Bryan Lewis (from Whitehead 2018)

How could the model predictions have been so wrong?! The model assumed that public health interventions would remain strong and consistent, quelling further transmission, whereas in reality, parts of the DCR were experiencing violent civil unrest that disrupted public health efforts. For example, Ebola care facilities were destroyed, and healthcare workers were murdered. As a result, the local effective reproductive number of the epidemic was higher in areas that experienced violent

events, and these areas likely kept the epidemic going far longer than predicted (Wannier et al. 2019). This local variation in transmission violates the assumptions of mass action and homogeneity that we first discussed in Section 5.3 (King et al. 2015) and serves as an important cautionary tale; when the assumptions used to build mathematical models are too simple, so too will be the model's predictions.

## 5.13 Summary

All mathematical models are simplifications of complex systems, and the simplifying assumptions (e.g. how contact rates are influenced by host density) affect our ability to understand and predict disease dynamics. Dynamic mathematical models include the SIR compartmental models described in this chapter, which can be deterministic or stochastic. These models can be used to describe or predict disease dynamics, and predictions may be qualitative or quantitative, depending on the goals of the model. For example, $R_0$ and $R_e$ can be used to predict whether a pathogen can invade a host population (and the host densities and vaccination thresholds that prevent or allow invasion) and whether a pathogen can persist in a host population over the long term. Predictions like these can be critical for controlling infectious diseases, as we will see in Chapter 14.

## 5.14 References

Anderson, R.M., May, R.M. (1991). Infectious Diseases of Humans. Oxford University Press Inc.

Antonovics, J., Iwasa, Y., Hassell, M.P. (1995). A generalized model of parasitoid, venereal, and vector-based transmission processes. The American Naturalist, 145, 661–75.

Aruna, A., Mbala, P., Minikulu, L., et al. (2019). Ebola Virus Disease Outbreak—Democratic Republic of the Congo, August 2018–November 2019. Morbidity and Mortality Weekly Report (MMWR), 68, 1162–5.

Barlow, N.D. (1996). The ecology of wildlife disease control: simple models revisited. Journal of Applied Ecology, 33, 303–14.

Barnett, K.M., Civitello, D.J. (2020). Ecological and evolutionary challenges for wildlife vaccination. Trends in Parasitology, 36, 970–78.

Bartlett, M.S. (1957). Measles periodicity and community size. Journal of the Royal Statistical Society Series A, 120, 48–70.

Begon, M., Bennett, M., Bowers, R.G., et al. (2002). A clarification of transmission terms in host-microparasite models: numbers, densities and areas. Epidemiology & Infection, 129, 147–53.

Bergstrom, C.T., Dean, N. (2020). What the proponents of 'natural' herd immunity don't say. The New York Times, 1 May: https://www.nytimes.com/2020/05/01/opinion/sunday/coronavirus-herd-immunity.html, accessed 5/8/2021.

Black, F.L. (1966). Measles endemicity in insular populations: critical community size and its evolutional implication. Journal of Theoretical Biology, 11, 207–11.

Borremans, B., Reijniers, J., Hens, N., Leirs, H. (2017a). The shape of the contact-density function matters when modelling parasite transmission in fluctuating populations. Royal Society Open Science, 4, 171308.

Borremans, B., Reijniers, J., Hughes, N.K., et al. (2017b). Nonlinear scaling of foraging contacts with rodent population density. Oikos, 126, 792–800.

Box, G.E.P. (1976). Science and statistics. Journal of the American Statistical Association, 71, 791–9.

De Castro, F., Bolker, B. (2005). Mechanisms of disease-induced extinction. Ecology Letters, 8, 117–26.

Diekmann, O., Heesterbeek, J.A.P., Metz, J.A.J. (1990). On the definition and the computation of the basic reproduction ratio R0 in models for infectious diseases in heterogeneous populations. Journal of Mathematical Biology, 28, 365–82.

Diekmann, O., Heesterbeek, J.A., Roberts, M.G. (2010). The construction of next-generation matrices for compartmental epidemic models. Journal of the Royal Society Interface, 7, 873–85.

Dietz, K. (1993). The estimation of the basic reproduction number for infectious diseases. Statistical Methods in Medical Research, 2, 23–41.

Dobson, A., Meagher, M. (1996). The population dynamics of brucellosis in the Yellowstone National Park. Ecology, 77, 1026–36.

Evershed, N., Ball, A. (2021). How coronavirus spreads through a population and how we can beat it. The Guardian: https://www.theguardian.com/world/datablog/ng-interactive/2021/sep/09/how-contagious-delta-variant-covid-19-r0-r-factor-value-number-explainer-see-how-coronavirus-spread-infectious-flatten-the-curve, accessed 9/9/2021.

Guerra, F.M., Bolotin, S., Lim, G., et al. (2017). The basic reproduction number (R0) of measles: a systematic review. The Lancet Infectious Diseases, 17, 420–28.

Gurley, E.S., Montgomery, J.M., Hossain, M.J., et al. (2007). Person-to-person transmission of Nipah virus in a Bangladeshi community. Emerging Infectious Diseases, 13, 1031–7.

Hamer, W.H. (1906). Epidemic disease in England - the evidence of variability and of persistence. The Lancet, 167, 733–8.

Heesterbeek, J.A.P. (2002). A brief history of R0 and a recipe for its calculation. Acta Biotheoretica, 50, 189–204.

Hopkins, S.R., Fleming-Davies, A.E., Belden, L.K., Wojdak, J.M. (2020). Systematic review of modelling assumptions and empirical evidence: does parasite transmission increase nonlinearly with host density? Methods in Ecology and Evolution, 11, 476–86.

Jones, D., Helmreich, S. (2020). A history of herd immunity. Lancet, 396, 810–11.

Jones, G. (2021). [Twitter] 3 February. Available at: https://twitter.com/ecologyofgavin/status/1357007679919591424?s=20, accessed 8/2/2021.

Kermack, W.O., McKendrick, A.G. (1927). A contribution to the mathematical theory of epidemics. Proceedings of the Royal Society of London, A: Mathematical, Physical, and Engineering Sciences, 115, 700–21.

King, A.A., Domenech de Cellès, M., Magpantay, F.M.G., Rohani, P. (2015). Avoidable errors in the modeling of outbreaks of emerging pathogens, with special reference to Ebola. Proceedings of the Royal Society of London, B, 282, 20150347.

Lafferty, K.D., DeLeo, G., Briggs, C.J., et al. (2015). A general consumer-resource population model. Science, 349, 854–7.

Lankester, M.W. (2010). Understanding the impact of meningeal worm, *Parelaphostrongylus tenuis*, on moose populations. Alces, 46, 53–70.

Lankester, M.W. (2018). Considering weather-enhanced transmission of meningeal worm, *Parelaphostrongylus tenuis*, and moose declines. Alces, 54, 1–13.

Levin, S.A. (1992). The problem of pattern and scale in ecology: the Robert H. MacArthur Award Lecture. Ecology, 73, 1943–67.

Lloyd-Smith, J.O., Getz, W.M., Westerhoff, H.V. (2004). Frequency-dependent incidence in models of sexually transmitted diseases: portrayal of pair-based transmission and effects of illness on contact behaviour. Proceedings of the Royal Society of London B: Biological Sciences, 271, 625–34.

McCallum, H., Barlow, N., Hone, J. (2001). How should pathogen transmission be modelled? Trends in Ecology & Evolution, 16, 295–300.

McCandless, D. (2014). Visualised: how Ebola compares to other infectious diseases. The Guardian: https://www.theguardian.com/news/datablog/ng-interactive/2014/oct/15/visualised-how-ebola-compares-to-other-infectious-diseases, accessed 9/9/2021.

Rodpothong, P., Auewarakul, P. (2012). Viral evolution and transmission effectiveness. World Journal of Virology, 1, 131–4.

Ross, R. (1905). The logical basis of the sanitary policy of mosquito reduction. Science, 22, 689–99.

Smith, D.L., Battle, K.E., Hay, S.I., et al. (2012). Ross, Macdonald, and a theory for the dynamics and control of mosquito-transmitted pathogens. PLoS Pathogens, 8, e1002588.

Smith, K.F., Sax, D.F., Lafferty, K.D. (2006). Evidence for the role of infectious disease in species extinction and endangerment. Conservation Biology, 20, 1349–57.

Wannier, S.R., Worden, L., Hoff, N.A., et al. (2019). Estimating the impact of violent events on transmission in Ebola virus disease outbreak, Democratic Republic of the Congo, 2018-2019. Epidemics, 28, 100353.

Whitehead, N. (2018). An encouraging prediction about the Ebola outbreak, NPR Goats and Soda, 8 June. [online]: https://www.npr.org/sections/goatsandsoda/2018/06/08/618012443/an-encouraging-prediction-about-the-ebola-outbreak, accessed 26/8/2021.

# SECTION 2

# Pathogen Sources

Some pathogens infect a single host species, and never spend time in the environment outside of that host species—an ideal disease control scenario because you can target that host species. But most pathogens spend time in the **environment** or other **reservoir host species**. Here (Chapters 6–8), we introduce several flavours of reservoirs and illus-trate how difficult it can be to accurately describe disease transmission dynamics among the sources, especially when an infectious disease is first emerg-ing. Read on for adventures with bat faeces, sheep faeces, cat faeces, deep faeces; and the healing power of a rooster's anus.

Red-winged blackbirds (*Agelaius phoeniceus*) have been linked to outbreaks of histoplasmosis.
Photo by Rainer Lutz Bauer, used with permission.

# CHAPTER 6

# The environment as a pathogen reservoir

## 6.1 Introduction

*New York, United States, Winter 2006–2007*—Clad in helmets, headlamps, and clothes warm enough to survive frigid conditions, biologists from the New York State Department of Environmental Conservation (DEC) headed towards a cave to conduct a routine bat survey. To navigate through the cave to reach the bats, the biologists had to crawl submerged to their chests in 7°C (45°F) water, wearing neoprene waders and rubber gloves to prevent hypothermia (which can be deadly). And while they had made the trip many times before, this time something would be horribly wrong.

The cave they were visiting was a known **hibernaculum**: a location where temperate insectivorous bats spend several months during the winter in **torpor**, with their body temperatures and metabolic rates reduced to conserve energy until spring arrives and insect availability returns. Hibernating bats usually arouse every few weeks throughout the winter, raising their body temperatures and flying around for a few hours before returning to their torpor state. But as the biologists approached the cave and went inside, they noticed that *many* bats were simultaneously awake and flying around inside and outside of the cave. And thousands of bats were dead.

The DEC biologists observed that roughly half the bats had fuzzy white growths on their wings, ears, and muzzles (Fig. 6.1), which they had never seen before. White growth on bats and dead bats were soon observed in several other caves in two counties in New York. As news of these events spread, information from the previous year also surfaced: a

hydrologist (who was also a recreational caver) had visited Howe Cave in New York in February 2006, and he had seen some dead bats there. By chance, he had taken some photographs that included bats, and returning to those photographs, he noticed that they too portrayed white growth on some bats. This remains the earliest documented case of what would become known as 'white nose syndrome' in North America.

As is often the case during the first outbreak of an emerging infectious disease, bat biologists were not sure if an infectious disease was the cause of the unusual bat behaviours and mortality observed during the winter of 2006–2007. But histological examinations soon showed that bats had skin lesions that were associated with a fungus—the same fungus that caused the white fuzzy growth (Blehert et al. 2009). Sequencing and phylogenetic analyses led to the description of a new species: *Pseudogymnoascus destructans*. (Actually, first it was called *Geomyces destructans*, and later the name was revised to reflect a whole new genus.) Later, experiments would show that healthy bats inoculated with *Pseudogymnoascus destructans* would develop disease symptoms during hibernation (see Koch's postulates, Chapter 3).

An immediate priority was figuring out how the fungus was transmitted among bats, because armed with that knowledge, interventions might be designed to disrupt transmission and prevent further spread. Experiments confirmed that the fungus grew and reproduced on bat epidermal tissue and could be transmitted from bat to bat. They also found that *Pseudogymnoascus destructans* has a dormant stage (conidia) that contaminates substrates

*Emerging Zoonotic and Wildlife Pathogens*. Dan Salkeld, Skylar Hopkins, and David Hayman, Oxford University Press.
© Oxford University Press (2023). DOI: 10.1093/oso/9780198825920.003.0006

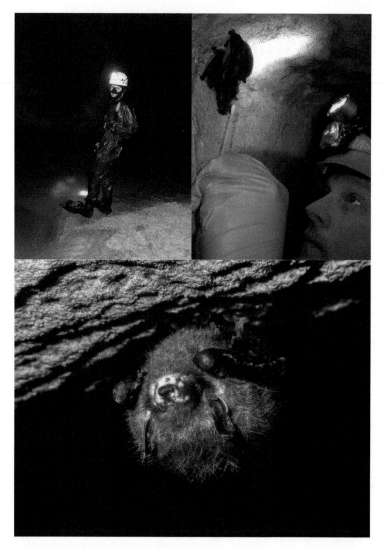

**Figure 6.1** (Top left) A biologist preparing to snorkel in a cold cave stream. Photo taken by the National Park Service and published in the public domain. (Top right) Dr Matt Niemiller swabbing hibernating little brown bats (*Myotis lucifugus*) for future qPCR analyses to test for the presence of *Pseudogymnoascus destructans*. Note that swabbing bats requires special wildlife sampling permits. (Bottom) A hibernating tri-coloured bat (*Perimyotis subflavus*) with visible *Pseudogymnoascus destructans* growing on its muzzle, ears and wings—symptoms of severe white nose syndrome.
Photograph taken in a cave in Jackson County, Alabama in January 2020. Photo by Matthew L. Niemiller.

where infected bats roost, and uninfected bats can become infected when they contact these infectious surfaces (Gargas et al. 2009, Lorch et al. 2011, 2013). Conidia can survive for long periods in hibernacula, including when bats are absent during non-winter months (Lorch et al. 2013). As we will see below, this persistent environmental stage creates a huge challenge for bat conservation.

Over time, researchers pieced together the typical infection cycle (Fig. 6.2). Uninfected bats return to hibernacula in the autumn to mate and then begin hibernation. They become exposed when they contact contaminated substrates or infected bats. When bats are newly infected, they do not show signs of disease, but as the fungus proliferates, it causes skin lesions. Bat skin is important for hibernation

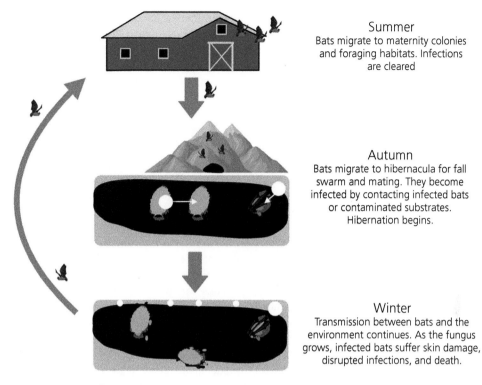

**Summer**
Bats migrate to maternity colonies and foraging habitats. Infections are cleared

**Autumn**
Bats migrate to hibernacula for fall swarm and mating. They become infected by contacting infected bats or contaminated substrates. Hibernation begins.

**Winter**
Transmission between bats and the environment continues. As the fungus grows, infected bats suffer skin damage, disrupted infections, and death.

**Figure 6.2** The typical annual infection cycle whereby bats become infected by and spread *Pseudogymnoascus destructans* and develop white nose syndrome.
Made using BioRender.com with bat cartoons from Skylar Hopkins.

physiology (e.g. bats absorb water through their skin in high humidity cave environments), so skin lesions caused by fungal infections appear to disrupt homeostasis and torpor, causing infected bats to rouse more frequently (Reeder et al. 2012, Warnecke et al. 2013). Though some arousal from torpor is normal for healthy bats, over-frequent arousal can kill infected bats; each arousal event uses substantial energy, and bats that use up too much energy starve before winter is over.

If bats can survive their infections until spring, they can clear their infections when their bodies return to temperatures too high for *Pseudogymnoascus destructans* growth. Surviving bats move on to their summer habitats and activities. (However, some bats may still die after hibernation due to sudden massive immune responses to fungal infection (Meteyer et al. 2012, Field et al. 2015).) Critically, surviving bats do not have strong long-term immunity; they can become reinfected

again the next autumn when they contact infected substrates or bats.

Within a given hibernaculum, different species may be more or less likely to get exposed to *Pseudogymnoascus destructans*. To demonstrate this, researchers covered a few bats of several species with fluorescent colour powder and released them back into their hibernacula (Hoyt et al. 2018). When they came back months later with a blacklight, they could determine which bats had contacted each other or contaminated surfaces based on residual powder marks—a method that was more accurate for predicting fungus transmission within and between bat species than just doing snapshot surveys to see which bats were contacting at any given time. The researchers found that powder trail evidence corresponded well to infection data. For example, northern long-eared bats (*Myotis septentrionalis*) were usually seen roosting alone, but the powder revealed that they often directly or indirectly contacted individuals of the same and other species; correspondingly, most northern long-eared bats were infected by *Pseudogymnoascus destructans* by the end of the hibernation season. In contrast, tri-coloured bats (*Perimyotis subflavus*) were rarely seen contacting *and* were rarely contaminated by powder from other bats, suggesting that they are solitary and spatially segregated from other bats during hibernation. This could explain why so few tri-coloured bats were infected by *Pseudogymnoascus destructans* by the end of the hibernation season.

When *Pseudogymnoascus destructans* is spread broadly throughout a hibernaculum, bats will become infected earlier during hibernations, experience more weeks of disrupted torpor, and be more likely to die from their infections. For example, imagine that an infected bat enters a fungus-free hibernaculum in early November and picks a nice spot to roost and go into torpor. The tiny bat (~8 g, if it's a little brown bat) contaminates a tiny patch of substrate. In a few weeks, the bat wakes up for a few hours, and chooses a new roost; this time, it cuddles up to another bat, infecting that bat and

contaminating another tiny patch of substrate. At this rate, most bats in the hibernacula either won't be exposed during that hibernation season, or they will be exposed so late that they don't experience major disease symptoms. But as this process repeats the next year and the year after that, *Pseudogymnoascus destructans* may be so widespread in the hibernaculum that most bats become infected as soon as they choose their first roost in autumn, forcing them to spend the entire winter battling their infections. This process can be seen in time series data for bat population declines: there is usually a ~2-year lag between when researchers first detect *Pseudogymnoascus destructans* on bats or substrates in a hibernaculum and when that hibernaculum experiences major population declines (Hoyt et al. 2020, Fig. 6.3).

Thus far, we have focused on *Pseudogymnoascus destructans* spread within a single hibernaculum, but of course, the fungus also spreads between hibernacula. By 2010, just a few years after first detection in New York, the fungus had made it all the way to the south-eastern United States (Fig. 6.4)—hundreds of kilometres! For the

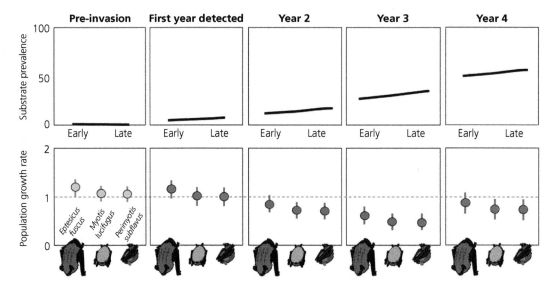

**Figure 6.3** Starting the first year that *Pseudogymnoascus destructans* is introduced to a hibernaculum, bats contaminate more and more of the substrate over time (top panels). As the substrate is increasingly contaminated, bats are more likely to become infected early during hibernation, leading to longer periods battling infections and increased mortality. This might explain why there is a time lag between when the fungus is first introduced and when bat population declines are observed (bottom panels). The dashed line shows a population growth rate of 1, indicating a constant population size.
Data are averages for North American bats and hibernacula and were published in Hoyt et al. (2020).

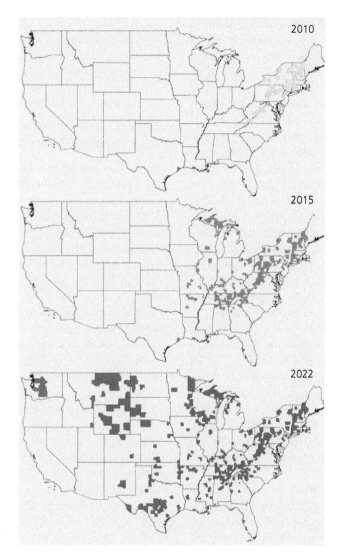

**Figure 6.4** A map time series of counties or districts that had PCR-confirmed white nose syndrome (WNS) cases in the United States by 2010, 2015 and 2022. This map does not include any information about geology, but counties with infected bats tend to be those with many caves and mines where bats hibernate. Note that Canada is not included even though WNS also spread there.
The data used to make this map came from whitenosesyndrome.org.

most part, it seemed that bats were responsible for spreading the fungus far and wide; bats fly long distances and may visit multiple hibernacula during fall swarm or even switch hibernacula mid-winter. All of these 'site switches' are opportunities for infected bats to carry the fungus to a new hibernaculum and establish a persistent environmental reservoir.

Humans can also spread the fungus from place to place if they don't decontaminate their gear between visiting different caves—a process that involves mechanically removing mud and dirt and then using chemicals to kill any remaining *Pseudogymnoascus destructans*. Therefore, part of the national white nose syndrome response was to restrict access to caves and suggest decontamination procedures that would reduce human contributions to the rapid spread. In practice, this means that biologists and responsible recreational cavers must strip down to their underwear on the side of the road after visiting caves, even if temperatures are below freezing, and separate their dirty gear into bags for future decontamination.

Despite these valiant efforts to prevent human spread, *Pseudogymnoascus destructans* made some long-distance jumps across North America. For example, in 2016, infected and diseased bats were found in Washington State (Lorch et al. 2016) and in California in 2019. Humans likely played some role in these long-distance transmission events because bats don't usually migrate from the east to west coast of the United States. Humans were probably also the cause of the initial introduction of *Pseudogymnoascus destructans* to New York. Somehow, a strain that originated in Eurasia (Puechmaille et al. 2010, Wibbelt et al. 2010, Hoyt et al. 2016) made it to Howe Cave, which happens to be connected to a commercial cave that does cave tours for the public and sells cave-aged cheese. Did an international tourist visiting New York accidentally bring *Pseudogymnoascus destructans* on their boots? Did soil from Eurasia get imported to make better cave-aged cheese? Cross-continent introductions like this one may be increasingly common in our globalized world (Chapter 9).

Before we end this case study, let's consider why we do not simply treat infected bats and sterilize hibernacula substrates to eliminate the fungus. This *can* work on a small scale, such as if government agencies focus on mediating disease impacts in a single priority hibernaculum for an endangered bat species. For example, individual bats can be treated with fungicides and researchers are testing the efficacy of treating hibernacula substrates with fungicides in a few mines. But there is no fungicide that targets only *Pseudogymnoascus destructans*, and cave managers and researchers worry that broadly killing fungi in sensitive cave environments could lead to different environmental catastrophes. Furthermore, treating bats and hibernacula requires immense human effort (remember that crawling in frigid water?), so it would be impossible to treat the thousands of hibernacula in the United States and Canada each year. Without effective control options,

some bat populations have declined towards extinction, while the force of infection (Chapter 5) from the environmental reservoir to surviving bats has remained high. We will return to the conservation implications of this disease in Chapter 12.

## 6.2 Five questions to define 'environmental reservoirs'

Clearly, *Pseudogymnoascus destructans* has a persistent **environmental reservoir**. But what, exactly, is an environmental reservoir? The term 'environment' has many possible definitions, which can be as broad as 'the natural world'. We will be more specific in our definition of the environment as it pertains to environmental reservoirs. We can build that specificity by answering five questions: who, what, when (and how), where, and why?

*Whose environment are we considering?* From the pathogen's perspective, an environmental reservoir is the surroundings or conditions where a pathogen exists when it is outside of a host—the pathogen's habitat during that life stage. From the host's perspective, an environmental reservoir is the surroundings or conditions where the host is exposed to the pathogen; it might be the host's entire habitat or just a small piece of the host's overall habitat.

*What distinguishes the environmental reservoir from other parts of the disease system?* Different professionals may look at the same disease system and make different decisions about which parts should be considered 'the environmental reservoir'. For example, from a medical perspective, malaria may be said to have an 'environmental reservoir' outside of the human host in the form of mosquito vectors. In contrast, someone who researches vector-borne diseases will likely consider mosquitoes to be separate from abiotic environments; they might say that malaria is a vector-borne disease with no environmental reservoir. One must decide whether to include *only* the abiotic, non-living environment (e.g. water, soil), or whether to also include living components of the system like vectors and reservoir hosts (Chapter 7). For simplicity, in this book, our definition of environmental reservoir excludes vectors and reservoir hosts.

However, note that 'non-living' does not mean that environmental reservoirs are dead vacuums for pathogens; they might feature critical biotic interactions between the pathogen and other organisms that are not vectors or reservoir hosts. For example, the fungus that causes snake fungal disease (ophidiomycosis) can survive in sterilized soils in the laboratory, such that soil is an environmental reservoir. However, the fungus does not appear to be able to survive or replicate for long in non-sterile soils where other fungi and bacteria exist; those biological interactions may prevent most soils from serving as environmental reservoirs (Campbell et al. 2021).

*When and how do pathogens spend time in environmental reservoirs?* Some pathogens can only survive for seconds outside of a host, whilst others can persist in the environment for decades. Some pathogens simply survive where they are deposited in abiotic environments, others can move around searching for hosts, and still others can eat and reproduce without a host indefinitely. There is no agreed time cut-off (e.g. 10 minutes) or life history requirement (e.g. pathogen must reproduce in environment) for categorizing environmental reservoirs or environmental transmission. Therefore, we can somewhat vaguely define environmental reservoirs as spaces where pathogens spend a proportion of their life cycle, and which contribute to a portion of transmission events. Alternatively, we can say that the amount of time spent and number of transmission events from the environmental reservoir is *not negligible*—we cannot ignore the environmental reservoir if we want to understand, predict, or control disease dynamics.

*Where do environmental reservoirs exist?* An environmental reservoir can be anything from a scattering of mouse faeces to an ocean to a dust storm. Environmental reservoirs are often categorized as soil, water, faeces, or plant/food surfaces, but there is often overlap among these categories, and indeed a disturbingly large overlap between faeces and the other categories.

*Why do pathogens have environmental reservoirs?* Some pathogens have little to no environmental transmission and yet still spread effectively, like HIV, which caused ~1.5 million new infections globally in 2021. If pathogens can spread so widely without environmental transmission, why has environmental transmission evolved? For some pathogens, the answer is that interactions between

infectious and susceptible hosts are rare enough (and transmission rates are low enough) that any genetic variants that could survive in the environment and thereby increase interaction potential with susceptible hosts had a **fitness advantage**—they would be better able to survive or reproduce. This helps explain why so many pathogens are predominantly transmitted through 'direct contacts' but also persist in the environment after their hosts expel bodily fluids into the environment by sneezing, vomiting, defaecating or urinating. Other pathogens might have started out as free-living species that lived full time in the environment, which then evolved to opportunistically infect hosts because it occasionally provided a method for rapid reproduction and dispersal. In fact, some pathogens are still simply opportunistic and host infections are incidental (see Section 6.3).

To wrap up, **environmental reservoirs** are nonliving habitats (e.g. soil, water, faeces, or plant/food surfaces) where pathogens spend a portion of their life cycle outside of hosts and vectors and from which they can be transmitted to new hosts.

## 6.3 Soil and plants as environmental reservoirs

Some of the most lethal and difficult to control pathogens are **sapronotic disease agents**: pathogens that can persist indefinitely without a host in the abiotic environment and cause a group of diseases called **sapronoses** (Kuris et al. 2014). Examples of sapronoses include brain-eating amoeba (*Naegleria fowleri*), tetanus (*Clostridium tetani*) and cholera (*Vibrio cholerae*). Because sapronotic disease agents can persist long-term in the environment, preventing transmission and infection may be limited to few options, including preventative medicine (e.g. tetanus vaccination) or eliminating contact with environments that contain the sapronotic agent (e.g. banning swimming in contaminated water bodies). These pathogens are often highly lethal because there is no 'lost transmission cost' to killing the host; if hosts are not a major source of dispersal, it does not matter if the pathogen is so virulent that it quickly kills the infected host. In fact, replicating as much as possible inside the host and causing subsequent host death might be highly advantageous.

Even for pathogens that are not fully sapronotic, long-term environmental persistence is often associated with high virulence. For example, anthrax, caused by the spore-forming bacterium *Bacillus anthracis*, is an obligately lethal pathogen where the host must die before subsequent transmission can occur. Animals are most likely infected by ingesting anthrax spores while grazing at anthrax carcass sites where animals have previously died from anthrax and decomposed. Transmission can also occur away from carcass sites when spores are transported by flies, vultures, water, etc. (Huang et al. 2022). The pathogen is a generalist that can infect most mammals (including humans) and some bird species (Easterday et al. 2020), but the main hosts of anthrax are large ungulates (e.g. zebra, bison, livestock).

In Etosha National Park, Namibia, the most common host for *Bacillus anthracis* is the plains zebra (*Equus quagga*, Fig. 6.5), though blue wildebeest (*Connochaetes taurinus*), springbok (*Antidorcas marsupialis*), and elephants (*Loxodonta africana*) are also vulnerable (Turner et al. 2014). These herbivores tend to avoid fresh carcass sites, which likely protects them from acquiring anthrax from recently dead animals. However, as carcasses decay in these relatively low nutrient environments, the increase in phosphorus and nitrogen can cause 'green-ups'. This lush vegetation attracts herbivores, who may then become infected when they unknowingly consume the anthrax bacterium on the vegetation (Ganz et al. 2014). Grazing at old carcass sites is especially likely to cause infection in new hosts in the first two years after host death. After that, the likelihood of exposure declines (Turner et al. 2014, Carlson et al. 2018), but infection is still possible if herbivores ingest soil in addition to plants.

And we mean that exposure and infection are still possible after a very, very long time. In far-northern Siberia, an outbreak of anthrax occurred that devastated reindeer (*Rangifer tarandus*) herds in 2016, more than *70 years* after the last known case. This outbreak wasn't due to imported anthrax,

**Figure 6.5** A plains zebra (*Equus quagga*) that died of anthrax on an active dust-bath site in Etosha National Park, Namibia. Zebra become infected by ingesting spores while grazing, dying within a week or so. Photo credit: Gabriella Flacke.
From Barandongo et al. (2018).

but followed a regional heat wave that caused the permafrost to thaw. The permafrost contained ancient reindeer carcasses that harboured the persistent bacterial spores that seeded the outbreak. In addition to reindeer, people were also infected; over 100 people were hospitalized, and one boy died (Doucleff 2016). In response, the government euthanized more than 2300 reindeer. This is yet another example of how global climate change and infectious diseases interact and impact human and animal health (Chapter 10).

As this example illustrates, when we refer to 'soil', we are imagining all kinds of earthy substrate (Fig. 6.6). This includes the permafrost combination that is soil, sand, gravel, ice, and (in this case) frozen reindeer carcasses. It also includes the thin veneer of mud on limestone rocks that can harbour *Pseudogymnoascus destructans* in caves. And of course, it also includes rich soil where plants grow, which is why soil-borne and food-borne transmission so often overlap.

*United States, 2013*—From 13 September 2013 to 3 May 2016, the same strain of *Listeria monocytogenes* was found in nine people who were seeking healthcare from four states in the US, three of whom died of listeriosis or other causes; most were elderly patients (CDC 2016). These cases were all eventually linked by DNA sequencing to frozen vegetables of several brands; all were associated with the conglomerate frozen food company that subsequently recalled many products. *Listeria* is a particularly hardy example of a pathogen with an environmental reservoir because it can survive and even grow during refrigeration and freezing—methods that humans have developed specifically to increase food longevity and reduce disease risk due to spoiling. *Listeria* outbreaks have even been linked to contaminated ice cream, proving that nothing in this world is sacred.

In many cases, soil and plant/food reservoirs are established after those substrates are contaminated by water or faecal reservoirs, the topics for the next two sections. Below, we will focus specifically on examples of zoonotic pathogens that are transmitted through water or faecal reservoirs, but we want to note that many pathogens that contaminate our food do not come from wildlife. Instead, they arise from human agriculture and livestock practices that allow faeces to mix with other reservoirs. Approximately 62% of mammalian biomass is now livestock (Bar-On et al. 2018), producing a *lot* of faeces, and humans often use that manure to fertilize crops or allow grazing animals to defecate near crops or water bodies (Phiri et al. 2020). This is a good reminder to wash your vegetables and avoid drinking water that hasn't been treated. This is also a reminder that hundreds of millions of people do not have reliable access to safe and clean water for drinking and washing their produce.

## 6.4 Faeces as an environmental reservoir

By way of caution: avoid the presence of ravens and other carrion eaters, for they are Death's envoys, tasked to deliver souls from this world unto the next.
—Benjamin Fielding (1743)

*North-eastern Arkansas, USA, 2011*—Two siblings, an eight-year-old boy and a five-year-old girl, attended a health clinic due to recently developed dry coughing, vomiting (technically: 'non-bloody emesis'), and vague abdominal discomfort (Haselow et al. 2014, Kroll 2014). They tested positive for streptococcal antigens (*Streptococcus* is a genus of bacteria), so the children were prescribed a ten-day course of antibiotics. However, the antibiotics were ineffective. Within the week, the children were suffering high-grade fevers and worsening coughs, so they were admitted to the hospital. They were diagnosed with pneumonia and treated with alternative antibiotics.

Further investigation into the potential circumstances behind this mysterious illness revealed that the children had attended a family gathering eight days prior to the symptom onset. While there, the children had made a bamboo fort, built a bamboo bonfire to roast hot dogs, raked up leaf litter and ash, and were generally playing in the dirt. Meanwhile, similar symptoms were also emerging in other family members; indeed, 18/19 attendees developed probable or confirmed cases of the disease, and seven cases were hospitalized.

Critically, someone mentioned that the location of the family gathering was also the site of a red-winged blackbird (*Agelaius phoeniceus*) roost. Migrating red-winged blackbirds can reach high densities in autumn: 'so many birds as to darken

**Figure 6.6** Dust bathing by African elephants (*Loxodonta africana*) and giraffes (*Giraffa camelopardalis*) caught up in a dust devil, Etosha National Park, Namibia. Dust clouds potentially expose animals to aerosolized soil-borne pathogens (although dust bathing is not considered a likely transmission mode for inhalational anthrax infections (Barandongo et al. 2018)).
Photos by Yathin Krishnappa.

the sky' (Haselow et al. 2014). This new information led the medical investigators to test the patients for *Histoplasma capsulatum*, a fungus associated with **guano** (accumulated bird or bat faeces) that causes histoplasmosis or 'caver's disease' (Fig. 6.7). Histoplasmosis was confirmed, so the medical team changed the course of treatment from antibiotics (for killing bacteria) to injections of antifungal drugs. Symptoms improved within 48 hours (Kroll 2014, Haselow et al. 2014), and thankfully, all patients recovered.

One must admire the fact that someone was knowledgeable about *Histoplasma capsulatum* and blackbird guano as a potential source. Histoplasmosis is quite rare in the United States, but it has prior history in and around Arkansas. In Arkansas's

*Histoplasma capsulatum* (fungus)—agent of histoplasmosis

 Airborne

Contact with either infected soil and/or guano can aerosolize *H. capsulatum* microconidia—the reproductive fungal spores—which are then inhaled and can develop into yeasts in the lungs' alveoli.

 Enters humans via inhalation. Presumably the same for bats & birds.

Presumably, fungal spores transit through the gastrointestinal tract in bat guano.

World map estimating regions most likely to have histoplasmosis: hyperendemic regions (darkest purple), regular cases (purple), and areas with reported, locally acquired cases (light purple). Histoplasmosis has a worldwide distribution and cases will also occur outside these zones. From Ashraf et al. (2020)

Environmental reservoir: bird & bat guano. Animal reservoir: potentially bats, birds poorly investigated

Bats are potential reservoirs as they can be infected and can harbour *H. capsulatum* in thelungs and gastrointestinal tract (e.g. Hoff and Bigler 1981, González-González et al. 2012, da Paz et al. 2018) Bat excrement cancontain the fungus, and after drying, thepathogen can be transmitted in dust orreproduce in nitrogen-rich guano (July et al. 2008). The role of birds is poorlyinvestigated; blackbird excreta have beenused as sources for molecular studies of thefungus, but it is unknown whether the birdshad the infection or whether guano providedan ideal growth medium (Taylor et al. 2000).

Red-winged blackbird (*Agelaius phoeniceus*). Photo: Laura Schoenle Thomas.

Although fungal isolation is the gold standard for demonstrating a fungal infection, it has low sensitivity because a high fungal burden is needed in tissue samples to obtain a successful fungal isolation. Contamination with other fast-growing microorganisms  can also interfere with the isolation process *H. capsulatum* isolation in nature is an unusual finding and, ingeneral, the rate of fungal isolation from bat droppings and/or other contaminated soil, as well as from infected bats, is low (González-González et al. 2012).

 Infections are often subclinical or oligosymptomatic in immunocompetent individuals. Coughing, headache, fever, chills, fatigue, chest pain and shortness of breath occur in symptomatic cases. Rarely (1 per 2000 cases), lung disease can be severe and disseminated histoplasmosis can be fatal. Mean incubation period is 10 days (range 3–25 days) (Valdez and Salata 1999).

 Influenced by actual exposure and immune system status. Cavers are at higher risk of exposure (Hoff and Bigler 1981, Valdez and Salata 1999)

**Figure 6.7** Anatomy of histoplasmosis.

neighbouring state, Louisiana, workers bulldozed a sugar cane field that was also a blackbird roosting site and thus covered with several inches of bird droppings. The workers subsequently experienced a histoplasmosis outbreak (Storch et al. 1980). Also, while working on an Arkansas courthouse in 1975, workers cleared pigeon guano from the courthouse tower and roof—'covered by old bird droppings a foot deep interspersed with bones and feathers'— and unfortunately deposited the bird droppings near an air conditioner intake. In the following two weeks, more than half (44/84) of the courthouse

employees developed histoplasmosis, as well as two of the workers involved in the excavations and nearly a quarter of the courthouse visitors. Investigators used an oil-droplet generator to recreate the path of airflow through the courthouse building, and the match between airflow and infection patterns suggested that infectious *Histoplasma capsulatum* spores had indeed circulated through the building (Dean et al. 1978). These examples and many others suggest that bird guano provides a rich growth medium for the fungus (Diaz 2018), which is especially risky to humans when it is aerosolized.

Limerick interlude, by Dr Skylar Hopkins:

> The courthouse was covered in poop,
> which had to be moved by the scoop.
> In front of the vent
> of the AC it went,
> and histo-exposed the whole group.

End of interlude.

Several species of bat also support fungal growth in cave roosts by provisioning the environment with guano and inoculating the guano with infective *Histoplasma capsulatum* spores (Diaz 2018). When recreational cavers visit guano-laden areas, they may stir up and inhale the spores, thus contracting the 'caver's disease'. This risk is relatively well-known among frequent cavers and researchers who work in caves, who all avoid guano, like two of the (intrepid) co-authors of this book. *Histoplasma* may pose a bigger health risk to infrequent and unwary visitors at commercial tourist caves (e.g. Julg et al. 2008). For example, the third co-author of this book (the once-wiry, Celtic one) can personally attest to this, after sporting a small round opacity approximately 1 cm in diameter in the upper zone of their chest x-ray. The abnormality resembled a granuloma (a collection of macrophage immune cells) and was originally suspected to be either tuberculosis or a cancerous growth. However, a positive result to a histoplasmin skin test suggested that the granuloma was almost certainly a histoplasmoma. The author likely acquired it during a stay in South America, during which he naively swam in subterranean lakes where bats were roosting overhead.

Many other parasites and pathogens enter the environment via faeces and can persist there for some time. For example, *Shigella* spp. bacteria, which infect >100 million people annually and cause shigellosis, can survive up to 12 days in faeces. The agent of typhoid fever, *Salmonella typhi*, which infects >10 million people per year, can persist in faecal environments for >48 days. And *Escherichia coli*, which infects >100 million people per year, can survive for months in faeces. There are also many larger parasite species, like the eggs and larvae of parasitic worms, that exit their hosts in faeces and remain there until future transmission (e.g. *Baylisascaris procyonis*, or raccoon roundworm). Clearly, faecal–oral (and food-borne) transmission of pathogens creates a huge global burden of human infectious diseases. This might explain why many animals (including humans) are so disgusted by faecal matter. As Weinstein et al. (2018) describe it: 'A rancid meal, a moist handshake, a pile of feces: these phenomena elicit disgust and avoidance that protect humans from our most pervasive consumer—infectious agents'.

## 6.5 Water as an environmental reservoir

Contrary to some popular beliefs, water can serve as an environmental reservoir for parasites and pathogens whether it is moving (e.g. streams) or stagnant (e.g. puddles, cisterns). For example, rainfall runoff often carries pathogens deposited on land (e.g. through animal faeces) to streams, rivers, ponds, lakes, and oceans, where they may survive until future encounters with susceptible host species. And the consequences of drinking contaminated water are dire. In fact, diarrheal disease is the second leading cause of global childhood mortality for children under five years old. This is especially true for poor children who drink pathogen-laden water and do not have access to life-saving rehydration therapies.

*Havelock North, New Zealand, 2016*—On 5–6 August, Havelock North experienced heavy rains. Then, between 7 and 24 August, 225 confirmed and 728 probable campylobacteriosis cases were reported to health authorities. (You might remember *Campylobacter* from Chapter 1, from the example where shepherds acquired infections from lambs.) A household cross-sectional study estimated that over one-third (39%) of residents and over half (56%) of households developed symptoms of gastroenteritis. The outbreak caused a total of 7570 (95% CI 6850–8320) illnesses, including 42 hospitalizations, three cases of Guillain-Barré syndrome, and four cases where *Campylobacter* was a contributive cause of death. This constitutes the largest ever recorded *Campylobacter* outbreak (Gilpin et al. 2020).

Subsequent investigations searched for and sequenced *Campylobacter* species from infected humans, the unchlorinated town water supply (which came from wells or 'bores'), and sheep paddocks near the bores. There were at least 12 genotypes of *Campylobacter* among outbreak cases, but 80% of the isolates were one of just three genotypes. And sure enough, these isolates were also found in the town water supply and sheep faeces (sheep again?!). The most likely pathway of contamination was from the sheep faeces to a stream that was hydrologically connected to the bore intake by the higher waters caused by heavy rainfalls.

The identification of the *Campylobacter* source and the description of the outbreak were possible because of careful surveillance and allowed for a prompt 'boil water notice' and initiation of water chlorination to prevent further transmission (Fig. 6.8).

*California coast*—*Toxoplasma gondii* is a protozoan parasite that sexually reproduces in felids (i.e. felids are the **definitive hosts**). Definitive hosts include wild cats, like bobcats (*Felis rufus*) and mountain lions (*Puma concolor*), and also domestic cats. Infected felids shed *Toxoplamsa gondii* **oocysts**—a hardy environmental stage that can survive for months—in their faeces. From faeces, the oocysts can be further spread to soil, water, and plants. These environmental reservoirs may be sources of infection for future felids that directly consume oocysts (e.g. by drinking contaminated water). But more often, oocysts are first consumed by

warm-blooded **intermediate hosts** (hosts in which sexual reproduction of the parasite does not occur), e.g. rats, that are then later consumed by felids. Inside the intermediate host mammal or bird, *Toxoplasma gondii* turns into a motile stage called a tachyzoite, reproduces asexually, and spreads throughout the intermediate host's body to encyst in tissues, including the muscles and brain. Intriguingly, some studies suggest that *Toxoplasma gondii* is a 'mind controlling' parasite that manipulates prey behaviour, such as by making rodents less neophobic. This makes them more brash and therefore less likely to avoid felid predators (Webster 2001). Note that pet cats would therefore be less likely to become infected (and spread those infections) if they were kept inside or on leash outside, such that they cannot consume oocysts or infected wildlife.

Of course, there is no guarantee that an oocyst spending time in an environmental reservoir will end up inside a prey animal that is likely to be consumed by a felid host. For example, humans are rarely eaten by felids, yet humans are often infected by *Toxoplasma gondii*. Infection prevalence varies by country depending on the propensity for humans to undertake certain behaviours, like eating raw meat and associating with outdoor cats,

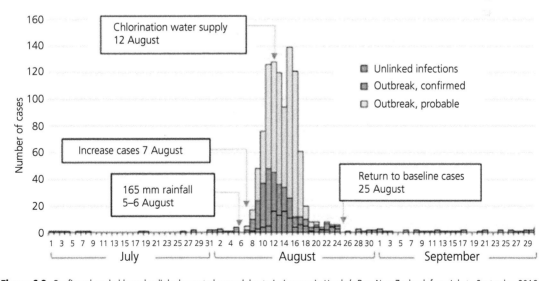

**Figure 6.8** Confirmed, probable and unlinked reported campylobacteriosis cases in Hawke's Bay, New Zealand, from July to September 2016. From Gilpin et al. (2020).

and climatic conditions that influence oocyst survival (Tenter et al. 2000, Lafferty 2006); prevalence is as high as ~12% in the United States and ~45% in France, so there is a good chance that you are infected if you are reading this!

Most people are not notably impacted by their *Toxoplasma gondii* infections. However, *Toxoplasma gondii* is hypothesized to have some side effects on the human brain (Flegr et al. 1996) and country-level infection prevalence may explain some variability in the frequency of personality traits, e.g. neuroticism (Lafferty 2006). Furthermore, infected humans with weak immune systems (e.g. infants, pregnant women, and people with HIV/AIDs) may develop serious and sometimes fatal disease, called toxoplasmosis. Immunocompromised people are therefore warned by product labels to avoid cleaning kitty litter, lest they should be exposed to an

environmental reservoir of oocysts originating from their pet cats.

Long-lived *Toxoplasma gondii* oocysts do not always stay in the local environmental reservoirs where they were first deposited in cat faeces. They can be flushed down toilets or transported by runoff into waterways, and even advanced sewage treatment processes do not kill oocysts (i.e. secondary treatment plants do not, but tertiary treatment plants do).

If oocysts reach the ocean, they can be ingested by marine mammals, potentially leading to serious disease (Fig. 6.9). For example, sea otter (*Enhydra lutris*) populations have increased off the California coast (Fig. 6.10) due to legal protections following near extinction due to hunting for the fur trade. However, *Toxoplasma gondii* and other infectious diseases may be preventing the sea otter population from

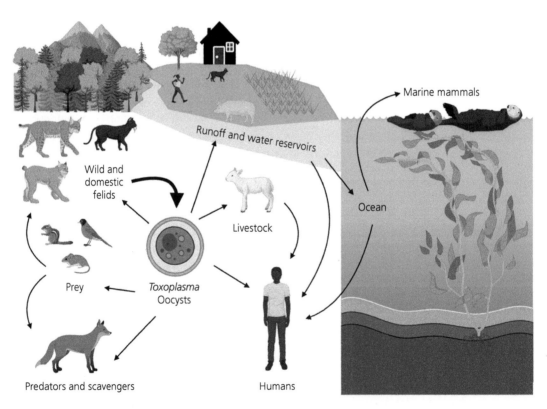

**Figure 6.9** *Toxoplasma gondii* has hardy oocysts that can survive in the environment for long periods, allowing them to be transported widely by runoff and ocean currents. These oocysts first enter environmental reservoirs through felid faeces, and accidental consumption of contaminated water, plants, or prey items leads to subsequent infection in felids or dead-end hosts, like sea otters and people. Made with BioRender.com.

**Figure 6.10** (Top) A champion sea otter (*Enhydra lutris*) at Elkhorn Slough, California, USA. (Bottom) Prime sea otter habitat, Big Sur, California.

(Top) Photo by mana5280 on Unsplash. (Bottom) Photo by Y_S on Unsplash.

recovering to the point of removal from the Endangered Species Act (Lafferty and Gerber 2002, Lafferty 2015). *Toxoplasma* is also known to contribute to deaths of other marine mammals, like Māui dolphins (*Cephalorhynchus hectori maui*) off the coast of New Zealand.

These conservation impacts are not constrained to local coasts, because *Toxoplasma* can survive long-distance dispersal by ocean currents or migratory intermediate hosts, like birds. One study in Antarctica found that 13% of pinnipeds (e.g. seals) had *Toxoplasma gondii* antibodies, even though there are no endemic felids that could be sources of infection (Rengifo-Herrera et al. 2012). This is just one example of how long-lived environmental stages of pathogens can persist in environmental reservoirs and rapidly spread across our increasingly globalized planet (Chapter 9).

## 6.6 Summary

Environmental reservoirs are non-living habitats where pathogens spend a portion of their life cycle outside of hosts and vectors and from which they can be transmitted to new hosts. Diseases caused by pathogens with environmental reservoirs are a major source of morbidity and mortality in humans (e.g. major diarrhoeal diseases) and wildlife (e.g. white nose syndrome in bats). A single pathogen may use multiple environmental reservoirs, such as anthrax persisting and being transmitted through soil, bones, plants, and water. And many pathogens may be maintained in both environmental *and* living reservoirs, as we shall see in the following chapter. When pathogens have persistent environmental reservoirs, it complicates disease control efforts (Chapters 12 and 14).

## 6.7 References

Ashraf, N., Kubat, R.C., Poplin, V., et al. (2020). Redrawing the maps for endemic mycoses. Micropathological, 185, 843–65.

Barandongo, Z.R., Mfune, J.K.E., Turner, W.C. (2018). Dust-bathing behaviors of African herbivores and the potential risk of inhalational anthrax. Journal of Wildlife Diseases, 54, 34–44.

Bar-On, Y.M., Phillips, R., Milo, R. (2018). The biomass distribution on Earth. Proceedings of the National Academy of Sciences, 115(25), 6506–11.

Blehert, D.S., Hicks, A.C., Behr, M., et al. (2009). Bat white-nose syndrome: an emerging fungal pathogen? Science, 323, 227.

Campbell, L.J., Burger, J., Zappalorti, R.T., et al. (2021). Soil reservoir dynamics of *Ophidiomyces ophidiicola*, the causative agent of snake fungal disease. Journal of Fungi, 7, 461.

Carlson, C.J., Getz, W.M., Kausrud, K.L., et al. (2018). Spores and soil from six sides: interdisciplinarity and the environmental biology of anthrax (*Bacillus anthracis*). Biological Reviews, 93, 1813–31.

CDC. (2016). Multistate outbreak of listeriosis linked to frozen vegetables. https://www.cdc.gov/listeria/outbreaks/frozen-vegetables-05-16/index.html, accessed 17/1/2023.

da Paz, G.S., Adorno, B.M.V., Richini-Pereira, V.B., et al. (2018). Infection by *Histoplasma capsulatum*, *Cryptococcus* spp. and *Paracoccidioides brasiliensis* in bats collected in urban areas. Transboundary and Emerging Diseases, 65, 1797–805.

Dean, A.G., Bates, J.H., Sorrels, C., et al. (1978). An outbreak of histoplasmosis at an Arkansas courthouse, with five cases of probable infection. American Journal of Epidemiology, 108, 36–48.

Diaz, J.H. (2018). Environmental and wilderness-related risk factors for histoplasmosis: more than bats in caves. Wilderness and Environmental Medicine, 29, 531–40.

Doucleff, M. (2016). Killing reindeer to stop anthrax could snuff out a nomadic culture. NPR: https://www.npr.org/sections/goatsandsoda/2016/10/12/496568291/killing-reindeer-to-stop-anthrax-could-snuff-out-a-nomadic-culture, accessed 16/10/20.

Easterday, W.R., Ponciano, J.M., Gomez, J.P., et al. (2020). Coalescence modeling of intrainfection *Bacillus anthracis* populations allows estimation of infection parameters in wild populations. Proceedings of the National Academy of Sciences USA, 117, 4273–80.

Field, K.A., Johnson, J.S., Lilley, T.M., et al. (2015). The white-nose syndrome transcriptome: activation of antifungal host responses in wing tissue of hibernating little brown myotis. PLoS Pathogens, 11, e1005168.

Fielding, B. (1743). Midwifery: A Pocket-Companion for Women in their Conception. London. Cited in 'To the Bright Edge of the World' by Eowyn Ivey.

Flegr, J., Zitková, Š., Kodym, P., Frynta, D. (1996). Induction of changes in human behaviour by the parasitic protozoan *Toxoplasma gondii*. Parasitology, 113, 49–54.

Ganz, H.H., Turner, W.C., Brodie, E.L., et al. (2014). Interactions between *Bacillus anthracis* and plants may promote anthrax transmission. PLoS Neglected Tropical Diseases, 8, e2903.

Gargas, A., Trest, M.T., Christensen, M., et al. (2009). *Geomyces destructans* sp. nov., associated with bat white-nose syndrome. Mycotaxon, 108, 147–54.

Gilpin, B.J., Walker, T., Paine, S., et al. (2020). A large scale waterborne Campylobacteriosis outbreak, Havelock North, New Zealand. Journal of Infection, 81, 390–95.

González-González, A.E., Aliouat-Denis, C.M., Carreto-Binaghi, L.E., et al. (2012). An Hcp100 gene fragment reveals *Histoplasma capsulatum* presence in lungs of *Tadarida brasiliensis* migratory bats. Epidemiology & Infection, 140, 1955–63.

Haselow, D.T., Safi, H., Holcomb, D., et al. (2014). Histoplasmosis associated with a bamboo bonfire – Arkansas, October 2011. Morbidity and Mortality Weekly Report (MMWR), 63, 165–8.

Hoff, G.L., Bigler, W.J. (1981). The role of bats in the propagation and spread of histoplasmosis: a review. Journal of Wildlife Diseases, 17, 191–6.

Hoyt, J.R., Langwig, K.E., Sun, K., et al. (2016). Host persistence or extinction from emerging infectious disease: insights from white-nose syndrome in endemic and invading regions. Proceedings of the Royal Society, B: Biological Sciences, 283, 20152861.

Hoyt, J.R., Langwig, K.E., Sun, K., et al. (2020). Environmental reservoir dynamics predict global infection patterns and population impacts for the fungal disease white-nose syndrome. Proceedings of the National Academy of Sciences, 117, 7255–62.

Hoyt, J.R., Langwig, K.E., White, J.P., et al. (2018). Cryptic connections illuminate pathogen transmission within community networks. Nature, 563, 710–13.

Huang, Y.-H., Kausrud, K., Hassim, A., et al. (2022). Environmental drivers of biseasonal anthrax outbreak dynamics in two multihost savanna systems. Ecological Monographs, 92, e1526.

Jülg, B., Elias, J., Zahn, A., et al. (2008). Bat-associated histoplasmosis can be transmitted at entrances of bat caves and not only inside the caves. Journal of Travel Medicine, 15, 133–6.

Kroll, D. (2014). Solving the mystery of 'Bamboo Bonfire' lung disease. https://www.forbes.com/sites/davidkroll/2014/03/06/solving-the-mystery-of-bamboo-bonfire-lung-disease/#131238676c26, accessed 20/2/2019.

Kuris, A.M., Lafferty, K.D., Sokolow, S.H. (2014). Sapronosis: a distinctive type of infectious agent. Trends in Parasitology, 30, 386–93.

Lafferty, K.D. (2006). Can the common brain parasite, *Toxoplasma gondii*, influence human culture? Proceedings of the Royal Society, B: Biological Sciences, 273, 2749–55.

Lafferty, K.D. (2015). Sea otter health: challenging a pet hypothesis. International Journal for Parasitology: Parasites and Wildlife, 4, 291–4.

Lafferty, K.D., Gerber, L.R. (2002). Good medicine for conservation biology: the intersection of epidemiology and conservation theory. Conservation Biology, 16, 593–604.

Lorch, J.M., Meteyer, C.U., Behr, M.J., et al. (2011). Experimental infection of bats with *Geomyces destructans* causes white-nose syndrome. Nature, 480, 376–8.

Lorch, J.M., Muller, L.K., Russell, R.E., et al. (2013). Distribution and environmental persistence of the causative agent of white-nose syndrome, *Geomyces destructans*, in bat hibernacula of the eastern United States. Applied and Environmental Microbiology, 79, 1293–301.

Lorch, J.M., Palmer, J.M., Lindner, D.L., et al. (2016). First detection of bat white-nose syndrome in western North America. mSphere, 1, e00148–16.

Meteyer, C.U., Barber, D., Mandl, J.N. (2012). Pathology in euthermic bats with white nose syndrome suggests a natural manifestation of immune reconstitution inflammatory syndrome. Virulence, 3, 583–8.

Phiri, B.J., Pita, A.B., Hayman, D.T.S., et al. (2020). Does land use affect pathogen presence in New Zealand drinking water supplies? Water Research, 185, 116229.

Puechmaille, S.J., Verdeyroux, P., Fuller, H., et al. (2010). White-nose syndrome fungus (*Geomyces destructans*) in bat, France. Emerging Infectious Diseases, 16, 290–93.

Reeder, D.M., Frank, C.L., Turner, G.G., et al. (2012). Frequent arousal from hibernation linked to severity of infection and mortality in bats with white-nose syndrome. PloS ONE, 7, e38920.

Rengifo-Herrera, C., Ortega-Mora, L.M., Álvarez-García, G., et al. (2012). Detection of *Toxoplasma gondii* antibodies in Antarctic pinnipeds. Veterinary Parasitology, 190, 259–62.

Storch, G., Burford, J.G., George, R.B., et al. (1980). Acute histoplasmosis. Description of an outbreak in northern Louisiana. Chest, 77, 38–42.

Taylor, M.L., Chávez-Tapia, C.B., Reyes-Montes, M.R. (2000). Molecular typing of *Histoplasma capsulatum* isolated from infected bats, captured in Mexico. Fungal Genetics and Biology, 30, 207–12.

Tenter, A.M., Heckeroth, A.R., Weiss, L.M. (2000). *Toxoplasma gondii*: from animals to humans. International Journal for Parasitology, 30, 1217–58.

Turner W.C., Kausrud, K.L., Krishnappa, Y.S. (2014). Fatal attraction: vegetation responses to nutrient inputs attract herbivores to infectious anthrax carcass sites. Proceedings of the Royal Society: Biological Sciences, 281, 20141785.

Valdez, H., Salata, R.A. (1999). Bat-associated histoplasmosis in returning travelers: case presentation and description of a cluster. Journal of Travel Medicine, 6, 258–60.

Warnecke, L., Turner, J.M., Bollinger, T.K., et al. (2013). Pathophysiology of white-nose syndrome in bats: a mechanistic model linking wing damage to mortality. Biology Letters, 9, 20130177.

Webster, J.P. (2001). Rats, cats, people and parasites: the impact of latent toxoplasmosis on behaviour. Microbes and Infection, 3, 1037–145.

Weinstein, S.B., Buck, J.C., Young, H.S. (2018). A landscape of disgust. Science, 359, 1213–14.

Wibbelt, G., Kurth, A., Hellmann, D., et al. (2010). White-nose syndrome fungus (*Geomyces destructans*) in bats, Europe. Emerging Infectious Diseases, 16, 1237–43.

# CHAPTER 7

# Reservoir hosts

## 7.1 Introduction

Reservoir hosts are populations, species or ecological communities that drive disease dynamics because they can become infected by a disease agent *and* sustain transmission long enough to serve as a source of infection for uninfected hosts (Lane et al. 2005).

Some host species may serve only as **spillover hosts** (also called **dead-end hosts**)—hosts that suffer infections and disease, but do not transmit the disease agent to subsequent hosts (see Chapter 1). For cases of spillover, the spillover host cases primarily result from cases in the reservoir host population (Cleaveland and Dye 1995, Haydon et al. 2002, Salkeld and Stapp 2006, Viana et al. 2014, Chapter 1). For example, *Borrelia burgdorferi* is maintained in *Ixodes* ticks and vertebrate wildlife host reservoirs, and humans are usually a dead-end host for the pathogen; there are hundreds of thousands of new cases of Lyme disease in people in the United States each year (Nelson et al. 2015, Kugeler et al. 2021), but humans do not contribute to transmission to other hosts. In systems like this it is critical to treat infected people, but that intervention has no effect on the force of infection from wildlife or vectors. To reduce human Lyme disease incidence, we must instead target the vectors and reservoir hosts (or the ways that humans interact with them). The same goes for *Campylobacter* infections, or histoplasmosis.

Zoonoses can infect more than one host species; otherwise, they wouldn't be able to infect animals *and* humans! Many pathogens persist in **multi-host communities**—they infect and are transmitted by multiple host species. Multi-host communities can make disease control efforts seem daunting—how can we understand or intervene in such complex systems? Often, though, just one or a few **reservoir host species** are usually responsible for maintaining pathogen transmission in broader host communities.

> We define a reservoir as one or more epidemiologically connected populations or environments in which the pathogen can be permanently maintained and from which infection is transmitted to the defined target population.
>
> Haydon et al. (2002)

## 7.2 Spillover from a single host reservoir: armadillos and leprosy

At its simplest, a zoonotic disease system can be represented as a single source maintenance population (the animal reservoir host) that transmits to a single non-maintenance target population (humans).

Leprosy, also known as Hansen's disease, is caused by the bacterium *Mycobacterium leprae*. Disease symptoms include skin lesions, nerve damage, disfigurement, and disability. Leprosy has long been socially stigmatized, including banishing infected people to leper colonies. This isolation practice started in biblical times because people realized that leprosy was contagious. Transmission is mainly human-to-human, via coughing and sneezing, and typically involves living in close contact with an untreated infected individual for an extended period. Fortunately, in the present day, leprosy can be cured using antibiotics if infected individuals have access to affordable, quality healthcare. But leprosy is still a prevalent and problematic disease: over 200,000 new cases were reported worldwide in 2018, mostly in the tropics and semi-tropics (Truman et al. 2011, Araujo et al.

*Emerging Zoonotic and Wildlife Pathogens*. Dan Salkeld, Skylar Hopkins, and David Hayman, Oxford University Press.
© Oxford University Press (2023). DOI: 10.1093/oso/9780198825920.003.0007

2016, Spencer 2018, WHO 2023; see Box 11.2 in Chapter 11).

Leprosy cases are infrequent in the US—approximately 150 new cases annually—and most cases occur in people that have previously worked or lived in leprosy-endemic areas outside of the United States. However, about a third of cases in the US report no prior travel history and have had no known contact with a person known to have had leprosy, suggesting an alternate source of infection (Truman et al. 2011).

*Mycobacterium leprae* also occurs naturally in wild nine-banded armadillos (*Dasypus novemcinctus*) in the southern states of the US where it is established. Interestingly, the nine-banded armadillo is the *only* known non-human animal host for *Mycobacterium leprae* in the Americas and is the primary animal model for experimental studies of leprosy. (Red squirrels, *Sciurus vulgaris*, have been observed infected with *Mycobacterium leprae* in the United Kingdom (Avanzi et al. 2016), and wild chimpanzees have recently been found with leprosy in West Africa (Hockings et al. 2021), but there are few other identified animal reservoir hosts.) Leprosy was absent from the Americas before the arrival of Columbus, and armadillo populations are thought to have originally become infected by people (anthroponotic spillover), possibly by digging in contaminated soil (Truman et al. 2011, Spencer 2018). In the US, armadillos interact with people in multiple ways: as pets, purses, boots, and something to grill on a barbecue (Spencer 2018; Fig. 7.1), suggesting a potential role as a reservoir host (Fig. 7.2).

In one study of armadillos' reservoir host potential, wild armadillos were captured from five southern US states and screened for *Mycobacterium leprae* (using liver, spleen, or lymph-node tissue) (Truman et al. 2011). Genome sequences from these *Mycobacterium leprae* samples were compared to *Mycobacterium leprae* genome sequences from frozen skin-biopsy specimens from human patients of the National Hansen's Disease Program outpatient clinic in Baton Rouge, Louisiana, USA. Among the human patients, there were several strains of *Mycobacterium leprae*, and many matched *Mycobacterium leprae* samples from regions outside the United States where the patients had previously lived.

autochthonous [ɔ:ˈtɑːkθənəs]—acquired by local transmission, i.e. not travel-associated.

However, a high proportion of the **autochthonous** human leprosy cases were infected with the same unique *Mycobacterium leprae* strain found in the local wild armadillos (Truman et al. 2011, Sharma et al. 2015). The study's authors, Truman et al. (2011), carefully warned that: 'Though it is difficult to establish specific causality, when these data are taken together, they strongly implicate armadillos as a source of infection. Therefore, leprosy appears to be a zoonosis in the southern United States.'

Furthermore, though leprosy has a long history in the United States, it also appears to be an emerging infectious disease because it is increasing in geographic range. Prior to 2009, cases were apparently absent from Florida, Georgia, Mississippi, and Alabama, but *Mycobacterium leprae* has now been observed in populations of nine-banded armadillos from these eastern states (Sharma et al. 2015). This likely has something to do with armadillos expanding their ranges in the United States in the past few decades.

And of course, armadillos are also present in South America. Near Belterra in western Pará state in the Brazilian Amazon, more than 60% (10/16) of armadillos have antibodies indicating past *Mycobacterium leprae* infection, and people frequently contact armadillos: in the year prior to a survey-questionnaire, roughly 65% of people had either hunted armadillos, cleaned the meat for cooking, or eaten them. Some of these interactions are prolonged, because hunters who capture wild armadillos in the surrounding forest don't always kill the animal at first encounter; sometimes the armadillos are brought home alive and kept in enclosures inside the house to fatten them up. And some of the interactions are frequent; amongst the surveyed people who dined on armadillo more than once a month, antibody titres were 50% higher than other survey participants. Overall, seven of the 146 people surveyed were currently diagnosed with leprosy or had previously been diagnosed and received treatment, and 63% (92/146) of people

**Figure 7.1** Armadillo, smooth on the inside, crunchy on the outside, being grilled somewhere between San Antonio and San Ignacio, Belize. 'Nice. Grilled with salt, pepper and some lime, from what I remember,' says one of the book authors (the grizzled, wise one). Two out of three of this book's authors have eaten armadillo; none of the authors have contracted leprosy.
Photos by David Hayman.

had antibody titres indicative of prior exposure (da Silva et al. 2018).

Given the many ways that humans and armadillos interact in the Amazon, there may be multiple transmission routes through which *Mycobacterium leprae* spreads from armadillos to people. When infected animals are kept in the home, household members could be exposed via airborne transmission. Alternately, handling, butchering, and cleaning the animals could result in direct transmission. Cooking armadillo meat should kill *Mycobacterium leprae*, but a locally prepared dish of raw armadillo

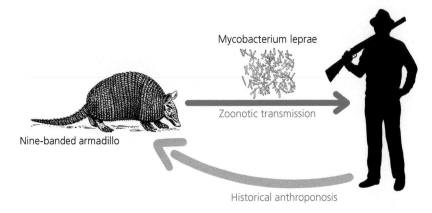

Mycobacterium leprae

Zoonotic transmission

Nine-banded armadillo

Historical anthroponosis

**Figure 7.2**  In the United States, nine-banded armadillos are reservoir hosts of *Mycobacterium leprae*, the disease agent that causes leprosy. Humans can be infected by strains currently circulating in armadillo populations and were likely the original source of infection to armadillos (anthroponosis) in the Americas.
Created by Hanna D. Kiryluk using Canva.com.

liver and onion ceviche could be another source of infection, especially as the liver is one of the main organs where *Mycobacterium leprae* growth is highest (da Silva et al. 2018).

## 7.3  Multiple reservoir hosts: rabies, dogs, and wildlife

Rabies virus has long been feared by humans due to the nearly 100% fatality rate when untreated and the terrifying symptoms that precede death. These symptoms may even have given rise to our legends of vampires and werewolves (Wasik and Murphy 2012). After a bite from a rabid animal, it can take weeks to months for the virus to travel to the human brain. At that point, the infected person may experience flu-like symptoms and pain at the bite wound (now possibly healed). Soon thereafter, the infected person experiences hydrophobia (a fear of water so strong that they may convulse or gag when presented with a glass of water) and/or sialorrhoea (drooling or excessive saliva). During the final stages of the disease, patients often experience fever, paraparesis (partial paralysis of both legs), tetraparesis (partial loss of voluntary motor function in all limbs), throat spasms, hallucinations, hypersexual behaviour (involuntary erections and orgasms), and finally, death.

*France, 1885*—Joseph Meister, age nine, was viciously attacked by the grocer's dog. Fearing the worst—hydrophobia, now called rabies—Joseph was rushed to the Pasteur Institute in Paris within a few days, hoping to become the first (publicized) recipient of the postexposure rabies vaccine. Joseph was injected with tissue from the spinal cord of a rabid rabbit that had previously been desiccated in a jar for 15 days to attenuate the rabies virus. Joseph received many more injections over the next 10 days, each one containing tissue from rabid rabbit spinal cords desiccated for shorter and shorter periods, and thus containing rabies virus that had been progressively less weakened. To confirm that Joseph was truly immune, he was finally inoculated with fresh spinal tissue taken from a dog infected with a particularly fast-killing rabies strain (see below). To Pasteur's (and presumably, the Meisters') great relief, Joseph Meister survived. He grew up and became the caretaker of the Pasteur Institute until his death in 1940, during World War II (Wasik and Murphy 2012).

Before Louis Pasteur developed the postexposure vaccine, many treatments were proposed to cure patients bitten by rabid animals, none of which worked. These included putting salves, poultices, brine pickles, a linen soaked with the oestral blood of a female dog, or the brains or hair from a rabid dog on the bite wound; the last remedy is where the saying 'the hair of the dog that bit you' comes from (Wasik and Murphy 2012). Even more dubious treatments existed, such as causing the patient to sweat profusely in a steam bath or using a live rooster's anus to suck the 'poison' out of the bite wound; a similar remedy has been used to treat venomous snake bites (Kubab 1928).

Having no antivenene available, I applied an indigenous treatment which is much in vogue in the Ratnagiri district [western India]. The fang marks were well incised and chickens, one after the other, with their anuses well stretched were applied to the site of the bite. The first few chickens dropped down dead within a few minutes. From the 42nd chicken onwards, the patient stated that he could distinctly feel the aspirating action of the chickens. In all 74 chickens died, 12 more were half-dead but recovered in about six hours, and the last 6 lost consciousness but recovered speedily; in all 96 chickens were used.
—Kubab 1928

In the late 1800s, the connection between rabid dog bites and human rabies was clear, but debates over the cause of rabies in dogs preceded the widespread adoption of the **germ theory of disease**—the theory that microorganisms or other infectious agents cause disease. Rather than being caused by an infectious agent, rabies was often proposed to be caused by sexual frustration in dogs, mistreatment of dogs, dehydration of dogs, or other triggers seemingly related to the peculiar symptoms of rabies infection (Wasik and Murphy 2012). These hypotheses were all examples of the theory of spontaneous generation—the idea that diseases appear due to bad air, unbalanced bodily humours, and so on. It was in this sceptical culture that Louis Pasteur created the first attenuated vaccines for chicken cholera and anthrax, and then turned his attention to rabies.

Pasteur was confident that rabies was caused by an infectious agent, but unlike chicken cholera and anthrax (both bacteria), Pasteur could not see or culture the comparatively tiny rabies virus. Therefore, he could not satisfy Koch's postulates (Chapter 3) and present proof of transmission. Furthermore, to develop a vaccine for the unculturable virus, Pasteur's team had to keep a constant supply of infected animals in the laboratory.

Pasteur's team first acquired saliva or infected brain tissue from rabid dogs, which they used to infect rabbits. Then they used brain tissue from infected rabbits to inoculate susceptible rabbits, serially infecting 21 batches of rabbits until the **incubation period**—the time from inoculation until symptom onset—decreased from an unpredictable period of weeks or months to a predictable eight days. When this new, lab-maintained rabies strain was injected into a dog, it was far more virulent than the typical rabies seen in village dogs; this strain was the final inoculation that 9-year-old Joseph Meister received after many days of injections of attenuated rabies-infected rabbit tissues (Wasik and Murphy 2012), proving that he had developed immunity to the rabies virus.

In the 21st century, rabies is a vaccine-preventable disease and postexposure rabies prophylaxis (PEP) is also available, though the treatment is still expensive and can be difficult to obtain (Knobel et al. 2005); in poor rural settings, there might not be any postexposure vaccines within easy travel of a person who has been bitten by an animal, travel may require time away from work and lost wages, and the treatment itself might cost 4–6% of a person's gross annual income (Wasik and Murphy 2012). Left untreated, rabies is fatal. Thus, there are more than 50,000 recorded deaths from rabies globally each year, mostly children in Asia and Africa (Shwiff et al. 2013). The true incidence may be 20–160 times higher, because many deaths likely go unreported (Lankester et al. 2014). Because medical treatment alone has failed to protect these most vulnerable populations, the gold standard for rabies control is preventing rabid bites to humans by controlling the reservoir hosts for rabies (Knobel et al. 2005).

More than 95% of human rabies cases are caused by rabid dog bites. Therefore, the most effective way to protect humans is to keep free-ranging dog populations under control as much as possible, and most importantly, to vaccinate dogs against rabies. Dog vaccinations are considered a cost-effective rabies control method, though sporadic funding can make it difficult to achieve dog vaccination targets (Lembo et al. 2010, Lankester et al. 2014, Rysava et al. 2020). Elimination of domestic dog rabies has, however, been achieved on islands and throughout very large regions, including in Mexico and Brazil (Rysava et al. 2020).

Once a dog population is rabies free, efforts must be constantly maintained to prevent imported cases of rabies. (This is why travelling internationally with pet dogs is so complicated and often requires rabies certificates and quarantines.) This raises the question of where infections in dog populations come from. Do domestic dogs get infected from wildlife or is it the other way around?

## 7.4 Interactions between domestic and wildlife reservoirs

*The Serengeti, Tanzania*—In the endless plains of the Serengeti, rabies affects a broad suite of wildlife hosts, including charismatic creatures like the

bat-eared fox (*Otocyon megalotis*), white-tailed mongoose (*Ichneumia albicauda*), spotted hyena (*Crocuta crocuta*), and nonchalant honey badger (*Mellivora capensis*) (Lembo et al. 2008). Domestic dogs (*Canis lupus familiaris*) are also affected, and in much greater numbers (though this observation may be biased because surveillance can more easily observe human–dog interactions and often neglects wildlife populations; (Lembo et al. 2008)). Though all these species are infected by rabies, they are not all necessarily reservoirs, just as humans are not rabies reservoirs even though humans can become infected. Does the virus circulate through wildlife species—in a so-called **sylvatic cycle**—sporadically infecting dogs (Fig. 7.3)? Or do dog populations primarily maintain the virus and act as the spillover source to wildlife? Or does the virus simply persist in an intricate web of transmission among and between all these mammalian components?

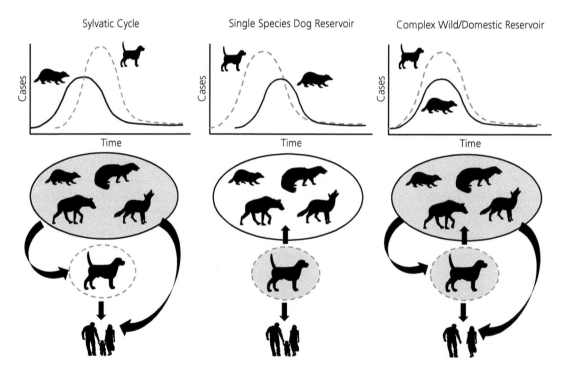

**Figure 7.3** In East Africa, multiple wild carnivore species, dogs and people are susceptible to rabies virus, but not all are likely reservoirs for the virus (Lembo et al. 2008). Most human rabies cases are caused by infected dog bites, but it has been difficult to determine whether rabies periodically spills over into dogs from wildlife, whether dogs are the main reservoir, or whether wildlife and dogs constitute one large multi-species reservoir. Grey-shaded circles infer the potential reservoir host community; arrows represent spillover.

Sampling host species in the Serengeti revealed that all (57/57) rabies virus specimens from dogs and a range of wildlife species belonged to the Africa 1b group of canid-associated rabies virus (Lembo et al. 2008). There was no phylogenetic clustering by host species, suggesting that viral transmission was occurring frequently and across all the examined host species. However, spatiotemporal analyses revealed that rabies cases in the domestic dog populations preceded and predicted cases in other wildlife (Fig. 7.4), and that the majority of

human and livestock rabies exposures were from domestic dogs. Thus, the available evidence suggests that the dog population is driving rabies transmission patterns (Cleaveland and Dye 1995, Lembo et al. 2008).

In south-east Tanzania, most rabies cases (57%) were attributable to domestic dogs but black-backed jackals (*Canis mesomelas*) (Fig. 7.5) accounted for 38% of human rabies exposures (a total of 688 probable human rabies exposures and 47 deaths from 2011–2019 were identified) (Lushasi et al. 2021).

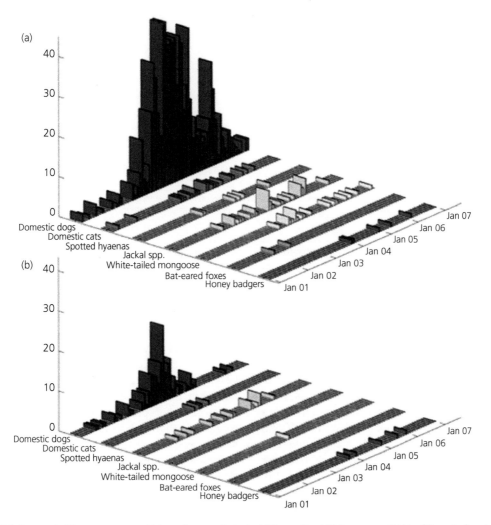

**Figure 7.4** Suspected rabies cases among multiple carnivore species in the (a) Serengeti and (b) Ngorongoro districts of Tanzania, from January 2001 to January 2007.
From Lembo et al. (2008). DOI: (10.1111/j.1365-2664.2008.01468.x)

**Figure 7.5** Black-backed jackal (*Canis mesomelas*) feeding on a springbok carcass in Etosha National Park, Namibia.
Photo by Yathin S. Krishnappa—Own work, CC BY-SA 3.0, https://commons.wikimedia.org/w/index.php?curid=21278579

There, wildlife-to-wildlife rabies transmission is more important than in the Serengeti and Ngorongoro systems, highlighting an important complication: reservoir host species for a given pathogen may vary geographically and over time.

Furthermore, different pathogens circulating in the same host communities may rely on different reservoir hosts (Box 7.1). For example, canine distemper virus (CDV) also infects dogs and multiple wildlife species in the Serengeti ecosystem, including lions. Decades ago, it appeared that periodic CDV epidemics in lions followed CDV epidemics in dogs (Viana et al. 2015), suggesting that dogs were the primary reservoir host. But over time, CDV epidemics in lions became more frequent, and no longer followed dog CDV outbreaks, suggesting that lion populations might now independently maintain CDV transmission (Viana et al. 2015). Comparing how rabies and CDV circulate in the same region among the same carnivore species

illustrates how difficult it can be to infer reservoir dynamics from observational time series in complex multi-host systems.

When possible, interventions that manipulate a suspected reservoir species might yield critical information for disease control. For example, vaccinating dogs may yield insights into whether dogs are really driving disease dynamics for rabies and CDV. In the Serengeti, vaccinating dog populations has reduced rabies incidence in dogs and rabies risks for people (Cleaveland et al. 2003), confirming that human rabies risks are predominantly caused by dog reservoirs. In contrast, CDV vaccination has failed to prevent CDV transmission to lions (Viana et al. 2015), further suggesting that dynamics in lions are separate from dynamics in dogs.

Of course, vaccinating dogs is not easy in rural settings where medical and veterinary resources are limited. Due to these logistical and funding constraints, vaccination targets in dog populations can be difficult to reach, and thus vaccinating dogs might not be sufficient to prevent rabies transmission from dogs to wild carnivores. This is especially true if infected dogs from outside the vaccination region continue to immigrate into the local dog population. Therefore, to protect rapidly declining populations of endangered species like the African

wild dog and Ethiopian wolf, vaccination strategies have focused on vaccinating both dogs and the endangered wild carnivores (Prager et al. 2011). This combined vaccination strategy has successfully reduced rabies mortality in Ethiopian wolves (Randall et al. 2006). We'll further discuss these endangered carnivores and their infectious diseases in Chapter 12.

---

### Box 7.1 Reservoir hosts' contributions to $R_0$

Not all species that contribute to pathogen transmission are 'reservoir host species' (also known as 'maintenance host species'; Haydon et al. 2002). Some are dead-end hosts and have no role in pathogen persistence. And some viable hosts simply don't drive transmission cycles—they cause a negligible number of new cases. Here we will explore a mathematical definition of what it means to be a reservoir or maintenance host versus some other kind of host.

Consider a simple mathematical model that represents a system comprised of a single pathogen that infects two host species: Host Species 1 and Host Species 2. (This model could be expanded to include as many host species as needed, but it is easiest to look at just two, $I_1$ and $I_2$.) We assume that the pathogen is environmentally transmitted, and that both host species can become infected when they are exposed (at rates $\beta_1$ and $\beta_2$). We also assume that both host species can shed pathogens into the environment when they are infected

(at rates $\lambda_1$ and $\lambda_2$). Finally, we assume that hosts can die ($d_i$) and pathogens in the environment can die ($\gamma$). We will use this simple and specific model (Fig. 7.6) to explore the reservoir potential of Hosts 1 and 2, but remember that we could do a similar exercise with a different model, e.g. one that uses direct transmission instead of environmental transmission.

Using the mathematical model implied by our compartmental diagram, we can derive an equation for $R_{0,tot}$, the 'total' basic reproductive rate of the pathogen. As you'll remember from Chapter 5, $R_0$ points to an important threshold for transmission dynamics: when $R_0 > 1$, epidemics tend to occur, whereas when $R_0 < 1$, initial chains of infection tend to stutter to a halt, and an invading pathogen does not successfully establish itself in the host community. For our current model, the $R_{0,tot}$ equation combines the contributions of Host Species 1 and Host Species 2 to pathogen

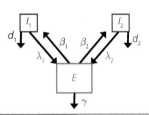

$$\frac{dP_i}{dt} = (1 - P_i)\beta_i E - d_i P_i$$

$$\frac{dE}{dt} = \sum_{i=1}^{n} \lambda_i P_i H_i - \gamma E$$

$$R_{0,i} = \frac{\beta_i \lambda_i H_i}{\gamma d_i}$$

$$R_{0,tot} = \sqrt{\sum_{i=1}^{n} R_{0,i}}$$

| | | | |
|---|---|---|---|
| $H_i$ | Abundance of host species $i$ (out of $n$ species) | $\beta_i$ | Transmission rate to host species $i$ |
| $I_i$ | Abundance of infectious individuals of host species $i$ | $d_i$ | Overall death rate of host species $i$ |
| $P_i$ | Prevalence of infection in host species $i$ | $\lambda_i$ | Rate that host species $i$ produces infectious stages |
| $E$ | Abundance of infectious stages in the environment | $\gamma$ | Mortality rate of environmental infectious stages |

**Figure 7.6** Stock and flow diagram and mathematical equations for a simple system where two host species share an environmentally transmitted pathogen.
Remade from Fenton et al. 2015.

**Box 7.1** *Continued*

transmission in the two-host community ($R_{0,1}$ and $R_{0,2}$). In fact, Fenton et al. (2015) showed that the overall basic reproductive rate, $R_{0,tot}$, is the square root of the summed ability for each host to maintain the pathogen alone; that is, if we calculate the basic reproductive rates for the pathogen when only one of the host species is present, $R_{0,1}$ and $R_{0,2}$, add them together, and take the square root, we'll get the $R_{0,tot}$. It turns out that these details—$R_{0,tot}$, $R_{0,1}$ and $R_{0,2}$—are all we need to classify each host species as a reservoir host or not! (Though getting that data is another question...)

The diagram in Fig. 7.7 is divided into four quadrants based on whether $R_{0,1}$ is greater than or less than 1 and whether $R_{0,2}$ is greater than or less than 1. When they are both less than 1, there are two possible outcomes: (1) if their sum ($R_{0,tot}$) is less than 1, the pathogen cannot be maintained in the host community and it will go extinct; or (2) if their sum is greater than 1, the pathogen can be maintained by both host species together; in this case, we have an **obligate multi-host system**, where the pathogen cannot persist in either host species alone.

When $R_{0,1}$ is greater than 1 and $R_{0,2}$ is less than 1, Host Species 1 can maintain the pathogen on its own, whereas Host Species 2 cannot. In this case, Host Species 1 is a reservoir host and Host Species 2 is a spillover host. Similarly, when $R_{0,2}$ is greater than 1 and $R_{0,1}$ is less than 1, Host Species 2 is a reservoir host and Host Species 1 is a spillover host.

And finally, we have the case where $R_{0,1}$ and $R_{0,2}$ are both greater than 1. Here, both species are reservoir hosts, and we have a **facultative multi-host community**—the pathogen circulates among both species, but it could also persist if one of the host species disappeared.

Having precise mathematical definitions can be helpful, but of course, it's always nice to see concepts illustrated with real-world data. We can make the same four quadrant diagram (Fig. 7.7), but with data from two rodent species—the white-footed mouse (*Peromyscus leucopus*) and the eastern deer mouse (*Peromyscus maniculatus*) sampled for a variety of parasites and pathogens in the eastern US (Streicker et al. 2013, Fenton et al. 2015). *Hymenolepis* (a tapeworm) requires both host species for persistence; *Eimeria spp.* (an apicomplexan) is a facultative multi-host pathogen that could use either host species; *Eimeria delicata* (another apicomplexan) requires white-footed mice as a reservoir host; and none of these pathogens are specific only to eastern deer mice as a reservoir host (Fenton et al. 2015).

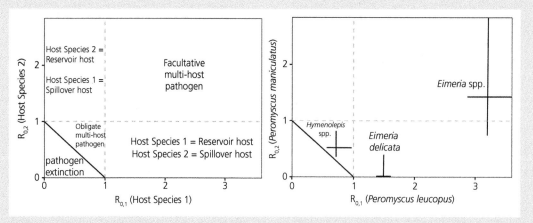

**Figure 7.7** On the left, values of $R_{0,1}$ and $R_{0,2}$ can be compared to the threshold value of 1 to determine whether Host Species 1 and 2 are reservoir hosts. On the right, real data from three pathogens in two rodent species are overlain on the same four-quadrant diagram. The crosses show the 2.5% to 97.5% quantiles for the estimated mean $R_{0,1}$ and $R_{0,2}$.

These diagrams were modified from those published in Fenton et al. (2015), and data from more parasite species than the three used here can be found in that paper.

## 7.5  Spillover from multiple reservoirs: Lyme disease

Lyme disease often begins innocuously—a barely noticeable bite from a nymphal *Ixodes* tick that is the size of a poppy seed. If the tick is infected and injects the bacterium that causes Lyme disease—*Borrelia burgdorferi*—during feeding, a symptomatic rash—*erythema migrans*—may develop as the bacteria disseminate. As the infection spreads, a slew of other flu-like symptoms follow: headache, fever, chills, myalgia, and aching joints. Without timely diagnosis and antibiotic treatment, disease progression is dire, including arthritis, facial palsy, neurological damage, and cardiac complications (Stanek et al. 2012, Forrester et al. 2014). With an estimated 300,000–450,000 new cases every year, Lyme disease is the most frequently contracted vector-borne disease in the United States and constitutes a massive public health burden (Nelson et al. 2015, Kugeler et al. 2021). Human cases of Lyme disease and the black-legged tick vector occur across the eastern US (Arsnoe et al. 2015, Eisen et al. 2016), but the disease is most prevalent in two geographical foci: the north-east and the upper Midwest regions (Eisen and Eisen 2018).

As a reminder (Chapter 4), black-legged ticks (*Ixodes scapularis*) are born uninfected. They can acquire *Borrelia burgdorferi* during their first blood-meal as **larvae** and transmit the bacterium (or acquire it if still uninfected) when they feed during the subsequent **nymphal stage**. Nymphs moult and become **adults**, which is the final life-stage of these Ixodid ticks. This life cycle is important because the ticks feed on many different host species and even switch host species between life stages, complicating our ability to understand and control reservoir dynamics and thereby protect people.

White-footed mice (*Peromyscus leucopus*) are commonly infected with *Borrelia burgdorferi*, and experimental investigations (xenodiagnosis—see Box 7.2) have demonstrated that the mice are able to infect a good proportion of feeding larvae (Donahue et al. 1987, LoGiudice et al. 2003). White-footed mice are therefore a potential reservoir host for Lyme disease, and targeting mice could be an effective way to reduce Lyme disease incidence in human populations.

To test this possibility, Tsao et al. (2004) ingeniously vaccinated white-footed mice against *Borrelia burgdorferi* in Connecticut woodlands. The vaccine they used was brilliant, because they found a way to cure infection in ticks by vaccinating mice! *Borrelia burgdorferi* commonly expresses an antigen called outer surface protein A (OspA) in tick hosts, but *not* in mouse hosts. Consequently, OspA-vaccinated mice, even infected ones, develop antibodies that kill the Lyme disease spirochaete *in the tick* whilst it feeds. The year after mice were immunized, *Borrelia burgdorferi* prevalence in nymphal black-legged ticks was reduced compared to control sites (where a sham vaccine was administered); vaccination caused fewer larvae to become exposed and/or infected by the spirochaetes when feeding on mice. Since mouse vaccination had such a large impact on infection prevalence in vectors, we can infer that mice are important reservoir hosts of *Borrelia burgdorferi*.

However, even though infection prevalence was reduced in nymphal ticks, there were still large numbers of infected ticks (low prevalence in a big population still means many infected individuals). Therefore, the vaccination intervention revealed that *other animals* were also acting as infectious sources for ticks: 'nonmouse hosts contributed more to infecting ticks than previously expected' (Tsao et al. 2004). But the relative importance of mice versus other mammals differed across sites; when mice were more abundant, they played a proportionately larger role in infecting larval ticks. This brings us to a critical observation about the Lyme disease system: there is spatial variation in potential reservoir host communities (e.g. some sites have abundant mice, some sites have fewer mice and perhaps more diverse host communities), and this heterogeneity among sites can influence local disease transmission and control intervention efficacy.

While Tsao et al. (2004) were vaccinating mice, another group of scientists were examining the reservoir competence of species inhabiting oak woodlands in nearby New York State. For *Borrelia burgdorferi* and other vector-borne diseases, **reservoir competence** is the ability of an infected host to infect ticks that are feeding on it. LoGiudice et al. (2003) trapped many species of mammals and birds and then transported them to laboratory cages for three days—enough time for any attached larvae to enjoy their bloodmeal to completion. Engorged larvae would then drop off their hosts into water pans that were beneath the animal cages, where they could be collected by the researchers, maintained whilst they moulted, and then examined for evidence of *Borrelia burgdorferi* infection. This superb investigation allowed description of: (1) the average larval loads on each species (i.e. number of larvae per host); (2) rates of successful tick moulting; and (3) reservoir competence of each host species (proportion of nymphal ticks infected). In other words, we can disentangle how much each vertebrate host contributes to the tick population (by feeding and moulting larval ticks) versus how much each host contributes to infection dynamics.

By capturing diverse woodland denizens—from raccoons to shrews to mice to ovenbirds—the researchers found that a broad suite of animals are competent reservoirs for *Borrelia burgdorferi* (LoGiudice et al. 2003). White-footed mice were again confirmed as important players in Lyme disease ecology: on average, an individual mouse hosted nearly 30 larvae (42% moulted) and infected >90% of them! But other animals were also important co-stars. Eastern chipmunks (*Tamias striatus*) had higher average larval infestations than mice (36 versus 28) and lower reservoir competence, infecting 55% of larvae. Shrews (*Blarina brevicauda* and *Sorex* spp.) were also putting a shift in, and infected 50% of their average 56 larvae. Indeed, genetic and ecological data suggest that inconspicuous shrews feed 35% of all ticks and 55% of *B. burgdorferi*-infected ticks (Brisson et al. 2008), whereas mice 'only' feed 10% of all ticks and 25% of infected ticks. So, surprisingly, shrews are the hosts with the mosts.

The take-home message here is that Lyme disease bacteria persist in multi-host systems, where different wildlife host species contribute in different ways to local disease ecology and force of infection. As we will see below, these complications have led to much debate and confusion regarding how best to control Lyme disease risk for humans.

---

### Box 7.2 Xenodiagnosis

Just like observing a virus or bacterium or parasite in a diseased host doesn't prove that the pathogen caused the disease (Koch's postulates; it could be something else), simply observing evidence of exposure or infection in a purported reservoir host does not prove that the species is a reservoir host. For example, exposure or infection may only indicate dead-end spillover or transient exposure of the pathogen (Estrada-Peña and de la Fuente, 2014). In addition to being infected, reservoir hosts must also be demonstrably infectious—they can transmit the pathogen to other hosts—and they must play a substantial role in maintaining the pathogen in a host community.

**Xenodiagnosis** is a diagnostic method for identifying hosts for vector-borne diseases (Fig. 7.8). The word derives from Greek: *xenos* for foreign, and *diagnosis*. To use this method, an infected host is exposed to blood-feeding vectors, and then the vector is examined for the disease agent. If exposed or infected vertebrate hosts can infect the vectors, then they are *infectious*. After vectors have been infected, they can be fed on uninfected vertebrate hosts to demonstrate whether the vector too can be infectious. These experiments can be logistically challenging.

Before serology and PCR were broadly available, xenodiagnosis was the primary technique used to determine infection status in case-patients! For example, body lice (*Pediculus humanus*) from people who were suspected to have trench fever were examined to demonstrate that *Bartonella quintana* was the etiologic agent of the disease. Similarly, the xenodiagnosis was integral in demonstrating the transmission of Chagas disease trypanosomes (*Trypanosoma cruzi*) by triatomine bugs (Meiser and Schaub 2011). Of course, there were drawbacks to relying heavily on xenodiagnosis, such as waiting for the pathogen to multiply in the vector, allergic reactions to the vector's bite, etc. (Meiser and Schaub 2011).

In the present day, xenodiagnoses can still be very useful. Tests using vectors may be more sensitive than those using host species. For example, naive western black-legged ticks (*Ixodes pacificus*) feeding on western gray squirrels (*Sciurus griseus*) were more likely to show evidence of *Borrelia burgdorferi* infection than PCR tests of tissue biopsies from the very same squirrels (Salkeld et al. 2008).

Additionally, now that genetic analyses are cheaper and more widespread, pathogens inside vectors have become

| **Box 7.2** *Continued* |
| --- |

a critical source of surveillance or **xenosurveillance**. For example, one can collect recently fed vectors from the wild and examine their bloodmeals for the presence of multiple pathogens (Fauver et al. 2017). In many cases, surveying many invertebrate vectors will be easier than surveying many vertebrate animals, so xenosurveillance can be a relatively easy way to discover which pathogens are circulating in a region and which host species vectors feed on the most.

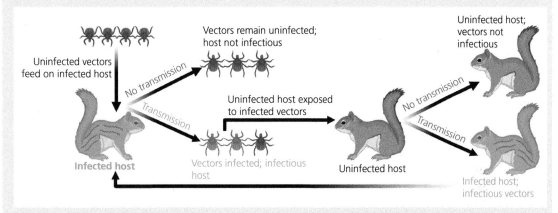

**Figure 7.8** Conceptual diagram of the xenodiagnosis framework used to identify reservoir hosts and vectors for vector-borne diseases. First, uninfected potential vectors are fed on infected hosts. Then the vectors are examined for the disease agent. Pathogen-positive vectors demonstrate that the host is capable of infecting vectors; 100% uninfected vectors suggest that the infected host is not infectious, or the vector species cannot acquire the pathogen. To compare reservoir competence among host species, this process could be repeated for multiple host species and the proportion of vectors that became infected can be quantified and compared. Vector competence can be demonstrated by allowing infected vectors to feed on uninfected reservoir hosts; infection of the naïve hosts demonstrates that the vectors can transmit the pathogen to this host species.
Figure adapted from Estrada-Peña and de la Fuente (2014); created using BioRender.com.

## 7.6 Reservoir hosts and the dilution effect

Lyme disease incidence in the north-east US has increased steadily since the disease was described in the early 1980s—a worrying trend for local communities. Given how debilitating Lyme disease can be, people are very interested in figuring out how to intervene and reduce potential (interacting) drivers (see Chapters 9–11), including land use change, climate change, and increasing abundance of deer that host adult ticks, as well as increased case detection due to media awareness and improved surveillance.

One potential cause of increased disease risk stems from the **nested-subset hypothesis**, which suggests that land use affects the structure or composition of wildlife reservoir communities in predictable ways. For example, as forests are carved up by housing developments (see forest fragmentation in Chapter 11), the resident wildlife community will likely be affected; some species might become rarer or disappear altogether, like the southern flying squirrel, whereas other species might remain and even increase in abundance, like white-footed mice. The nested-subset hypothesis specifically suggests that these changes to host communities, or species 'assemblages', are predictable and non-random, because a subset of species are generally more resilient whereas other species are more sensitive to land use change (Nupp and Swihart 2000), which is intriguing because we already know how different species act as *Borrelia burgdorferi* reservoir hosts.

However, first, we need to decide how to define Lyme disease risk for humans. At first glance, we could measure human disease risk as 'nymphal infection prevalence' (Chapter 4)—the proportion of nymphs infected with the bacterium. A nymphal infection prevalence (NIP) of 80% suggests a higher risk than an NIP of, say, 12%. But what if we're looking at two people, one of whom has 10 ticks on them with 80% infection prevalence and the other has 67 ticks on them with a prevalence of 12%. They both have eight infected ticks on them! Infection prevalence is important, but it does not tell the whole story.

An alternative measure for disease risk is the *density* of infected nymphs (DIN)—the number of infected nymphs in an area. For example, if you're walking a trail where there are 200 questing nymphs along 100 metres where infection prevalence is 20%, the density of infected nymphs is 40/100 m. If the infection prevalence is the same (20%) but there are only 10 nymphs in that 100 metre stretch of trail, the density of infected nymphs is 2/100 m. Which trail would you prefer to walk down? And this scenario shows that we should consider multiple measures of disease risk.

So... is there a predictable impact of land use change on local Lyme disease risk? One way to answer this question is to develop a mechanistic or process-based model (Chapter 5) that represents the Lyme disease system in different host community scenarios to predict how those scenarios will influence the density and prevalence of infected nymphal ticks. Such a model can be parameterized using data on larval tick loads, species' reservoir competence, host density for each reservoir species, and other important system details. This 'host community model' can describe Lyme disease risk in intact woodlands with all potential reservoir species present and compare it to circumstances when host species are subsequently lost until only resilient mice and chipmunks persist (LoGiudice et al. 2003).

We can do some back-of-the-envelope calculations here with our own simple mechanistic model. The goal of our model is to estimate the magnitude of infected ticks contributed by each host species. We assume that the total density of ticks (infected + uninfected) is a function of the density of the host species (more hosts mean more ticks), which can be estimated using field data or published literature, multiplied by the average larval tick load feeding on that host species, multiplied by the tick moulting percentage for that host species. So, for

eastern chipmunks: 25 chipmunks/ha × 36 average tick larvae on each host × 0.412 (the proportion of larvae that successfully moult to become nymphs) = 371 total nymphs/ha. The reservoir competence of chipmunks is 55%, so chipmunks contribute 371 × 0.55 = 205 infected nymphs/ha. If you do this for each host species and sum the results, you can create the predicted nymphal infection density and prevalence for each host community scenario (Fig. 7.9).

When we apply the model across imagined host communities that decline in species richness, the prevalence (infected proportion) of *Borrelia burgdorferi* infection increases. Therefore, forests with more 'intact' ecological communities will have lower nymphal infection prevalence (and reduced human disease risk) compared to forest patches where white-footed mice are abundant and the sole source of infected ticks. This relationship—decreasing infection prevalence in the tick population with increasing host diversity—is referred to as the **dilution effect**. The dilution effect is exciting because it suggests that diverse ecological communities may protect people from vector-borne pathogens, potentially uniting the goals of conservation and public health: if generalizable, intact ecosystems have lower disease risk.

However, the situation is more complicated than it may first appear. For example, when LoGiudice et al. (2003) first published their model, they carefully warned that the dilution effect may not be caused by host diversity per se. Instead, it may be mostly driven by the abundance of a few particularly important reservoir host species. Furthermore, if we return to our model diagram (Fig. 7.9), we see that although nymphal infection prevalence *decreases* with host diversity, the density of infected nymphal ticks *increases*. So, the model's predictions depend on which metric we use to represent human disease risk! Adopting different measures of disease risk changes the interpretation of whether intact biodiversity might be good or bad for human health.

Of course, this was a modelling approach, so we made simplifying assumptions to explore hypotheses. For example, the model used species-specific average tick burdens that do not fluctuate across time or space. The model assumed that host densities are constant across host communities (i.e. competition for resources or other interactions does not

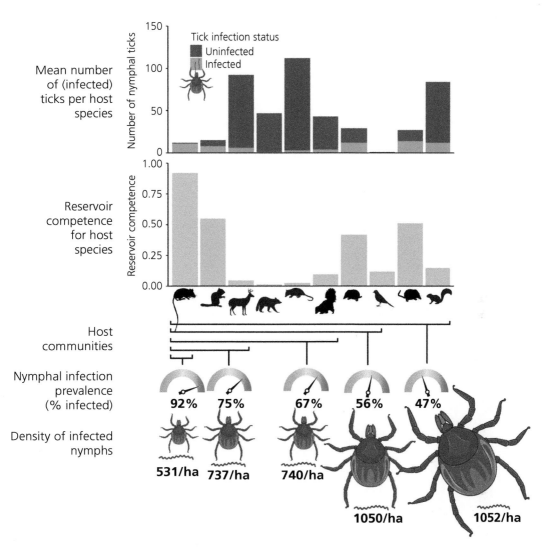

**Figure 7.9** Host community and *Borrelia burgdorferi* dynamics in the north-east US. Vertebrate host species vary in the number of ticks that they feed and infect and therefore how much they contribute to local disease risk. This can be estimated by nymphal infection prevalence (the proportion of feeding ticks that successfully moult and are infected; prevalence rainbows) or density of infected nymphs (conceptualized as the size of the tick icon). The host species are arranged on the x-axis in a 'nested' fashion, showing a gradient from the most diverse and intact host community (all species present) to a highly defaunated host community (e.g. only white-footed mice remain). Compared to just mice, more diverse host communities generate a lower nymphal infection prevalence but a higher density of infected nymphs.

This figure is based on data from LoGiudice et al. (2003) and imagining that white-footed mice and chipmunks have the same densities in all host communities (50 mice and 25 chipmunks/ha). Figure was created using BioRender.com.

change species densities in more/less diverse communities). The model also assumed that infection with *Borrelia burgdorferi* did not influence tick survival; that moulting success data observed in laboratory conditions were representative of moulting success in the wild; and that reservoir competence for each host species remains consistent across the tick activity season regardless of host immunity, previous tick exposures, or prior *Borrelia burgdorferi* infections. These assumptions might be reasonable, but we also know that these details can have important impacts on disease dynamics.

## 7.7 Multiple reservoir hosts and multiple pathogens: tick-borne diseases

Lyme disease is not the only tick-borne scourge of the north-eastern US. A bite from a black-legged tick can also potentially transmit *Borrelia miyamotoi*, *Babesia microti*, Powassan virus, or *Anaplasma phagocytophilum* to humans (to name just four).

*Anaplasma phagocytophilum* is a bacterium that causes human granulocytic anaplasmosis, which normally features the typical symptom list of fever, chills, headache, and myalgia, but can also result in more severe cases, leading to hospitalization and death (Jin et al. 2012). In the north-eastern United States, a single strain appears to infect humans. It's referred to as *Anaplasma phagocytophilum* human-active (*A. phagocytophilum*-ha) (Keesing et al. 2014). For ease of reading, we will refer to this strain as '*Anaplasma*' for the rest of this chapter.

*Babesia microti* is an apicomplexan parasite (like *Plasmodium* that causes malaria) that infects red blood cells and causes babesiosis in humans. Infections are often asymptomatic, or limited to flu-like symptoms, but can result in haemolytic anaemia (destruction of red blood cells), splenomegaly (an enlarged spleen), hepatomegaly (enlarged liver), renal failure, and even death. Most frequently, *Babesia microti* is transmitted by tick bites, but transmission can also be vertical (transplacentally from mother to foetus) or vehicle-borne (during blood transfusions) (Young et al. 2019).

Both *Anaplasma* and *Babesia* circulate among the same ecological vertebrate host community as *Borrelia burgdorferi*, but reservoir competence for the same host species can be quite different for the different pathogens (Fig. 7.10), e.g. raccoons are not competent hosts for *Borrelia* or *Anaplasma*, but they are highly competent hosts for *Babesia*.

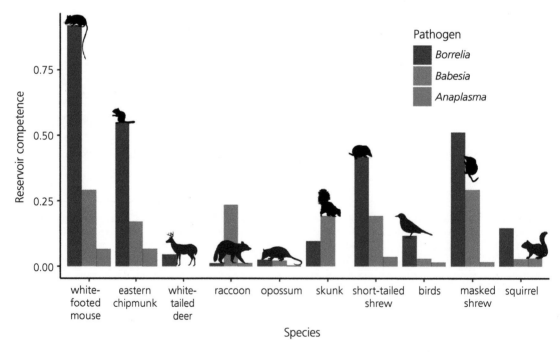

**Figure 7.10** Reservoir competence (average proportion of ticks infected with the disease agent by hosts) for three zoonotic tick-borne pathogens: *Borrelia burgdorferi*, *Babesia microti* and the human-infectious strain of *Anaplasma phagocytophilum* (LoGiudice et al. 2003, Hersh et al. 2012, Keesing et al. 2014). White-tailed deer are not competent reservoir hosts for *Babesia microti* (Piesman et al. 1979, Goethert et al. 2021) or for *Anaplasma phagocytophilum* (Massung et al. 2005, Reichard et al. 2009, Keesing et al. 2014). Data for larval tick loads and moulting success are from LoGiudice et al. (2003). Flying squirrels were not examined in LoGiudice et al. (2003) so were ignored here. Bird species and squirrel species were combined in LoGiudice et al. (2003), so realized reservoir competence for *Babesia* and *Anaplasma* for these groups was calculated by combining and calculating number of ticks infected/number of ticks tested across the species, i.e. 18/628 for *Babesia* for red and grey squirrels; 35/1161 for *Babesia* from birds; 21/655 for *Anaplasma* from red and grey squirrels; and 24/1585 for *Anaplasma* from birds (Hersh et al. 2012, Keesing et al. 2014).

Image created in BioRender.com using some icons from PhyloPic.org, including *Peromyscus leucopus* by Nina Skinner (CC by 3.0).

Understanding the role of human disease risk and ecological communities therefore needs to include multiple pathogens.

## 7.8 Idiosyncrasies of human behaviour and exposure to tick-borne pathogens

It can be difficult to figure out which host species contribute the most to pathogen transmission in host communities, and thus which host communities might have a higher risk of spillover to people (or other focal hosts). To further complicate matters, the way that humans interact with reservoir host communities can vary too! For example, do Lyme disease cases occur after exposure to ticks whilst hiking or hunting in natural areas, which presumably have relatively high host diversity (Fig. 7.11)? Or do more tick encounters and subsequent cases occur in suburban backyards, where host diversity is presumably lower and dominated by a few human-adapted species (Brownstein et al. 2005)? These two scenarios likely differ in both host communities and human behaviours relevant to exposure. These questions were explored using satellite imagery near Lyme, Connecticut, the disease's eponymous town. Interestingly, relationships between measures of 'risk' (density of ticks and prevalence of infection) and incidence of human cases showed conflicting trends when compared to land-use metrics (Brownstein et al. 2005). How could that be?! Data on human movement and

exposure are not easy to obtain, so perhaps the use of human case data averaged across a decade and aggregated by jurisdictional boundaries confounded the results (see Chapter 4). Or perhaps we do not understand enough about the processes driving local *Borrelia burgdorferi* dynamics in wildlife host communities and the exposure and transmission to humans.

Often, the best way to figure out what is happening in a complex system is to do a controlled manipulation and observe what happens next. One such rigorous manipulation—a four-year randomized, placebo-controlled, double-blinded study—found that tick-control interventions in residential areas (fungal sprays and insecticidal bait-boxes) resulted in fewer questing ticks and fewer ticks feeding on rodents, but had no demonstrable effect on incidence of tick-borne diseases in humans (Keesing et al. 2022). One possible explanation for these results is that people are not actually exposed in backyards; if most exposure occurs in natural areas, then tick interventions in suburban areas do not target transmission hot spots. Furthermore, spatial analyses linking human cases with the geographical factors linked to residences may be meaningless; if most exposures occur when people are recreating away from home, we do not need data on their backyards, but rather an improved understanding of how people move and where and when they are exposed to tick bites. For example, one could use mobile apps to reveal movements and activities at

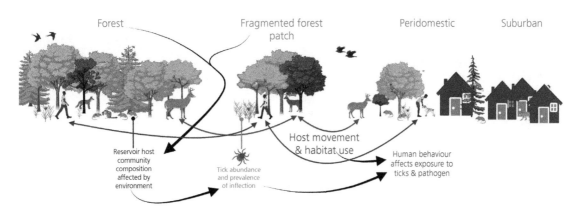

**Figure 7.11** Conceptual diagram illustrating the links between habitat, host community ecology, human behaviour and vector biology that affect exposure to tick bites and tick-borne pathogens.
This figure was inspired by Diuk-Wasser et al. 2021 and made with BioRender.com.

scales germane to tick exposures (Fernandez et al. 2019) and accounting for the fact that biting ticks may go unnoticed for several days.

Given how fascinatingly complex this and other multi-host systems are, the applicability of the dilution effect to human zoonoses remains elusive. The lack of consistent, demonstrable causality between biodiversity and Lyme disease incidence could be due to the difficulties of monitoring and science; flaws in the underlying hypothesis; the scale-dependent nature of disease ecology; or idiosyncrasies that are too difficult to capture.

## 7.9 Contributions of non-reservoir hosts to local disease ecology

The unambiguous moral of the tick-borne disease ecology story is that pathogens, tick-borne or not, can persist in complex multi-host and multi-pathogen systems. And, like politics, disease ecology is a local, idiosyncratic process—pathogen dynamics will vary both geographically and temporally. In fact, let's take one last look at Lyme disease, but in a totally different context.

Though it receives less lyme-light (sorry) than in the north-eastern US, Lyme disease is endemic in California. In north-western California specifically, the archetypal habitat for *Borrelia burgdorferi* dynamics is black oak woodland, and the vector is the western black-legged tick, *Ixodes pacificus* (Fig. 7.12). Rather than mice or shrews, western gray squirrels

(*Sciurus griseus*) are the predominant reservoir hosts for *Borrelia burgdorferi*; they infect nearly 90% of feeding larvae (Lane et al. 2005, Salkeld et al. 2008, Nieto et al. 2010, Roy et al. 2017).

However, a preponderance of ticks feed on two reptile hosts: western fence lizards (*Sceloporus occidentalis*) (Fig. 7.12) and alligator lizards (*Elgaria* spp.). These charismatic reptiles have been observed loaded with 18–35 larvae, and nearly 90% of lizards are infested (Eisen et al. 2004). Incredibly, the lizards' blood contains borreliacidal blood factors that destroy the Lyme disease bacterium. Therefore, lizard-fed ticks do not contribute to *Borrelia burgdorferi* transmission (Lane and Quistad 1998, Kuo et al. 2000, Salkeld and Lane 2010, Swei et al. 2011).

To adopt the host community model described above to host communities in California, it is important to include the role of tick hosts that do not contribute to *Borrelia burgdorferi* infections. For example, lizards provide bloodmeals to ticks and thus contribute to the tick population, but they do not transmit infections. Therefore, in Californian oak woodlands, nymphal infection prevalence is determined by the relative contributions of western gray squirrels and western fence lizards (Salkeld and Lane 2010).

And California isn't the only place where reptiles play a role in Lyme disease dynamics! Back in the eastern US, tick–host associations show a clear shift from mammals in the north to reptiles

**Figure 7.12** (a) Western fence lizard (*Sceloporus occidentalis*) infested with larval and nymphal western black-legged ticks (*Ixodes pacificus*) in the nuchal pocket and neck. (b) Adult female western black-legged ticks questing on vegetation.
Photos by Ervic Aquino.

in the south. Just like in California, south-eastern reptiles (e.g. skinks, anoles, snakes) host ticks but are not good *Borrelia burgdorferi* reservoirs (Ginsberg et al. 2021). This, and related factors (climate, host-seeking behaviour of ticks, and tick densities), likely explain why the south-eastern US has far fewer Lyme disease cases in humans than the north-eastern US. Once again, host communities are an important driver of geographical differences in the incidence of zoonoses.

## 7.10 Summary

It is almost a truism that ecology is complex and that community processes depend on local context, and this certainly applies to host community and disease ecology! In some cases, pathogens are maintained by a single wildlife host reservoir species or population, as we saw with leprosy and armadillos. In many other cases, pathogens are transmitted among multiple host species, and it can be difficult to determine which reservoir host species are the most important for maintaining transmission in the host community. Host community models that aim to describe the different roles of interacting species can be a valuable way to understand local pathogen dynamics, especially when considering the oft-overlooked interactions of non-reservoirs (Fig. 7.13). However, even armed with models and data, it can be difficult to identify reservoir host species. This will be the topic of the next chapter.

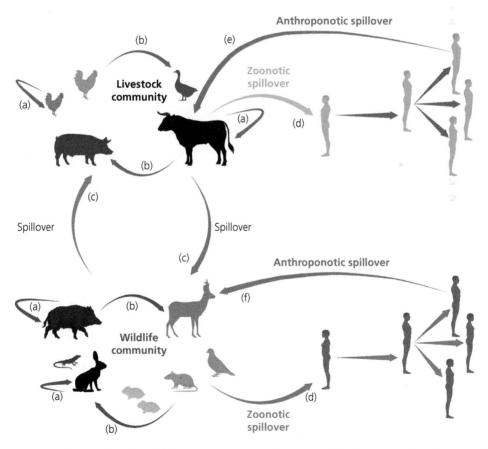

**Figure 7.13** Conceptual diagram illustrating multiple pathogen transmission cycles among wildlife, domestic animals and humans. Pathogens can be transmitted within a single host species (a) or between multiple host species (b). Spillover may occur between wildlife and domestic animals (c) and to humans (d). Humans may also act as a source of disease to animal populations, whether livestock (e) or wildlife (f).

## 7.11 References

Araujo, S., Freitas, L.O., Goulart, L.R., Goulart, I.M.B. (2016). Molecular evidence for the aerial route of infection of *Mycobacterium leprae* and the role of asymptomatic carriers in the persistence of leprosy. Clinical Infectious Diseases, 63, 1412–20.

Arsnoe, I.M., Hickling, G.J., Ginsberg, H.S., et al. (2015). Different populations of blacklegged tick nymphs exhibit differences in questing behavior that have implications for human Lyme disease risk. PLOS One, 10, e0127450.

Avanzi, C., Del-Pozo, J., Benjak, A. (2016) Red squirrels in the British Isles are infected with leprosy bacilli. Science, 354, 744–7.

Brisson, D., Dykhuizen, D.E., Ostfeld, R.S. (2008). Conspicuous impacts of inconspicuous hosts on the Lyme disease epidemic. Proceedings of the Royal Society B: Biological Sciences, 275, 227–35.

Brownstein, J.S., Skelly, D.K., Holford, T.R., Fish, D. (2005). Forest fragmentation predicts local scale heterogeneity of Lyme disease risk. Oecologia, 146, 469–75.

Cleaveland, S., Dye, C. (1995). Maintenance of a microparasite infecting several host species: rabies in the Serengeti. Parasitology, 111, S33–S47.

Cleaveland, S., Kaare, M., Tiringa, P., et al. (2003). A dog rabies vaccination campaign in rural Africa: impact on the incidence of dog rabies and human dog-bite injuries. Vaccine, 21, 1965–73.

da Silva, M.B., Portela, J.M., Li, W., et al. (2018). Evidence of zoonotic leprosy in Pará, Brazilian Amazon, and risks associated with human contact or consumption of armadillos. PLoS Neglected Tropical Diseases, 12, e0006532.

Diuk-Wasser, M.A., VanAcker, M.C., Fernandez, M.P. (2021). Impact of land use changes and habitat fragmentation on the eco-epidemiology of tick-borne diseases. Journal of Medical Entomology, 58, 1546–64.

Donahue, J.G., Piesman, J., Spielman, A. (1987). Reservoir competence of white-footed mice for Lyme disease spirochetes. American Journal of Tropical Medicine and Hygiene, 36, 92–6.

Eisen, L., Eisen, R.J., Lane, R.S. (2004). The roles of birds, lizards, and rodents as hosts for the western blacklegged tick *Ixodes pacificus*. Journal of Vector Ecology, 29, 295–308.

Eisen, R.J., Eisen, L. (2018). The blacklegged tick, *Ixodes scapularis*: an increasing public health concern. Trends in Parasitology, 34, 295–309.

Eisen, R.J., Eisen, L., Beard, C.B. (2016). County-scale distribution of *Ixodes scapularis* and *Ixodes pacificus* (Acari: Ixodidae) in the continental United States. Journal of Medical Entomology, 53, 349–86.

Estrada-Peña, A., de la Fuente, J. (2014). The ecology of ticks and epidemiology of tick-borne viral diseases. Antiviral Research, 108, 104–28.

Fauver, J.R., Gendernalik, A., Weger-Lucarelli, J., et al. (2017). The use of xenosurveillance to detect human bacteria, parasites, and viruses in mosquito bloodmeals. American Journal of Tropical Medicine and Hygiene, 97, 324–9.

Fenton, A., Streicker, D.G., Petchey, O.L., Pedersen, A.B. (2015). Are all hosts created equal? Partitioning host species contributions to parasite persistence in multihost communities. The American Naturalist, 186, 610–22.

Fernandez, M.P, Bron, G.M., Kache, P.A., et al. (2019). Usability and feasibility of a smartphone app to assess human behavioral factors associated with tick exposure (The Tick App): quantitative and qualitative study. JMIR mHealth uHealth, 7, e14769.

Forrester, J.D., Meiman, J., Mullins, J., et al. (2014). Update on Lyme carditis, groups at high risk, and frequency of associated sudden cardiac death – United States. Morbidity and Mortality Weekly Report (MMWR), 63, 982–3.

Ginsberg, H.S., Hickling, G.J., Burke, R.L., et al. (2021). Why Lyme disease is common in the northern US, but rare in the south: the roles of host choice, host-seeking behavior, and tick density. PLoS Biology, 19, e3001066.

Goethert, H.K., Mather, T.N., Buchthal, J., Telford III, S.R. (2021). Retrotransposon-based blood meal analysis of nymphal deer ticks demonstrates spatiotemporal diversity of *Borrelia burgdorferi* and *Babesia microti* reservoirs. Applied and Environmental Microbiology, 87, e02370–20.

Haydon, D.T., Cleaveland, S., Taylor, L.H., et al. (2002). Identifying reservoirs of infection: a conceptual and practical challenge. Emerging Infectious Diseases, 8, 1468–73.

Hersh, M.H., Tibbetts, M., Strauss, M., et al. (2012). Reservoir competence of wildlife host species for *Babesia microti*. Emerging Infectious Diseases, 18, 1951–7.

Hockings, K.J., Mubemba, B., Avanzi, C., et al. (2021). Leprosy in wild chimpanzees. Nature, 598, 652–6.

Jin, H., Wei, F., Liu, Q., Qian, J. (2012). Epidemiology and control of human granulocytic anaplasmosis: a systematic review. Vector Borne Zoonotic Diseases, 12, 269–74.

Keesing, F., McHenry, D.J., Hersh, M., et al. (2014). Prevalence of human-active and variant 1 strains of the tick-borne pathogen *Anaplasma phagocytophilum* in hosts and forests of eastern North America. American Journal of Tropical Medicine and Hygiene, 91, 302–09.

Keesing, F., Mowry, S., Bremer, W., et al. (2022). Effects of tick-control interventions on tick abundance, human encounters with ticks, and incidence of tickborne diseases in residential neighborhoods, New York, USA. Emerging Infectious Diseases, 28, 957–66.

Knobel, D.L., Cleaveland, S., Coleman, P.G., et al. (2005). Re-evaluating the burden of rabies in Africa and Asia. Bulletin World Health Organization, 83, 360–68.

Kubab, K.V. (1928). An indigenous treatment for snakebite. Indian Medical Gazette, 63, 491.

Kugeler, K.J., Schwartz, A.M., Delorey, M.J., et al. (2021). Estimating the frequency of Lyme disease diagnoses, United States, 2010-2018. Emerging Infectious Diseases, 27, 616–19.

Kuo, M.M., Lane, R.S., Giclas, P.C. (2000). A comparative study of mammalian and reptilian alternative pathway of complement-mediated killing of the Lyme disease spirochete (Borrelia burgdorferi). Journal of Parasitology, 86, 1223–8.

Lane, R.S., Mun, J., Eisen, R.J., Eisen, L. (2005). Western gray squirrel (Rodentia: Sciuridae): a primary reservoir host of Borrelia burgdorferi in Californian oak woodlands? Journal of Medical Entomology, 42, 388–96.

Lane, R.S., Quistad, G.B. (1998). Borreliacidal factor in the blood of the western fence lizard (Sceloporus occidentalis). Journal of Parasitology, 84, 29–34.

Lankester, F., Hampson, K., Lembo, T., et al. (2014). Implementing Pasteur's vision for rabies elimination. Science, 345, 1562–4.

Lembo, T., Hampson, K., Haydon, D.T., et al. (2008). Exploring reservoir dynamics: a case study of rabies in the Serengeti ecosystem. Journal of Applied Ecology, 45, 1246–57.

Lembo, T., Hampson, K., Kaare, M.T., et al. (2010). The feasibility of canine rabies elimination in Africa: dispelling doubts with data. PLOS Neglected Tropical Diseases, 4, e626.

LoGiudice, K., Ostfeld, R.S., Schmidt, K.A., Keesing, F. (2003). The ecology of infectious disease: effects of host diversity and community composition on Lyme disease risk. Proceedings of the National Academy of Sciences, USA, 100, 567–71.

Lushasi, K., Hayes, S., Ferguson, E.A., et al. (2021). Reservoir dynamics of rabies in south-east Tanzania and the roles of cross-species transmission and domestic dog vaccination. Journal of Applied Ecology, 58, 2673–85.

Massung, R.F., Courtney, J.W., Hiratzka, S.L., et al. (2005). Anaplasma phagocytophilum in white-tailed deer. Emerging Infectious Diseases, 11, 1604–06.

Meiser, C.K., Schaub, G.A. (2011). Xenodiagnosis. 273–99 in H. Mehlhorn (Ed.), Nature Helps. . . How Plants and Other Organisms Contribute to Solve Health Problems, Springer Berlin Heidelberg.

Nelson, C.A., Saha, S., Kugeler, K.J., et al. (2015). Incidence of clinician-diagnosed Lyme disease, United States, 2005–2010. Emerging Infectious Diseases, 21, 1625–31.

Nieto, N.C., Leonhard, S., Foley, J.E., Lane, R.S. (2010). Coinfection of western gray squirrel (Sciurus griseus) and other sciurid rodents with Borrelia burgdorferi sensu stricto and Anaplasma phagocytophilum in California. Journal of Wildlife Diseases, 46, 291–6.

Nupp, T.N., Swihart, R.K. (2000). Landscape-level correlates of small-mammal assemblages in forest fragments of farmland. Journal of Mammalogy, 81, 512–26.

Piesman, J., Spielman, A., Etkind, P., et al. (1979). Role of deer in the epizootiology of Babesia microti in Massachusetts, USA. Journal of Medical Entomology, 15, 537–40.

Prager, K.C., Woodroffe, R., Cameron, A., Haydon, D.T. (2011). Vaccination strategies to conserve the endangered African wild dog (Lycaon pictus). Biological Conservation, 144, 1940–48.

Randall, D.A., Marino, J., Haydon, D.T., et al. (2006). An integrated disease management strategy for the control of rabies in Ethiopian wolves. Biological Conservation, 131, 151–62.

Reichard, M.V., Roman, R.M., Kocan, K.M., et al. (2009). Inoculation of white-tailed deer (Odocoileus virginianus) with Ap-V1 or NY-18 strains of Anaplasma phagocytophilum and microscopic demonstration of Ap-V1 in Ixodes scapularis adults that acquired infection from deer as nymphs. Vector-Borne and Zoonotic Diseases, 9, 565–8.

Roy, A.N., Straub, M.H., Stephenson, N., et al. (2017). Distribution and diversity of Borrelia burgdorferi sensu lato group bacteria in sciurids of California. Vector-Borne and Zoonotic Diseases, 17, 735–42.

Rysava, K., Mancero, T., Caldas, E., et al. (2020). Towards the elimination of dog-mediated rabies: development and application of an evidence-based management tool. BMC Infectious Diseases, 20, 778.

Salkeld, D.J., Lane, R.S. (2010). Community ecology and disease risk: lizards, squirrels, and the Lyme disease spirochete in California, USA. Ecology, 91, 293–8.

Salkeld, D.J., Leonhard, S., Girard, Y.A., et al. (2008). Identifying the reservoir hosts of the Lyme disease spirochete Borrelia burgdorferi in California: the role of western gray squirrels (Sciurus griseus). American Journal of Tropical Medicine & Hygiene, 79, 535–40.

Salkeld, D.J., Stapp, P. (2006). Seroprevalence rates and transmission of plague (Yersinia pestis) in mammalian carnivores. Vector Borne and Zoonotic Diseases, 6, 231–9.

Sharma, R., Singh, P., Loughry, W.J., et al. (2015). Zoonotic leprosy in the southeastern United States. Emerging Infectious Diseases, 21, 2127–34.

Shwiff, S., Hampson, K., Anderson, A. (2013). Potential economic benefits of eliminating canine rabies. Antiviral Research, 98, 352–6.

Spencer, J.S. (2018). Humans gave leprosy to armadillos – now they are giving it back to us. The Conversation: https://theconversation.com/humans-gave-leprosy-to-armadillos-now-they-are-giving-it-back-to-us-99915#comment_1686381, accessed 17/7/2023.

Stanek, G., Wormser, G.P., Strle, F. (2012). Lyme borreliosis. Lancet, 379, 461–73.

Streicker, D.G., Fenton, A., Pedersen, A.B. (2013). Differential sources of host species heterogeneity influence the transmission and control of multihost parasites. Ecology Letters, 16, 975–84.

Swei, A., Ostfeld, R.S., Lane, R.S., Briggs, C.J. (2011). Impact of the experimental removal of lizards on Lyme disease risk. Proceedings of the Royal Society B: Biological Sciences, 278, 2970–78.

Truman, R.W., Singh, P., Sharma, R., et al. (2011). Probable zoonotic leprosy in the southern United States. New England Journal of Medicine, 364, 1626–33.

Tsao, J.I., Wootton, J.T., Bunikis, J., et al. (2004). An ecological approach to preventing human infection: vaccinating wild mouse reservoirs intervenes in the Lyme disease cycle. Proceedings of the National Academy of Sciences, USA, 101, 18,159–64.

Viana, M., Cleaveland, S., Matthiopoulos, J., et al. (2015). Dynamics of a morbillivirus at the domestic–wildlife interface: canine distemper virus in domestic dogs and lions. Proceedings of the National Academy of Sciences, USA, 112, 1464–9.

Viana, M., Mancy, R., Biek, R., et al. (2014). Assembling evidence for identifying reservoirs of infection. Trends in Ecology & Evolution, 29, 270–79.

Wasik, B., Murphy, M. (2012). Rabid: A Cultural History of the World's Most Diabolical Virus. Viking Press.

WHO (World Health Organization). (2023). Leprosy. https://www.who.int/news-room/fact-sheets/detail/leprosy, accessed 7/17/2020.

Young, K.M., Corrin, T., Wilhelm, B., et al. (2019). Zoonotic *Babesia*: a scoping review of the global evidence. PLoS ONE, 14, e0226781.

# CHAPTER 8

# Identifying animal reservoirs during an epidemic

Definitive identification of infection reservoirs is extremely challenging and rarely possible.
**—Lembo et al. (2008)**

## 8.1 Evidence of infection

Identifying reservoir hosts is no easy matter. Ide-ally, one must be able to demonstrate all the proper-ties of the reservoir host (i.e. hosts can be infectious, population can maintain the pathogen), whilst, at the same time, considering additional species or host populations so one doesn't falsely rule out other sources of infection.

A good example of the search for a pathogen reservoir is that of the 'original' **severe acute respiratory syndrome virus**, otherwise known as the SARS coronavirus, or SARS-CoV. SARS-CoV is one of the posterchildren of emerging zoonotic viruses in the 21st century, though it is now somewhat superseded by its relative, SARS-CoV-2, the disease agent that causes COVID-19. However, SARS-CoV provides an excellent model system for our pur-poses here.

The original SARS emerged in Guangdong province in China in late 2002, though the outbreak really took off in 2003, eventually resulting in 8096 recorded cases, including 774 deaths (9.6% case fatality ratio) (Peiris et al. 2004). The virus was new to science, and the source was unknown. However, early cases had been associated with restaurant workers handling wild animals as exotic food, so an investigation started in a live animal retail market in Shenzhen, a major city in Guangdong province. Nasal and faecal swabs from 25 individual animals were tested for SARS Co-V viral nucleic acid, using reverse transcription–polymerase chain reaction

(RT-PCR), electron microscopy, virus isolation and serology (Guan et al. 2003).

... Guangdong. It's a province of ravenous, unsqueamish carnivores, where the list of animals considered delectable could be mistaken for the inventory of a pet store or a zoo. ... a wildcat fellow named Guan Yi, with the instincts of an epidemiologist and the balls of a brass macaque, had crossed into China and, with cooperation from some local officials, taken swabs from the throats, the anuses, and the cloacae of animals on sale in the biggest live market in Shenzhen.

—David Quammen

Swabs from four of six Himalayan palm civets were positive using the RT-PCR assay (Table 8.1). Four of six palm civets were also positive using virus iso-lation, but two of these weren't the same animals that tested positive by RT-PCR; in total, six out of six Himalayan palm civets had test results indicating current infections (virus present) or prior exposure (antibodies present) (Fig. 8.1). The single raccoon dog that was sampled also tested positive for the virus and antibodies (Table 8.1). In six other species sampled, no virus was detectable (no active infec-tions), but sera from one Chinese ferret-badger had neutralizing antibody to the coronavirus.

How should we interpret these data? The first inference is simply that SARS-CoV can infect three species: Himalayan palm civets, raccoon dogs and Chinese ferret-badgers. Furthermore, the preva-lence of the virus in these three potential reservoir species in the market was high (50–100%). Infec-tions in these species in the market are consistent

*Emerging Zoonotic and Wildlife Pathogens.* Dan Salkeld, Skylar Hopkins, and David Hayman, Oxford University Press.
© Oxford University Press (2023). DOI: 10.1093/oso/9780198825920.003.0008

**Table 8.1** Animal species tested for coronavirus detection in a live animal retail market in Shenzhen, China, in 2003. Data from Guan et al. (2003).

| Animal species | Evidence for current infection—PCR (% prevalence) | Evidence for current infection—virus isolation (% prevalence) | Evidence for prior infection—neutralizing antibody (% prevalence) |
|---|---|---|---|
| Beaver (*Castor fiber*) | 0/3 | | 0/3 |
| Chinese ferret-badger (*Melogale moschata*) | 0/2 | | 1/2 (50%) |
| Chinese hare (*Lepus sinensis*) | 0/4 | | 0/3 |
| Chinese muntjac (*Muntiacus reevesi*) | 0/2 | 0/2 | 0/2 |
| Domestic cat (*Felis catus*) | 0/4 | | 0/3 |
| Hog-badger (*Arctonyx collaris*) | 0/3 | | 0/1 |
| Himalayan palm civet (*Paguma larvata*) | 4/6 (67%) | 4/6 (67%) | 3/4 (75%) |
| Raccoon dog (*Nyctereutes procyonoides*) | 1/1 (100%) | 1/1 (100%) | 1/1 (100%) |

with the hypothesis that the virus spillover into humans was from a wild animal source in the Chinese food markets. There was also evidence of higher rates of virus exposure in wild-animal traders (8/20, 40% seropositive) and workers who slaughter these animals (3/15, 20% seropositive) compared to vegetable traders (1/20, 5%) or control sera from patients admitted to a Guangdong hospital for non-respiratory diseases (0/60) (Guan et al. 2003).

However, there are also caveats to be recognized.

The sample sizes for examined individual animals, especially for each species, is very small. If just one more beaver had been examined and happened to be seropositive, then prevalence would be an intriguing 25% for that species. Conversely (and hypothetically), if seroprevalence was approximately 10% in beavers (1/10 sampled hosts), then you would need to have a larger sample size to reliably observe evidence of infection in beavers. This is why statistics is so important! When one estimates an infection prevalence from a population sample (e.g. from three beavers), there is uncertainty around that estimate due to chance, and the uncertainty is larger for smaller sample sizes (Figs. 8.1, 8.2). Of course, it is easy to disparage these small sample sizes, but we must also recognize that larger sample sizes may not have been possible; investigators were sampling animals in a food market where the vendors might be reluctant, and even quite hostile, to an investigation that could paint them as perpetrators of a pandemic zoonosis.

Ahhh, if I had a nickel for every time I wish I'd sampled just one more beaver.

—Anon.

Given these sampling constraints, the data may not paint a full or accurate picture due to the sampling design. The prevalence values could be inflated if the sampled animals weren't independent. For example, they could be clusters of animals all exposed from a similar source, such as if all six Himalayan palm civets had been kept in a single cage, or nearby cages, and all simultaneously exposed to some other source of SARS-CoV. This further reminds us that these data do not tell us anything of the direction of transmission. These samples were taken after people were infected, so it is feasible that infected people transmitted their viruses to the wildlife; genomic analyses can help provide more information on this (see below). Also, we don't have a full picture of the animals that were present but not sampled. Were the eight species examined here the most common species? If not, the animals that were sampled here might not be representative of the whole panoply of exotic food sources, and the 'true' reservoir species could have been missed. It was 'not clear whether any one or more of these animals are the natural reservoir in the wild. It is conceivable that civets, raccoon dog (Fig. 8.3), and ferret badgers were all infected from another, as yet unknown, animal source, which is in fact the true reservoir in nature. However, because of the culinary practices of southern China, these market animals may

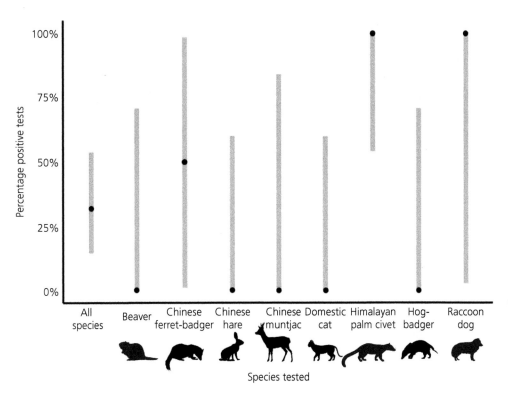

**Figure 8.1** The combined serological and infection prevalence of animals infected with SARS coronavirus in the market in China as measured by Guan et al. (2003). These data demonstrate proof of principle that wildlife were infected with the SARS coronavirus, but small sample sizes have large two-sided binomial confidence intervals (grey bars) and thus lots of uncertainty regarding the actual prevalence of SARS infection and exposure. Typically, seropositive and virus positive data should *not* be combined, but here we combine them to illustrate the importance of judiciously interpreting information from small sample sizes.

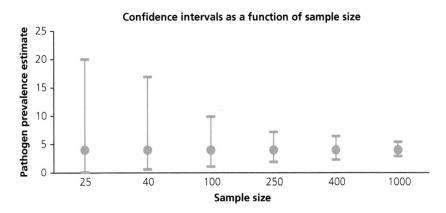

**Figure 8.2** The impact of sample size on 95% confidence intervals. Here, prevalence is a constant 4%, but small sample sizes have large confidence intervals. Larger sample sizes improve interpretation of pathogen prevalence, as long as aggregation of data is not misleading. From Salkeld et al. (2021).

**Figure 8.3** (a) Hog-badger (*Arctonyx collaris*) and (b) raccoon dog (*Nyctereutes procyonoides*).
Photo (a) by Rushenb, CC BY-SA 4.0, https://commons.wikimedia.org/wiki/File:Arctonyx-collaris-hog-badger.jpg, and photo (b) by Ryzhkov Sergey, CC BY-SA 4.0, https://commons.wikimedia.org/w/index.php?curid=79109907

be intermediate hosts that increase the opportunity for transmission of infection to humans. Further extensive surveillance on animals will help to better understand the animal reservoir in nature and the interspecies transmission events that led to the origin of the SARS outbreak' (Guan et al. 2003).

The SARS epidemic (or pandemic) that began in late 2002 was declared over by the World Health Organization in July 2003. But four new cases of SARS were reported during December 2003 and January 2004 in Guangzhou, Guangdong province (Wang et al. 2005). One patient was a 20-year-old waiter from a restaurant that butchered and cooked palm civets. The waiter denied eating palm civets herself, but during her work she was often in the vicinity of the palm civets' cages, which were stacked at the front door of the restaurant so customers could select their desired palm civet for dinner. A second SARS case was a 40-year-old physician who had eaten at the restaurant on New Year's Eve; though his meal was not described, his dining table was within 5 metres of the civet cages. Antibodies against SARS-CoV were also detected in an additional two restaurant employees (of 39) with no reported history of illness or fever: the head waiter, who often assisted customers in choosing palm civets, and a cook. Analyses of the complete genome sequences of SARS-CoV from palm civets at the restaurant and human cases suggested that the humans had indeed been infected by the palm civets at the restaurant because they were in close proximity (the employees at the restaurant had not eaten palm civets) (Wang et al. 2005).

These new human cases suggested that palm civets were a reservoir host for SARS-CoV, and in response, an estimated 10,000 civets were culled in China (Parry 2004, Wang et al. 2005). The World Health Organization expressed serious concerns about the methods adopted to kill the civets (including drowning in disinfectant and clubbing), citing worries for the safety of the people involved. Additionally, Dr J. Hall, WHO's communicable diseases surveillance and response coordinator for China at the time, was concerned that palm civet culls would 'squander the opportunity for extensive sampling from civets and may lead to a false sense of security ... [as] they could be an intermediary carrier' (quoted in Parry 2004).

It was like a medieval pogrom against satanic cats.
—David Quammen

Fortunately, there was a survey in the Xinyuan animal market, Guangzhou, China in January 2004, prior to the civet culling (Kan et al. 2005). In this survey, 91 palm civets (*Paguma larvata*) and 15 raccoon dogs (*Nyctereutes procyonoides*) were sampled and every one of them tested positive for a SARS-CoV-like virus!

To try and determine the source of infections in the market animals, 1107 palm civets from 25 farms in 12 Chinese provinces were sampled during 2004. Not a single one of them tested positive for SARS-CoV-like virus! One farmer had supplied 17 palm civets to the Xinyuan market and all of them were coronavirus-positive at the market, even though none of the civets at his farm (n = 169) were positive.

How could this be? Perhaps civets, raccoon dogs and Chinese ferret-badgers were just temporary reservoir hosts, which were not infected in the wild but could acquire infections in the unnatural market ecosystem where many species of animals were presented in small wire cages piled atop one another (see Chapter 9) (Tu et al. 2004, Li et al. 2005). The zoological biodiversity at the Xinyuan animal market included live donkeys, calves, goats, sheep, piglets, American mink, raccoon dogs, farmed foxes, hog-badgers, porcupines, nutria, guinea pigs, rabbits, and birds.

If that was the case, then what was the original reservoir host for the animal infections in the market? In other words, what infected the civets, raccoon dog and Chinese ferret-badger in the first place?

From March to December of 2004, Li et al. (2005) sampled 408 bats in their native habitats (not in markets), including nine species from six genera from three families from four locations in China (Guangdong, Guangxi, Hubei, and Tianjin) (Table 8.2). Three species of horseshoe bats (Fig. 8.4) showed high SARS-CoV antibody prevalence: Pearson's horseshoe bat, the least horseshoe bat, and the big-eared horseshoe bat (Table 8.2). These serological findings were corroborated by PCR analyses of faecal samples that tested positive for SARS-related coronaviruses by a SARS-CoV RT-PCR in two of the same species, plus one sample from a greater horseshoe bat (Table 8.2). Based on these results, Li et al. (2005) advanced a plausible mechanism for the emergence of SARS coronavirus from bat populations to civets to humans: 'Fruit bats including *R. leschenaultii*, and less frequently insectivorous bats [e.g. the horseshoe bats, genus *Rhinolophus*], are found in markets in southern China. An infectious consignment of bats serendipitously juxtaposed with a susceptible amplifying species, such as *P. larvata* [the Himalayan palm civet], at some point in the wildlife supply chain could result in spillover and establishment of a market cycle while susceptible animals are available to maintain infection.'

Subsequent coronavirus analyses and the emergence of SARS-CoV-2 and the COVID-19 pandemic have all helped inform the relationship between coronaviruses. Betacoronaviruses (which include SARS and related viruses) all have their ancestors in bats. The SARS viruses identified in people and civets are in the *Sarbecovirus* subgenus and also appear to have evolved from ancestral viruses in

**Table 8.2** Detection of antibodies to SARS-CoV and PCR amplification of N and P gene fragments with SARS-CoV-specific primers from bat species sampled in four locations in China. Note that reporting multiple tests can help distinguish between historically (antibody positive) and actively infected animals (PCR) at a time and place, revealing whether some sites experienced more recent exposure than others. The ND designation indicates that prevalence was not determined because of poor sample quality or unavailability of specimens from individual animals. Recreated from Li et al. (2005).

| Sampling time | Sampling location | Bat species | Antibody test positive/total (%) | PCR analysis of faecal swabs: positive/total (%) | PCR analysis of respiratory swabs: positive/total (%) |
|---|---|---|---|---|---|
| March 2004 | Naning, Guangxi | *Rousettus leschenaultii* | 1/84 (1.2%) | 0/110 | ND |
| | Maoming, Guandong | *Rousettus leschenaultii* | 0/42 | 0/45 | ND |
| | | *Cynopterus sphinx* | 0/17 | 0/27 | ND |
| July 2004 | Naning, Guangxi | *Rousettus leschenaultii* | ND | 0/55 | 0/55 |
| | Tianjin | *Myotis ricketti* | ND | 0/21 | 0/21 |
| | | *Rhinolophus pusillus* | ND | 0/15 | ND |
| | | *Rhinolophus ferrumequinum* | 0/4 | 1/8 (12.5%) | ND |
| November 2004 | Yichang, Hubei | *Rhinolophus macrotis* | 5/7 (71%) | 1/8 (12.5%) | 0/3 |
| | | *Nyctalus plancyi* | 0/1 | 0/1 | ND |
| | | *Miniopterus schreibersi* | 0/1 | 0/1 | ND |
| | | *Myotis altarium* | 0/1 | 0/1 | ND |
| | | *Rousettus leschenaultii* | 1/58 (1.8%) | ND | ND |
| December 2004 | Nanning, Guangxi | *Rhinolophus pearsonii* | 13/46 (28.3%) | 3/30 (10%) | 0/11 |
| | | *Rhinolophus pussilus* | 2/6 (33.3%) | 0/6 | 0/2 |

**Figure 8.4** Horseshoe bats—the genus *Rhinolophus*—are a diverse group that comprise approximately 70 species across Europe, Asia, Africa and Australia. The 'horseshoe' relates to a large, obvious 'noseleaf', which emits echolocation calls that are used to locate their flying insect prey, as illustrated here in the rufous horseshoe bat (*Rhinolophus rouxii*). Some horseshoe bat species, though not this particular species, have been identified with coronavirus infections.
Photo by Aditya Joshi, CC BY-SA 3.0, https://commons.wikimedia.org/w/index.php?curid=20131071

bats. However, the precise timing of spillover, and whether there were other intermediate hosts, is still unknown and much debated.

## 8.2 Evidence of exposure is not evidence of reservoir competence

Screening for antibodies is distinct from isolating virus, just as a footprint is distinct from a shoe.
—David Quammen

The case of SARS-CoV, bats, and animal markets demonstrates that evidence of infection in a particular host or species is not necessarily evidence that the host or species maintains and sustains the disease agent population. Instead, these data show that

there was exposure and infection, but we cannot tell whether the host is a dead-end, a link in a stuttering chain of infection, or a sustaining reservoir host. 'Infected' is not the same as 'infectious'.

The interpretation of serology and/or evidence of current infection (detecting the actual disease agent) remains 'fraught with uncertainties', because: (1) serology tests can be cross-reactive (i.e. they can be evidence of a prior, different-but-related infection or even background reactivity unrelated to infection); (2) timing of the test can miss the infection, or occur when antibody titres are declining; and (3) tests may be subject to human errors of interpretation and cut-off thresholds (Viana et al. 2014, Chapter 3). These issues can be especially problematic for wild animals and epidemics of previously undescribed pathogens, because the tests to detect the pathogen are being developed as the epidemic unfolds and known negative and positive control sera are not yet available (Gilbert et al. 2013). Due to these complications, several lines of evidence are required before a *bona fide* identification of reservoir host status can be established.

## 8.3 Genomic analyses to identify reservoir sources

**Genomics** is a suite of DNA sequencing and bioinformatics methods used to sequence, assemble, and analyse the **whole genome** of an organism instead of examining just one gene or gene fragment. Genomics allows a more detailed and complete understanding of genetic structure and function, and of the evolutionary relationships among organisms.

Of course, genomic analyses are not simple. Nor do they always manage to sequence the whole genome of the target organism. However, most DNA sequencing techniques produce short fragments of sequences targeting smaller parts of a genome (often of known genes) using specific primers that bind to that target sequence and can be amplified using PCR. Sequencing methods called shotgun sequencing contrastingly generate thousands of small DNA fragments (often 50–300 base pair nucleotides long) from random places in the genome. Bioinformatic tools are then used to align these fragments and, hopefully, assemble them into a complete genome. This may be especially difficult for large genomes and genomes that can change through processes other than mutation and selection, like bacteria that can uptake DNA from the environment or other bacteria.

Genomic epidemiology is the application of genomic methods to epidemiology to understand pathogen transmission and evolution. For example, we often want to know the origin of emerging infectious agents and how they are being transmitted. We also often want to know which genetic changes have allowed a pathogen to escape host immune systems or become more pathogenic. And sometimes we want to know exactly who infected whom, so we can use tiny genetic changes (i.e. substitutions) that accumulate as a pathogen is transmitted

from one host to the next to reconstruct transmission chains among individuals and within communities. These days, we can answer these questions rapidly because integrating genomic, clinical, and epidemiological data can be done in near real-time, e.g. during Ebola virus, Zika virus, and SARS-CoV-2 pandemics (Worobey et al. 2002, 2014a, b, Rambaut et al. 2008, Baize et al. 2014, Ladner et al. 2019, Harvey et al. 2021).

The 2002–2003 SARS-CoV outbreak was one of the earliest and most prominent epidemics where genome sequencing was used. Because there were samples from multiple host species and multiple countries, genetic analyses were able to identify zoonotic transmission during the epidemic (Chinese SARS Molecular Epidemiology Consortium 2004, Kan et al. 2005, Lu et al. 2004, Ruan et al. 2003). SARS-CoV was identified by sequencing clinical samples from the index patient, which allowed the rapid development of diagnostic tests (Poon et al. 2003). Sequence and epidemiological data established that more than one independent zoonotic transmission event had caused human cases (Guan et al. 2003). Comparison of early, middle, and late outbreak human isolates to civet isolates also identified genetic changes that occurred throughout the outbreak, including changes shown to be essential for efficient receptor (ACE2) recognition and binding, and so human infection (Li et al. 2005, Pacciarini et al. 2008). Genetic analyses also helped epidemiologists understand the key role hospital super-spreading events were playing in transmission. Sequencing analyses, therefore, led to identification of the source of human infection, mutations that facilitated human infection, and key factors in the control of the outbreak.

During the Ebola virus epidemic in West Africa in 2013–2016, the power of genomic epidemiology became even more evident (Gire et al. 2014, Gardy and Loman 2018, Ladner et al. 2019). Large, near-real time, whole-genome sequencing studies that used portable sequencing platforms in the field were used to identify transmission chains and describe the epidemic dynamics. These genomic analyses established that the outbreak was initiated by a single spillover event with subsequent sustained human-to-human transmission (Holmes et al. 2016), and novel Ebola virus mutations were identified in human hosts (Diehl et al. 2016, Dietzel et al. 2017, Urbanowicz et al. 2016). Genomic analyses also revealed that transmission could occur via breast milk (Arias et al. 2016) and sexual intercourse with asymptomatic survivors (genomics even revealed a case of Ebola virus sexually transmitted 470 days after onset of symptoms in the initial case and which sparked a new disease cluster), which influenced public health recommendations (Christie et al. 2015, Diallo et al. 2016, Mate et al. 2015, Whitmer et al. 2018, Ladner et al. 2019, Dudas et al. 2017). Ebola virus dynamics in wildlife, and the drivers of emergence, are still uncertain (Hranac et al. 2019), but molecular and genomic analyses have also been used to identify surprising transmission pathways and hosts. For example, genomics revealed that pigs were discovered as the source of Reston Ebola virus after people fell sick (Barrette et al. 2009, WHO 2009). Similarly, genomic analyses of Lassa virus outbreaks have demonstrated that they occur because of multiple distinct spillover events from rodents, rather than extensive human-to-human transmission (Andersen et al. 2015, Siddle et al. 2018, Ladner et al. 2019).

Genomic sequencing helps the reconstruction and understanding of global spread and evolution of pathogens, e.g. chytrid fungi and Tasmanian devil facial tumour disease (O'Hanlon et al. 2018, Murchison et al. 2010, 2012). Along similar lines, genomics can also reveal how long pathogens have been circulating before their 'discovery'. From 1998 to 2012, an extended outbreak of *Salmonella enterica* serovar Typhimurium definitive type 160 (mercifully abbreviated to DT160) affected >3000 humans and killed wild birds in New Zealand (Bloomfield et al. 2017). Whole-genome sequencing of 109 DT160 isolates from sources throughout New Zealand showed that DT160 was introduced into the country around 1997 and rapidly spread, becoming more genetically diverse over time. Because there was no evidence of host group differentiation between isolates collected from human, poultry, bovid, and wild bird sources, transmission between these host groups was likely occurring throughout the 14 years.

Similar approaches have been adopted to understand the transmission chains of SARS-CoV-2. The release of the complete genome data just days after the virus was sequenced allowed the rapid development of both early specific diagnostic qRT-PCR assays (Corman et al. 2020) and the development of novel mRNA vaccines (Oliver et al. 2020, 2021). After SARS-CoV-2 emerged, the hunt to find the source rapidly identified related viruses in Rhinolophid bats and Malayan pangolins (Lam et al. 2020). However, genomic and molecular clock analyses applied to the genomic data suggest that neither is likely the direct relative; instead, SARS-CoV-2 likely diverged from the current known relatives several decades ago (Boni et al. 2020, Fig. 8.5) and, moreover, subsequent studies have found viruses with even greater similarity to SARS-CoV-2 in bats overall and in specific parts of the genome, with further support for several fragments of SARS-CoV-2 genome being from different donor strains due to recombination (Temman et al. 2022). Genomic approaches have also illustrated how SARS-CoV-2 evolves and spreads (GISAID 2020).

Genomic approaches have also been critical for vaccine development beyond SARS-CoV-2. For example, influenza viruses are constantly evolving, either by antigenic drift or antigenic shift. **Antigenic drift** occurs when viral mutations alter the virus's surface proteins, which are recognized by the immune system (e.g. haemagglutinin and neuraminidase for influenza). Over time, antigenic drift allows the virus to be different enough to escape the hosts' immune response, and this is why influenza vaccines must be updated every year. **Antigenic shift** is a more abrupt, major change due to reassortment of the viral genome, and these rare events can occur when a host is co-infected with more than one virus. For example, influenza viruses have segmented genomes, and antigenic shift can create new viral strains with the new haemagglutinin and/or neuraminidase protein combinations that are different enough from other circulating viral strains that it creates a new emerging infectious disease. This was the case in 1998, when an H3N2 virus with genes from North American and Eurasian pigs, people, and birds infected pigs and people, and later in 2009 when pig-origin viruses combined to make a new H1N1 virus that infected people—the so-called 'swine flu' pandemic (Smith et al. 2009). Without modern genomic approaches, it would be difficult to trace these origins and create updated vaccines that target the most important strains in circulation.

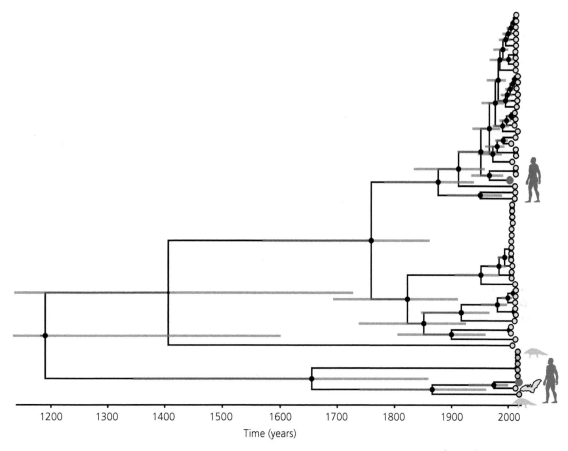

**Figure 8.5** Time-measured phylogenetic tree and divergence times for sarbecoviruses—a subgenus of the *Betacoronavirus* genus (details in Boni et al. 2020). Grey tips in the phylogenetic tree correspond to viruses obtained from bats, green to viruses obtained from pangolins, blue for SARS-CoV and red for SARS-CoV-2. 95% credible interval bars are shown for all internal node ages in blue, showing the uncertainty estimates. The divergence times mean that these are still relatively distantly related viruses if we are trying to discover who infected whom through viral spillover, so there are many events that could have taken place in the time between when viruses were detected in people and pangolins and when their ancestors were in bat hosts; as more viruses are discovered they may fill in some of these gaps.

Genomic analyses and phylogenetic trees can suffer **ascertainment bias** when the methods used to ascertain samples inadvertently bias the data. For example, if the reservoir host source of a pathogen is believed to be a bat, then research will naturally focus on capturing and sampling bats. Any newly discovered related pathogens then confirm, not surprisingly, that the nearest known relatives are indeed circulating in bats. Ideally, broader sampling of the wildlife community should take place to rule out or include additional hosts. Similarly, focusing sampling efforts on a particular place (e.g. central Asia) may confirm that a pathogen is most often found in central Asia! Sampling designs—not just for genetics or genomics—must be carefully designed to consider underlying biases.

## 8.4 Causal association

As epidemics unfold, how *can* we identify reservoir hosts? Or failing that, how can we identify reservoirs retrospectively, in the hopes of preventing future outbreaks? Experiments are often the gold standards for proving causal relationships, but they are not always possible, especially during the rapid response phase of an epidemic. For example, transmission experiments with wildlife are often difficult, and field manipulations are extremely challenging, but they are also the ideal approach for demonstrating the roles of reservoirs in pathogen transmission under natural circumstances. For example, vaccination of a host species to prevent its role in transmission will illustrate whether it has a central role in pathogen transmission, or whether it is part of a suite of reservoir hosts, as was the case for *Borrelia burgdorferi* in Connecticut, USA (Tsao et al. 2004). We often have to make do with imperfect data.

There is no single method that works in every situation because circumstances and data availability will vary. But there are ways to evaluate the strength of evidence (or design investigations to maximize evidence strength) using frameworks such as the **Bradford Hill criteria** for epidemiologic causal association (Table 8.3).

The august Sir Austin Bradford Hill was the epidemiologist who made the argument that lung cancer could be associated with smoking cigarettes. As Sir Hill (1965) phrased it: 'In what circumstances can we pass from [an] observed association to a verdict of causation?' He realized that in many cases, experiments are not possible, but available evidence can still be weighed to suggest possible causal relationships, test hypotheses, and guide further research. Sir Hill was also involved in the first randomized

clinical trial, which tested the efficacy of streptomycin in treating tuberculosis. (Sir Hill had himself experienced a serious tuberculosis infection.) In the next section, we use an example to illustrate the usefulness of the nine criteria proposed by Sir Hill (originally conveyed in Fredricks and Relman 1996).

What I do not believe—and this has been suggested—that we can usefully lay down some hard-and-fast rules of evidence that must be obeyed before we can accept cause and effect. None of my nine viewpoints can bring indisputable evidence for or against the cause-and-effect hypothesis and none can be required as a *sine qua non*. What they can do, with greater or less strength, is to help us to make up our minds on the fundamental question—is there any other way of explaining the set of facts before us, is there any other answer equally, or more, likely than cause and effect?

—Hill, 1965

## 8.5 Finding the reservoir for Hantavirus Pulmonary Syndrome

*Four Corners region, south-western USA, 1993*—A young Navajo man was rushed to a hospital in New Mexico suffering from shortness of breath. He died shortly after arriving, and case investigators discovered that the man's fiancée had also died just a few days previously, and also after experiencing breathlessness. The investigation team was prompted to review recent death records in the area and discovered five further cases of healthy young people who had died under similar circumstances. Dr James Cheek, then of the Indian Health Service, noted that: 'I think if it hadn't been for that initial pair of people that became sick within a week of each other, we never would have discovered the illness at all' (Barcott 2012). The 1993 outbreak in the Four Corners area (Colorado, New Mexico, Arizona, and Utah) of the south-western United States would ultimately claim 27 lives, with a case fatality ratio of 56%. This disease that caused the outbreak was called Sin Nombre virus (SNV)—Spanish for 'without a name'.

Serum samples from symptomatic patients were sent to the Centers for Disease Control and Prevention. The samples were positive for antibodies to several types of hantaviruses, but at that time,

**Table 8.3** Hill's epidemiologic criteria for causal association.

| Causal criterion | Causal association |
|---|---|
| Strength of association | What is the relative risk? |
| Consistency of association | Is there agreement among repeated observations in different places, at different times, using different methodology, by different researchers, under different circumstances? |
| Specificity of association | Is the outcome unique to the exposure? |
| Temporality | Does exposure precede the outcome variable? |
| Biological gradient | Is there evidence of a dose-response relationship? |
| Plausibility | Does the causal relationship make biological sense? |
| Coherence | Is the causal association compatible with present knowledge of the disease? |
| Experimentation | Does controlled manipulation of the exposure variable change the outcome? |
| Analogy | Does the causal relationship conform to a previously described relationship? |

hantaviruses were not recognized as a cause of pulmonary disease in humans. No other antibodies for known pulmonary pathogens were detected, suggesting that this was a new-to-science disease caused by an undescribed hantavirus. A specific RT-PCR assay confirmed the presence of this virus in 10 out of 10 patients studied, with evidence of multiple organ involvement (Fredricks and Relman 1996).

Rodents were known to be the usual hosts of previously characterized hantaviruses, prompting surveillance of potential rodent reservoirs in the Four Corners area. Deer mice (*Peromyscus maniculatus*) were seropositive for the hantavirus (33%) and positive for the disease agent by PCR (82%). In addition, the timing of the outbreak correlated with a local population boom of deer mice in the Four Corners region, and most of the victims had some degree of exposure to rodents. Furthermore, the virus genotypes found in patients closely matched the virus genotypes found in deer mice trapped in or around their homes, where genotypes in different geographic regions were distinct. Interestingly, investigators were able to propagate virus in culture from rodent cell sources, but not from human cell lines (Fredricks and Relman 1996). The disease is now referred to as Hantavirus Pulmonary Syndrome (HPS), and deer mice are considered the reservoir host of the virus.

However, the 1993 outbreak was not the first instance of this disease. The earliest known case of HPS confirmed by Western medicine was a 38-year-old Utah man in 1959. That case was discovered retrospectively by examining samples of preserved tissue belonging to people who had died of unexplained adult respiratory distress syndrome. And the disease had also been recognized by the Navajo Indians, who had observed the correlation of disease onset in humans when mice were abundant (Barcott 2012).

The evidence that Sin Nombre virus spilled over from deer mouse hosts and caused illness and death in exposed humans is compelling. Cases were likely to have been exposed to (infected) mouse populations; exposure to mice preceded disease—both in individual cases and with the recent boom in deer mouse populations; and the links were plausible,

because other hantaviruses use rodent reservoir hosts. However, at the time, hantaviruses were not thought to cause pulmonary disease, so the new findings were not compatible with the contemporary knowledge of the disease. An open-minded approach allows contemporary knowledge to be updated.

## 8.6 Outbreaks are not always caused by spillover from reservoirs

*Gouéké, Guinea, 2021*—A few years after the 2013–2016 Ebola virus outbreak severely affected Guinea, Sierra Leone, and Liberia, a new outbreak of Ebola virus was observed in Guinea in 2021. The outbreak began in Gouéké, a town roughly 200 km from the epicentre of the previous epidemic (Keita et al. 2021). The probable index case was a 51-year-old hospital nurse, who was misdiagnosed with malaria and salmonellosis and died in late January 2021. In the week after her death, her husband and other family members who attended her funeral also fell ill, and four of them died. Blood samples from two patients were confirmed positive for Ebola virus on 13 February. By 5 March, 14 confirmed cases and four probable cases of Ebola virus disease had been identified, causing nine deaths. No new cases were reported for 25 days, two cases occurred on 1 and 3 April, and then the outbreak was declared over on 19 June 2021.

How did the outbreak start? What was the spillover event from a reservoir or intermediate host (e.g. a bat, ape, or duiker)? Surprisingly, genomic characterization of the Ebola virus samples from 2021 suggested that this cluster was *not* the result of a new spillover event. Instead, it appears that this was a resurgence of a virus that had previously circulated in the 2013–2016 Ebola virus outbreak in West Africa. Maximum likelihood phylogenetic reconstruction positioned the genomes from the 2021 Guinea outbreak within a single cluster among the EBOV viruses from the 2013–2016 West Africa outbreak.

Given how little the virus had changed in six years, the virus either had a very slow long-term evolutionary rate, or more intriguingly, it persisted in a host with a long phase of dormant infection (Keita et al. 2021). If this mechanism is valid, and it appears that it is, then it explains how a seemingly rare disease can sporadically burst into outbreaks and then 'disappear' for long periods (Box 8.1, Fig. 8.6). In the case of Ebola virus disease, the virus was already known to be capable of persisting in the bodily fluids of survivors and relapsing. Indeed, in the large 2013–2016 outbreak, several transmission chains were observed that were associated with viral persistence in semen, even six months after infection. If relapses of illness after a long latent period can cause new transmission chains, it challenges the assumption that outbreaks are always the result of zoonotic spillover (from bats or apes or duikers) leading to human-to-human transmission.

Instead, new outbreaks could be the result of transmission from humans who were infected during a previous epidemic, even one occurring a half-decade previously.

This Guinean outbreak of Ebola virus reminds us that genome sequencing is and should be a major component of outbreak response because it allows insights into pathogen evolution and transmission. Additionally, we must always be humble and prepared to revisit our understanding of how disease systems work (Box 8.2). Even when there is a known reservoir host for a pathogen, we may discover that some outbreaks have nothing to do with that reservoir host.

---

### Box 8.1 Reservoir hosts and latent infections

This chapter contains the example of Ebola virus potentially hiding out in human bodies for long periods before re-emerging and causing new outbreaks. Long latency might also play a role in other disease systems when wildlife reservoir hosts can maintain pathogens for a long time. For example, a 45-year-old shepherd in China died from bubonic plague (*Yersinia pestis*) after excavating, skinning, and eating a hibernating marmot (*Marmota himalayana*), suggesting that the bacterium can viably persist in a marmot host during periods of torpor (Xi et al. 2022). Therefore, plague may become resurgent after the hosts' emergence from hibernation (Tang et al. 2022). Similarly, rabies virus can sometimes persist for long periods in big brown bats (*Eptesicus fuscus*), and this may be due to bat hibernation and pathogen dormancy (George et al. 2011). In fact, it might be very beneficial for pathogens to have ways to persist in their hosts for a long time, because pathogens can only be maintained in host populations over the long term if there is a way for the pathogen to be transmitted from one host generation to the next.

In temperate regions, many animals have seasonal population cycles where adults from multiple generations don't overlap and/or adults and their offspring don't temporally overlap. For example, two-spotted ladybird beetles (*Adalia bipunctata*) overwinter as adults, emerge in the spring, and in May–June they have lots of sex and lay lots of eggs, and then die. Also, at some point, the new cohort emerges from the eggs and individuals develop from larvae to sexually mature adults.

Two-spotted ladybirds also suffer from mite (*Coccipolipus hippodamiae*) infestations that feed on ladybird blood. Interestingly, not all ladybird populations have mites, and the presence or absence of mites in a population is consistent over time; populations in far north Sweden (north of 61°N) tend to be mite-free, and populations south of 61°N tend to have mites. Yet the northern ladybirds are susceptible to the mites when experimentally infested in the laboratory, so the geographical differences are not due to physiological or biological mechanisms (Pastok et al. 2016).

Instead, that division by latitude suggests that ecology (specifically, phenology) might play an important role in intergenerational mite transmission. Here's what happens: in August, southern mite populations have adults from both the overwintered ladybird beetle generation *and* the new ladybird beetle generation (Pastok et al. 2016). At least some of the new adults have already had sex when the old generation is still present, which means that sexual contacts between generations are happening. In contrast, in the northern populations, the overwintered ladybirds die sooner, and the new ladybirds mature later, so there is no overlap among generations (Pastok et al. 2016). That generation gap prevents sexually transmitted mites from persisting in those northern populations!

This is just one example of how phenology plays an important role in pathogen transmission. As host phenologies continue to change in response to the changing global climate (see Chapter 10), the role of host phenology and pathogen latent periods may become increasingly important determinants of disease dynamics.

**Box 8.1** *Continued*

**Figure 8.6** Two-spot ladybirds (*Adalia bipunctata*).
Photo by Lech Borowiec.

---

**Box 8.2 Pentastomes and pythons**

Burmese pythons (*Python molurus bivittatus*) are native to south-east Asia, but there is an invasive population of pythons in southern Florida, USA. The breeding population is established in the four southernmost counties, and occasional sightings of pythons further north are likely newly released pets. Like many invasive species, the Burmese python did not come to Florida alone.

*Raillietiella orientalis* is a pentastome parasite—a large crustacean that looks like a worm and lives inside snake lungs (Fig. 8.7). In its native range in Asia and Australia, *Raillietiella orientalis* is a host generalist; it infects snake species from multiple snake families, including Burmese pythons. A study published in 2018 found *Raillietiella orientalis* in pythons in Florida (Miller et al. 2018)—the first time this parasite species had ever been documented in the United States. The survey also found that *Raillietiella orientalis* was not limited to pythons in Florida; by dissecting native species that had been run over by cars, the investigators found that

nine native species also had the huge crustacean inside them (Miller et al. 2018).

Since *Raillietiella orientalis* had never been documented from the native species before, the most likely explanation seemed to be a recent spillover from pythons. Sequencing data supported this hypothesis (Miller et al. 2018). Furthermore, extensive sampling of native snake species (498 snakes from 26 species) that were roadkilled from 2012 to 2016 in three neighbouring US states (Florida, Alabama and Georgia) found that *Raillietiella orientalis* only occurred in native snake species within the range of the established python population. Thus, several lines of evidence pointed to pythons as the original spillover host and as a reservoir host.

As researchers started to study *Raillietiella orientalis* and its impacts on native snake species, concerning data began to accumulate. In 2018, *Raillietiella orientalis* was found in three native pygmy rattlesnakes (*Sistrurus miliarus*) in

**Box 8.2** *Continued*

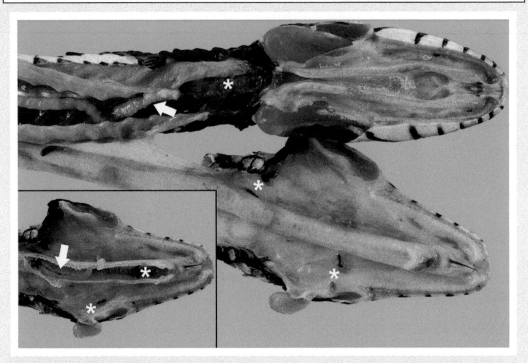

**Figure 8.7** A necropsied banded water snake that was infected with *Raillietiella orientalis* (shown by white arrows) and trematode parasites (denoted with white asterisks). The pentastomes are in the trachea and oesophagus.
Photo copyright: 2020 Walden, Iredale, Childress, Wellehan and Ossiboff; used under the terms of the Creative Commons Attribution License (CC BY).

central Florida—further north than the established range of Burmese pythons (Farrell et al. 2019). Later research showed that morbidity and mortality of pygmy rattlesnakes was caused by co-infections with *Raillietiella orientalis* and ophidiomycosis (a snake fungal disease) (Farrell et al. 2019).

Then, in 2019, someone saw a black racer (a native snake species) in Alachua County (north central Florida) regurgitate a banded water snake (another native species, *Nerodia fasciata*). The water snake seemed very unwell—understandably—and was brought to a rehabilitation facility. It was soon euthanized due to its extreme illness, and a necropsy revealed co-infection with *Raillietiella orientalis* and ophidiomycosis. This report placed the parasite in a native host species 340 km north of the established range of the Burmese python (Walden et al. 2020) (Fig. 8.8). These increasingly northward observations suggest that while *Raillietiella orientalis* originally spilled over from pythons, it is now established and spreading through native snake populations without pythons serving as the main reservoir host species.

While snakes are the definitive hosts for *Raillietiella orientalis* (the host where the adult parasites reproduce), they are not the only hosts. Like other pentastomes, *Raillietiella orientalis* has a complex life cycle, where eggs from the definitive host are released into the environment, the subsequent parasite larvae infect an intermediate host species, and consumption of infected intermediate hosts leads to infection in definitive host species (i.e. trophic transmission). This begs the question: what species are serving as intermediate hosts for *Raillietiella orientalis* in Florida?

As we write this textbook, one recent laboratory study has found that several species of insects, lizards and frogs can serve as intermediate hosts (Palmisano et al. 2022). This remains an active area of research, where fully understanding the life cycle feels urgent: since many wild species are collected in Florida and sold as pets or food for pets, there may be high risks of people accidentally transporting *Raillietiella orientalis* outside of Florida through the pet trade. This isn't just a hypothetical scenario, because a pet snake

**Box 8.2** *Continued*

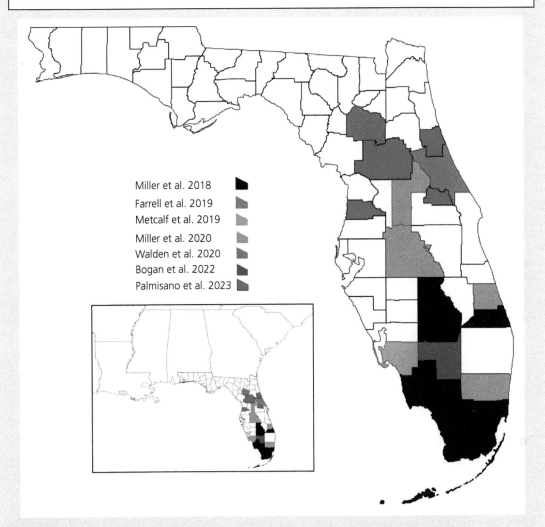

**Figure 8.8** Counties where pentastomes, *Raillietiella orientalis*, have been observed in native snake species in Florida, south-eastern USA. Up to 2016, the parasite had only been found in the southernmost counties where Burmese pythons are established, but subsequent studies have documented a northern range expansion of the parasite, suggesting that it is now maintained and spread by native species.

This figure was created by Jenna Palmisano as an adaptation of a figure published in Palmisano et al. (2023).

purchased in Michigan was recently found infected with this pentastome (Farrell et al. 2023).

This case study is a good example of using multiple lines of evidence to identify a reservoir host (the Burmese python). Surveillance in this case study was optimal because investigators were able to sample many hosts over a large region due to python culling programmes and the unfortunately large supply of snake roadkill.

This case study also demonstrates that after spillover occurs, the pathogen or parasite can adapt and adopt new hosts. As the parasite adapts to local circumstances, continued investigations may be necessary to predict and control the emerging infectious disease. Finally, this example illustrates how the pet trade, invasive species and our increasingly connected world can drive emerging infectious disease events—topics we will explore in detail in Chapter 9.

## 8.7 Notes on sources

Information on the discovery of Hantavirus Pulmonary Syndrome was from: https://www.cdc.gov/hantavirus/outbreaks/history.html (accessed 7/17/20).

## 8.8 References

Andersen, K.G., Shapiro, B.J., Matranga, C.B., et al. (2015). Clinical sequencing uncovers origins and evolution of Lassa virus. Cell, 162, 738–50.

Arias, A., Watson, S.J., Asogun, D., et al. (2016). Rapid outbreak sequencing of Ebola virus in Sierra Leone identifies transmission chains linked to sporadic cases. Virus Evolution, 2, vew016.

Baize, S., Pannetier, D., Oestereich, L., et al. (2014). Emergence of Zaire Ebola virus disease in Guinea. New England Journal of Medicine, 371, 1418–25.

Barcott, B. (2012). The story behind the Hantavirus outbreak at Yosemite. Outside: https://www.outsideonline.com/1930876/death-yosemite-story-behind-last-summers-hantavirus-outbreak, accessed 17/7/23.

Barrette, R.W., Metwally, S.A., Rowland, J.M., et al. (2009). Discovery of swine as a host for the Reston ebolavirus. Science, 325, 204–06.

Bloomfield, S.J., Benschop, J., Biggs, P.J., et al. (2017). Genomic analysis of Salmonella enterica serovar Typhimurium DT160 associated with a 14-year outbreak, New Zealand, 1998–2012. Emerging Infectious Diseases, 23, 906–13.

Boni, M.F., Lemey, P., Jiang, X., et al. (2020). Evolutionary origins of the SARS-CoV-2 sarbecovirus lineage responsible for the COVID-19 pandemic. Nature Microbiology, 5, 1408–17.

Chinese SARS Molecular Epidemiology Consortium. (2004). Molecular evolution of the SARS coronavirus during the course of the SARS epidemic in China. Science, 303, 1666–9.

Christie, A., Davies-Wayne, G.J., Cordier-Lasalle, T., et al. (2015). Possible sexual transmission of Ebola virus—Liberia, 2015. Morbidity and Mortality Weekly Report (MMWR), 64, 479–81.

Corman, V.M., Landt, O., Kaiser, M., et al. (2020). Detection of 2019 novel coronavirus (2019-nCoV) by real-time RT-PCR. Eurosurveillance, 25, 2000045.

Diallo, B., Sissoko, D., Loman, N.J., et al. (2016). Resurgence of Ebola virus disease in Guinea linked to a survivor with virus persistence in seminal fluid for more than 500 days. Clinical Infectious Diseases, 63, 1353–6.

Diehl, W.E., Lin, A.E., Grubaugh, N.D., et al. (2016). Ebola virus glycoprotein with increased infectivity dominated the 2013–2016 epidemic. Cell, 167, 1088–98.

Dietzel, E., Schudt, G., Krähling, V., et al. (2017). Functional characterization of adaptive mutations during the West African Ebola virus outbreak. Journal of Virology, 91, e01913–16.

Dudas, G., Carvalho, L.M., Bedford, T., et al. (2017). Virus genomes reveal factors that spread and sustained the Ebola epidemic. Nature, 544, 309–15.

Estrada-Peña, A., de la Fuente, J. (2014). The ecology of ticks and epidemiology of tick-borne viral diseases. Antiviral Research, 108, 104–28.

Farrell, T.M., Agugliaro, J., Walden, H.D.S., et al. (2019). Spillover of pentastome parasites from invasive Burmese pythons (Python bivittatus) extends beyond the geographic range of pythons in Florida. Herpetological Review, 50, 73–6.

Farrell, T.M., Walden, H.D., Ossiboff, R.J. (2023). The invasive pentastome Raillietiella orientalis in a banded water snake from the pet trade. Journal of Veterinary Diagnostic Investigation, 35, 201–03.

Fauver, J.R., Gendernalik, A., Weger-Lucarelli, J., et al. (2017). The use of xenosurveillance to detect human bacteria, parasites, and viruses in mosquito bloodmeals. American Journal of Tropical Medicine and Hygiene, 97, 324–9.

Fredricks, D.N., Relman, D.A. (1996). Sequence-based identification of microbial pathogens: a reconsideration of Koch's Postulates. Clinical Microbiology Reviews, 9, 18–33.

Gardy, J.L., Loman, N.J. (2018). Towards a genomics-informed, real-time, global pathogen surveillance system. Nature Reviews Genetics, 19, 9–20.

George, D.B., Webb, C.T., Farnsworth, M.L., et al. (2011). Host and viral ecology determine bat rabies seasonality and maintenance. Proceedings of the National Academy of Sciences, USA, 108, 10,208–13.

Gilbert, A.T., Fooks, A.R., Hayman, D.T.S., et al. (2013). Deciphering serology to understand the ecology of infectious diseases in wildlife. EcoHealth, 10, 298–313.

Gire, S.K., Goba, A., Andersen, K.G., et al. (2014). Genomic surveillance elucidates Ebola virus origin and transmission during the 2014 outbreak. Science, 345, 1369–72.

GISAID (Global Initiative on Sharing All Influenza Data). (2020). GISAID mission. https://www.gisaid.org/about-us/mission (accessed 24/6/2020).

Guan, Y., Zheng, B.J., He, Y.Q., et al. (2003). Isolation and characterization of viruses related to the SARS coronavirus from animals in southern China. Science, 302, 276–8.

Harvey, W.T., Carabelli, A.M., Jackson, B., et al. (2021). SARS-CoV-2 variants, spike mutations and immune escape. Nature Reviews Microbiology, 19, 409–24.

Hill, A.B. (1965). The environment and disease: association or causation? Proceedings of the Royal Society of Medicine, 58, 295–300.

Holmes, E.C., Dudas, G., Rambaut, A., Andersen, K.G. (2016). The evolution of Ebola virus: insights from the 2013–2016 epidemic. Nature, 538, 193–200.

Hranac, C.R., Marshall, J.C., Monadjem, A., Hayman, D.T. (2019). Predicting Ebola virus disease risk and the role of African bat birthing. Epidemics, 29, 100366.

Kan, B., Wang, M., Jing, H., et al. (2005). Molecular evolution analysis and geographic investigation of severe acute respiratory syndrome coronavirus-like virus in palm civets at an animal market and on farms. Journal of Virology, 79, 11,892–900.

Keita, A.K., Koundouno, F.R., Faye, M., et al. (2021). Resurgence of Ebola virus in 2021 in Guinea suggests a new paradigm for outbreaks. Nature, 597, 539–43.

Ladner, J.T., Grubaugh, N.D., Pybus, O.G., Andersen, K.G. (2019). Precision epidemiology for infectious disease control. Nature Medicine, 25, 206–11.

Lam, T.T.Y., Jia, N., Zhang, Y.W., et al. (2020). Identifying SARS-CoV-2-related coronaviruses in Malayan pangolins. Nature, 583, 282–5.

Lembo, T., Hampson, K., Haydon, D.T., et al. (2008). Exploring reservoir dynamics: a case study of rabies in the Serengeti ecosystem. Journal of Applied Ecology, 45, 1246–57.

Li, F., Li, W., Farzan, M., Harrison, S.C. (2005). Structure of SARS coronavirus spike receptor-binding domain complexed with receptor. Science, 309, 1864–8.

Li, W., Shi, Z., Yu, M., et al. (2005). Bats are natural reservoirs of SARS-like coronaviruses. Science, 310, 676–9.

Lu, H., Zhao, Y., Zhang, J., et al. (2004). Date of origin of the SARS coronavirus strains. BMC Infectious Diseases, 4, 1–6.

Mate, S.E., Kugelman, J.R., Nyenswah, T.G., et al. (2015). Molecular evidence of sexual transmission of Ebola virus. New England Journal of Medicine, 373, 2448–54.

Meiser, C.K., Schaub, G.A. (2011). Xenodiagnosis. 273–99 in H. Mehlhorn (Ed.), Nature Helps. . .: How Plants and Other Organisms Contribute to Solve Health Problems, Springer-Verlag.

Miller, M.A., Kinsella, J.M., Snow, R.W., et al. (2018). Parasite spillover: indirect effects of invasive Burmese pythons. Ecology and Evolution, 8, 830–40.

Murchison, E.P., Schulz-Trieglaff, O.B., Ning, Z., et al. (2012). Genome sequencing and analysis of the Tasmanian devil and its transmissible cancer. Cell, 148, 780–91.

Murchison, E.P., Tovar, C., Hsu, A., et al. (2010). The Tasmanian devil transcriptome reveals Schwann cell origins of a clonally transmissible cancer. Science, 327, 84–7.

O'Hanlon, S.J., Rieux, A., Farrer, R.A., et al. (2018). Recent Asian origin of chytrid fungi causing global amphibian declines. Science, 360, 621–7.

Oliver, S.E., Gargano, J.W., Marin, M., et al. (2020). The Advisory Committee on Immunization Practices' interim recommendation for use of Pfizer-BioNTech COVID-19 Vaccine - United States, December 2020. Morbidity and Mortality Weekly Report (MMWR), 69, 1922–4.

Oliver, S.E., Gargano, J.W., Marin M, et al. (2021). The Advisory Committee on Immunization Practices' interim recommendation for use of Moderna COVID-19 Vaccine - United States, December 2020. Morbidity and Mortality Weekly Report (MMWR), 69, 1653–6.

Pacciarini, F., Ghezzi, S., Canducci, F., et al. (2008). Persistent replication of severe acute respiratory syndrome coronavirus in human tubular kidney cells selects for adaptive mutations in the membrane protein. Journal of Virology, 82, 5137–44.

Palmisano, J.N., Bockoven, C., McPherson, S.M., et al. (2022). Infection experiments indicate that common Florida anurans and lizards may serve as intermediate hosts for the invasive pentastome parasite, *Raillietiella orientalis*. Journal of Herpetology, 56, 355–61.

Palmisano, J.N., Farrell, T.M., Hazelrig, C.M., Brennan, M.N. (2023). Documenting range expansion of the invasive pentastome parasite, *Raillietiella orientalis*, using southern black racer and eastern coachwhip road mortality. The Southeastern Naturalist, 22, N17–N22.

Parry, J. (2004). WHO queries culling of civet cats. BMJ, 328, 128.

Pastok, D., Hoare, M.J., Ryder, J.J., et al. (2016). The role of host phenology in determining the incidence of an insect sexually transmitted infection. Oikos, 125, 636–43.

Peiris, J.S.M., Guan, Y., Yuen, K.Y. (2004). Severe acute respiratory syndrome. Nature Medicine, 10, S88–97.

Poon, L.L., Chan, K.H., Wong, O.K., et al. (2003). Early diagnosis of SARS coronavirus infection by real time RT-PCR. Journal of Clinical Virology, 28, 233–8.

Quammen, D. (2012). Spillover: Animal Infections and the Next Human Pandemic. W.W. Norton.

Rambaut, A., Pybus, O.G., Nelson, M.I., et al. (2008). The genomic and epidemiological dynamics of human influenza A virus. Nature, 453, 615–19.

Ruan, Y., Wei, C.L., Ling, A.E., et al. (2003). Comparative full-length genome sequence analysis of 14 SARS coronavirus isolates and common mutations associated with putative origins of infection. The Lancet, 361, 1779–85.

Salkeld, D.J., Lagana, D.M., Wachara, J., et al. (2021). Examining prevalence and diversity of tick-borne pathogens in questing *Ixodes pacificus* ticks in California. Applied and Environmental Microbiology, 87, e00319–21.

Siddle, K.J., Eromon, P., Barnes, K.G., et al. (2018). Genomic analysis of Lassa virus during an increase in cases in Nigeria in 2018. New England Journal of Medicine, 379, 1745–53.

Smith, G., Vijaykrishna, D., Bahl, J., et al. (2009). Origins and evolutionary genomics of the 2009 swine-origin H1N1 influenza A epidemic. Nature, 459, 1122–5.

Tang, D., Duan, R., Chen, Y., et al. (2022). Plague outbreak of a *Marmota himalayana* family emerging from hibernation. Vector-Borne and Zoonotic Diseases, 22, 410–18.

Temmam, S., Vongphayloth, K., Baquero, E., et al. (2022). Bat coronaviruses related to SARS-CoV-2 and infectious for human cells. Nature, 604, 330–36.

Tsao, J.I., Wootton, J.T., Bunikis, J., et al. (2004). An ecological approach to preventing human infection: vaccinating wild mouse reservoirs intervenes in the Lyme disease cycle. Proceedings of the National Academy of Sciences, USA, 101, 18, 159–64.

Tu, C., Crameri, G., Kong, X., et al. (2004). Antibodies to SARS coronavirus in civets. Emerging Infectious Diseases, 10, 2244–8.

Urbanowicz, R.A., McClure, C.P., Sakuntabhai, A., et al. (2016). Human adaptation of Ebola virus during the West African outbreak. Cell, 167, 1079–87.

Viana, M., Mancy, R., Biek, R., et al. (2014). Assembling evidence for identifying reservoirs of infection. Trends in Ecology & Evolution, 29, 270–79.

Walden, H.D.S., Iredale, M.E., Childress, A., et al. (2020). Case report: invasive pentastomes, *Raillietiella orientalis* (Sambon, 1922), in a free-ranging banded water snake (*Nerodia fasciata*) in north central Florida, USA. Frontiers in Veterinary Science, 7, 467.

Wang, M., Yan, M., Xu, H., et al. (2005). SARS-CoV infection in a restaurant from palm civet. Emerging Infectious Diseases, 11, 1860–65.

Whitmer, S.L.M., Ladner, J.T., Wiley, M.R., et al. (2018). Active Ebola virus replication and heterogeneous evolutionary rates in EVD survivors. Cell Reports, 22, 1159–68.

World Health Organization (WHO). (2009). Ebola Reston in pigs and humans, Philippines. Weekly Epidemiological Record, 84(07), 49–50.

Worobey, M., Han, G.Z., Rambaut, A. (2014a). Genesis and pathogenesis of the 1918 pandemic H1N1 influenza A virus. Proceedings of the National Academy of Sciences, USA, 111, 8107–12.

Worobey, M., Han, G.Z., Rambaut, A. (2014b). A synchronized global sweep of the internal genes of modern avian influenza virus. Nature, 508, 254–7.

Worobey, M., Rambaut, A., Pybus, O.G., Robertson, D.L. (2002). Questioning the evidence for genetic recombination in the 1918 'Spanish flu' virus. Science, 296, 211.

Xi, J., Duan, R., He, Z., et al. (2022). First case report of human plague caused by excavation, skinning, and eating of a hibernating marmot (*Marmota himalayana*). Frontiers in Public Health, 10, 910872.

# SECTION 3

# Drivers of Infectious Disease Emergence

Infectious diseases *emerge* because of causal **drivers**. This section (Chapters 9–11) introduces three broad categories of driver: **globalization**, where travel and trade have enabled pathogens to spread further and faster than ever before; **climate change**, which is shifting the environment and interactions between hosts, vectors, and pathogens; and **land-use change**, like deforestation and urbanization, which change how humans and wildlife interact. These drivers are not independent of each other, and often many drivers are acting simultaneously. In many cases, an ecological driver like increasing temperature might be the **proximate** cause (the most recent or most responsible cause) of an emerging infectious

Field biologists recapture a large invasive Burmese python in Florida. This animal, 'Johnny', is used as a sentinel snake that leads researchers to catch and remove other Burmese pythons.
Photo by Gena Steffens.

disease event, whereas a more **ultimate** cause (the real deep cause) might be poverty, poorly designed political systems, or some other non-ecological driver. This sets up feedback loops where social injustices drive ecological change and emerging infectious diseases, and then ecological change and emerging infectious diseases further exacerbate social injustices. As we explore these complexities, Mr Charles Darwin reappears; moose ticks demonstrate how much climate change sucks; multiple squirrels will have a pox upon them; and we will ruminate on infectious diseases in Arctic ruminants.

# CHAPTER 9

# Emerging infectious diseases and globalization—travel, trade, and invasive species

During this period the physical condition of the dogs caused considerable anxiety. One after another of them fell sick and wasted away... The two doctors—young Macklin and McIlroy, the senior surgeon—performed post mortem on each dog and discovered that the majority of them suffered from huge red worms, often a foot or more long, in their intestines. Furthermore, there was nothing that could be done to cure the sick animals. One of the few items the expedition had failed to bring from England was worm powder.

**—Alfred Lansing, in *Endurance: Shackleton's Incredible Voyage*, recounting the fate of sled dogs fed raw seal meat and blubber when Shackleton's crew shipwrecked in Antarctica in 1915.**

## 9.1 Travel brings zoonotic infections to non-endemic areas

In 1881, it took more than a month to travel by ship from London to Hong Kong; today, it takes half a day by plane. Travel from far-flung areas can now occur within many more pathogens' incubation period (Chapter 2), exposing naive populations in new places to novel pathogens. We live in a world where we share pathogens! Globalization and the spread of infectious diseases is not an entirely new phenomenon, because diseases (e.g. tuberculosis, polio, smallpox, diphtheria, rinderpest) have travelled alongside colonization, exploration, slavery, war, travel, and trade-routes throughout human history (Baker et al. 2022). But the rate of spread may be rising in our increasingly connected world (Fig. 9.1).

*Maramagambo Forest, Queen Elizabeth National Park, Uganda, 2007–2008*—Volcanic caves filled with Egyptian fruit bats (*Rousettus aegyptiacus*), bat-eating pythons, and occasional forest cobras might

not be common travel bucket list items. But in 2008, a Dutch tourist travelling in Uganda to see mountain gorillas was intrigued enough to visit the 'python cave' in Maramagambo Forest. After her trip, she flew back to the Netherlands in good health, but subsequently became seriously ill, suffering with fever, liver failure, and severe haemorrhaging (Fig. 9.2). Three days after being placed in an induced coma, the woman died. She had become infected with Marburg virus, a filovirus related to the Ebola viruses, whilst in Uganda.

Marburg virus ... filamentous and twisty, like an anguished tapeworm.    —David Quammen, 'Spillover'

Eighteen months beforehand, on Christmas Day 2007, an American tourist had also been intrigued enough by the prospect of watching pythons hunt bats to visit the same Egyptian fruit bat roost (Fujita et al. 2009). The woman entered the cave for about 15 to 20 minutes wearing shorts and sandals, and reported that the bats '...were flying in and out

*Emerging Zoonotic and Wildlife Pathogens*. Dan Salkeld, Skylar Hopkins, and David Hayman, Oxford University Press.
© Oxford University Press (2023). DOI: 10.1093/oso/9780198825920.003.0009

**Figure 9.1** Global air travel routes demonstrate how connected the world is. Lines represent travel links between airports, coloured by passenger capacity in people per day (thousands: red, hundreds: yellow, tens: blue).
From Kilpatrick and Randolph (2012).

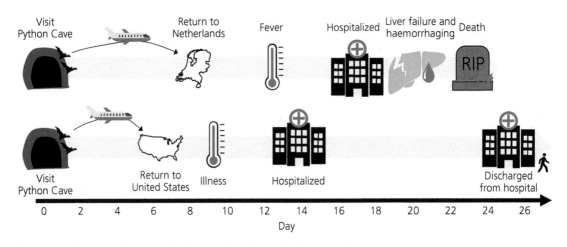

**Figure 9.2** Marburg virus infection timeline for two tourists visiting Python Cave, Uganda.

and screeching and making all sorts of noise... And the smell was super powerful. Everybody had their hand over their noses' (Barnes 2014).

A week later, on New Year's Eve 2007, the woman flew through London and then Iowa on her way home to Golden, Colorado. Signs of a nascent Marburg virus infection started during her return travels—headache, rash, nausea—but were not enough to discourage the woman from hanging out with family, visiting Starbucks, and shaking people's hands (Barnes 2014).

However, her health later deteriorated: she repeatedly consulted doctors and urgent care, and during her third doctor visit she collapsed and was rushed to the hospital. She was lucky and slowly recovered; the more-than-200 healthcare profession-als that interacted with her during her illness shared her luck and did not contract the virus. The cause of the illness in the American tourist was only diagnosed after she read about the case of the Dutch woman who had visited the same bat cave and become infected with the same virus.

Both women contracted Marburg virus in Africa and brought it home on a plane. They weren't quarantined and they interacted with hundreds of other people while potentially contagious. Though Marburg virus did not spread to other humans during these travels, these two cases vividly demonstrate that pathogens—like those lurking in bat caves in Uganda—have the potential to travel to different continents in a matter of days. (Sidenote: Though you may now be ludicrously inspired to visit the cave to see the Marburg virus spillover site, entry to the cave was closed to tourists in 2008.)

Exiting the cave at dusk, I had just taken off my respirator and glasses when I heard a tremendous noise above my head. I looked up, just for a second, and caught a dollop of fresh guano directly in my left eye. It was hot and it burned. I knew this was a 'wet contact', as bad as a bite.

Back at camp, I immediately called the Ugandan arm of the Centers for Disease Control to see what they knew, if anything, about those bats. There was a long pause on the other end of the line. 'You shouldn't have gone in there,' he said. 'Marburg circulates in that cave.'

—Joel Sartore

## 9.2 Travel drives stuttering outbreaks in non-endemic areas

*Senegal and USA, 2008*—Two researchers, Brian Foy and Kevin Kobylinski, were studying the potential of ivermectin to combat malaria transmission in Senegal. Because their work required catching mosquitoes, the researchers were reluctant to use insect repellent that would, umm, repel the target of their study. But the researchers wisely took malaria prophylactic drugs and were also vaccinated against some other local pathogens.

Roughly a week after returning home to Fort Collins, Colorado, the two researchers became sick (Foy et al. 2011). Foy suffered from swollen wrists and ankles, arthralgia (joint pain), maculopapular rash on his torso, extreme fatigue, headache, light-headedness, chills, aphthous ulcers, and, later on, prostatitis and haematospermia. If you need a translation for those terms: a maculopapular rash comprises both flat and raised skin lesions (macule refers to flat, discoloured skin lesions; papules are small, raised bumps); aphthous ulcers or canker sores are small ulcers that develop inside your mouth; prostatitis denotes a swollen, inflamed

prostate gland; and haematospermia is blood in the semen. TMI is too much information. Kobylinski suffered similar symptoms: swollen wrists and ankles, joint pain, maculopapular rash on torso, extreme fatigue, headache, and light-headedness. In short, they were both very ill. As an infectious disease researcher, Foy suspected that perhaps he and Kobylinski had been infected with a mosquito-borne virus (Box 9.1)—maybe dengue or Chikungunya?—so they collected blood samples to attempt a diagnosis.

However, then a third patient appeared. Foy was married with children, and 10 days after his triumphant return from Senegal, his wife, Joy Chilson Foy, became case patient 3, presenting swollen wrists and ankles, arthralgia, maculopapular rash on torso, extreme fatigue, headache, light-headedness, malaise, chills, photophobia, muscle pain, conjunctivitis, and aphthous ulcers. Joy Chilson Foy had not left northern Colorado for a year, so autochthonous transmission—i.e. acquired locally and not travel-associated—had to have been the cause. But how?

If the trio were indeed infected by mosquito-borne pathogen, then infected mosquitoes would have transmitted the pathogen between the Foys. However, mosquitoes were low in number near the Foys' residence, and Joy Chilson Foy's disease symptoms appeared faster than the normal extrinsic incubation period in mosquitoes. Furthermore, the Foy children were not infected, which would be unlikely if there were infected mosquitoes flying around or if the pathogen were transmitted by close contact.

The blood test results revealed that Foy and Kobylinski had been vaccinated or exposed to dengue and yellow fever (both flaviviruses), but Joy Chilson Foy's results had no positive results to dengue and yellow fever. Following those inconclusive results, all three patients recovered, and the blood samples were stored away while the cause of their illness remained a mystery.

A year later, Kobylinski returned to Senegal for further research. By happenstance, he chatted to a generously lamb-chopped Andrew Haddow, a researcher for the US Army Medical Research Institute of Infectious Diseases. Haddow's team was there to study obscure viruses that affected people in that region. When Kobylinski mentioned that he'd probably had one of those lesser-known viruses and described the test results and symptoms, Haddow suggested that the pathogen might have been Zika virus (Cohen 2016).

You might have heard of Zika virus now because there was a Zika virus pandemic in 2015, but back in 2008, Zika virus was not a well-known pathogen! At that time, it was believed to circulate in wildlife, and sporadic human infections did not seem to cause serious illness. It was not on most virologists' radars. However, Andrew Haddow knew about Zika virus because more than six decades earlier, Andrew's grandfather, Alexander Haddow, had discovered it!

While researching yellow fever in the Zika forest, Uganda, in 1947, Alexander Haddow and other researchers were putting Rhesus monkeys (*Macaca mulatta*) in cages on a tower in the forest to learn about mosquitoes and arboviruses in the forest canopy. One of these 'sentinel' monkeys, Rhesus 766, developed a fever and a blood sample from the monkey was injected into the brains of some Swiss albino mice, which developed sickness 10 days post inoculation. The researchers harvested the mouse brains and isolated a 'filterable transmissible agent'. Mosquitoes trapped at the same canopy platforms also harboured the infectious agent, which was called Zika virus after its forest namesake.

Armed with this new information, the 2008 blood samples from Foy and Kobylinski were re-examined, and they were indeed positive for antibodies for Zika virus. Top marks to Drs Haddow! But until that point, Zika virus was thought to be transmitted only by mosquito vectors. The curious incident in 2008 suggested that what was formerly regarded as a vector-borne disease could also be transmitted by direct contact.

In particular, direct transmission by sexual intercourse could explain Joy Chilson Foy's case and seems especially likely given that Brian Foy had prostitis and haematospermia. But the investigators were unable to isolate the *virus* (only antibodies) from their stored blood samples, and sexual transmission of a mosquito-borne virus had never been documented before. So, when they published an account of their own infections, they concluded that: 'If sexual transmission could be verified in subsequent studies, this would have major implications toward the epidemiology of ZIKV and possibly other arthropod-borne flaviviruses' (Foy et al. 2011).

These viral cases demonstrate that pathogens can infect people in one place and the infected people can travel to different places or continents before succumbing to illness. During travel or at the new destination, the infected person can expose others, who may or may not pass on their own infections. In the best-case scenarios, no additional people are infected (e.g. the Marburg virus cases) or very few additional people are infected as transmission stutters to a halt (e.g. the Colorado Zika virus cases).

---

### Box 9.1 Arboviruses

The term **arbovirus** is a contraction of **ar**thropod-**bo**rne **virus** and refers to viruses that are maintained in nature through biological transmission between a susceptible vertebrate host and a haematophagous (blood-feeding) arthropod (Musso and Gubler 2016) (Fig. 9.3). Often, humans are dead-end hosts of arboviruses (e.g. West Nile virus) but there are famous exceptions where infected humans transmit the pathogens to vectors, e.g. mosquito-borne dengue, Chikungunya, yellow fever, and Zika viruses.

Though arboviruses are often associated with mosquitoes, a variety of arthropods are capable of transmitting viruses. Sandflies can transmit Toscana, Sicilian, and Punta Toro viruses; midges (*Culicoides* spp.) can transmit Bluetongue, Akabana, Schmallenberg, and Oropouche viruses; and hard ticks are responsible for transmission of tick-borne encephalitis and Crimean-Congo haemorrhagic fever viruses.

Although they're called *arthropod-borne* viruses, arboviruses can have multiple transmission routes (see Table 9.1). For example, West Nile virus can be vehicle-borne during blood transfusions and Zika virus can be transmitted via vertical transmission or direct transmission via sexual contact. But in all cases, arthropods play a role, which means that when infected humans *or* infected arthropods show up in susceptible populations, new outbreaks can occur.

**Box 9.1** *Continued*

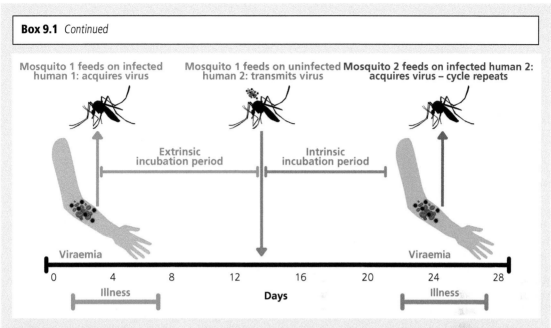

**Figure 9.3** Transmission of arboviruses occurs when a blood-feeding arthropod acquires the pathogen from an infectious host, and then transmits the virus to a new host in a subsequent bloodmeal. The **extrinsic incubation period** is the time between infection of the vector and when it becomes able to transmit the pathogen, often synonymous with the time required for the disease agent to disseminate from the midgut, where it arrives during the bloodmeal, to the salivary glands, where it can be expectorated during a bloodmeal and infect a new host. The **intrinsic incubation period** is the time between infection of the host and when the host is infectious to feeding arthropods, e.g. viraemia in the bloodstream.
Created using Canva.com by Hanna D. Kiryluk.

**Table 9.1** Non-exhaustive description of some arboviruses.

| Virus | Known arthropod vector(s) | Notes | Source |
|---|---|---|---|
| African swine fever virus (ASFV) Genus: *Asfivirus* Family: Asfarviridae | Soft ticks: *Ornithodoros moubata* in Africa; *Ornithodoros erraticus* in Europe | Direct or indirect contact with infected pigs, faeces, or body fluids. Does not cause illness in humans. Causes high mortality in domestic pigs, and consequently causes significant economic losses. First identified in Kenya in the 1920s but spread to Europe in the middle of the 20th century, and later to South America and the Caribbean. Eradicated from most of Europe in the 1990s. Unfortunately, in 2007, African swine fever virus was confirmed in Georgia and slowly spread to Armenia, Azerbaijan, Russia, and Belarus, affecting domestic pigs and wild boar. Re-occurrence in the European Union (EU) was reported in 2014. ASFV occurred in China in 2018, and by 2021 had spread to 16 Asian countries. In 2021, the disease reappeared in the Americas after an absence of almost 40 years. | Galindo and Alonso 2017; WOAH |

*continued*

**Table 9.1** *Continued*

| Virus | Known arthropod vector(s) | Notes | Source |
|---|---|---|---|
| Crimean-Congo haemorrhagic fever Genus: *Orthonairovirus* Family: Nairoviridae | Hard ticks, *Hyalomma* spp. Sexual, transovarial, and transstadial transmissions have been demonstrated in the ticks | Also transmitted by direct contact with infected fluids of animals and humans, e.g. nosocomial infections, slaughterhouses. Considered endemic in Africa (hence the Congo), the Balkans (hence Crimea), the Middle East, and Asian countries south of the 50th parallel. An emerging infectious disease in Spain. | Sánchez-Seco et al. 2022 |
| O'nyong-nyong virus Genus: *Alphavirus* Family: Togaviridae | Anopheline mosquitoes, e.g. *Anopheles funestus* and *Anopheles gambiae* (which are also malaria vectors) | Endemic to sub-Saharan Africa. The Acholi people of north-western Uganda called the disease 'o'nyong-nyong', meaning 'very painful weakening of joints'. In 1959–1962 an epidemic spanned eastern and western Africa and caused more than 2 million cases in eastern Africa alone. | Rezza et al. 2017 |
| Oropouche virus Genus: *Orthobunyavirus* Family: Peribunyaviridae | Biting midge *Culicoides paraensis* | First isolated in Vega de Oropouche, Trinidad. Outbreaks reported in Brazil, Panama, Peru, and Trinidad and Tobago. Reservoir host ecology not well understood, but evidence includes seropositive pale-throated three-toed sloths (*Bradypus tridactylus*), non-human primates, and birds. Urban outbreaks sparked by viraemic humans. The two largest epidemics of Oropouche fever (>100,000 people infected) were recorded in the Amazonian cities of Belém and Manaus in 1980–1981. | Sakkas et al. 2018 |
| Ross River virus (disease is 'Ross River fever' or 'epidemic polyarthritis') Genus: *Alphavirus* Family: Togaviridae | *Culex* and *Aedes* spp. mosquitoes | Endemic in Australia and Papua New Guinea. Caused a 1979–1980 outbreak that encompassed Fiji, New Caledonia, Samoa, and the Cook Islands. | Harley et al. 2001 |
| Toscana virus Genus: *Phlebovirus* Family: Bunyaviridae | Sandflies in Italy, e.g. *Phlebotomus perniciosus* and *Phlebotomus perfiliewi* | Important viral etiologic agent of summertime meningitis in Mediterranean countries. Toscana virus is possibly maintained in sandfly populations. Neither mammals nor birds have been recognized as reservoirs, but few studies have investigated this. The infectiousness of humans to sandflies is unknown. | Charrel et al. 2005 |
| Yellow fever virus Genus: *Flavivirus* Family: Flaviviridae | *Aedes aegypti*—the yellow fever mosquito | Yellow fever virus likely arose in east Africa, was imported into the Americas from West Africa during the period of the slave trade or first contact between Africa and the Americas, and then travelled westwards. 200,000 cases each year and 30,000 deaths. Sylvatic cycles are maintained by non-human primates. | Bryant et al. 2007 |

## 9.3 Travel drives pandemics

*Yap Island, Federated Republic of Micronesia, 2007—* In April and May 2007, Zika virus was observed far from the African tropical forests where it had been documented circulating in wildlife and occasionally infecting humans. Physicians on Yap Island—a remote Pacific Island archipelago east

of the Philippines—noticed an outbreak of cases exhibiting rash, conjunctivitis, fever, arthralgia, and arthritis. The outbreak couldn't be attributed to dengue or other arboviruses, and 10 of 71 samples (14%) were positive for Zika virus RNA, suggesting the cause of the illness. Further investigation suggested that approximately three-quarters of the Yap population had antibodies to Zika virus (i.e.

previous exposure). And 38% of the people with antibodies reported prior illness, though there were no deaths or hospitalizations attributed to the infections. The discrete timing of the outbreak suggested that this was a new introduction of Zika virus into Yap, either introduced by an infected person who acted as a source of infection for local *Aedes* mosquitoes or introduced by an infected mosquito that acted as a source of infection for local people (Duffy et al. 2009). However, Zika virus was not detected in local mosquitoes, possibly because infection rates were low in the mosquito population.

In 2013, a Zika outbreak hit French Polynesia, causing an estimated 19,000 suspected cases in a population of approximately 270,000 (Cao-Lormeau et al. 2014, Cohen 2016). Some of the cases developed serious neurological complications, including **Guillain-Barré syndrome**, an autoimmune disease that causes acute or subacute flaccid paralysis and, in rare instances, death (Cao-Lormeau et al. 2016, Cohen 2016). During the outbreak, a 44-year-old Tahitian experienced two bouts of Zika, and two weeks after recovering from his second episode, he noticed signs of haematospermia. Investigation revealed positive test results for Zika virus in his semen samples, and later in urine samples, even though his blood samples were negative (Musso et al. 2015).

Soon, outbreaks started to erupt throughout the Pacific: in New Caledonia, the Cook Islands, and Easter Island in 2014, and in Vanuatu, the Solomon Islands, Samoa, and Fiji in 2015 (Musso and Gubler 2016). And travel-associated Zika virus cases started appearing sporadically in far flung places: a Norwegian exposed in Tahiti, a Canadian who had travelled to Thailand, Japanese tourists who vacationed in Bora Bora, an Australian who explored the Cook Islands, etc. (Musso and Gubler 2016, Cohen 2016).

In early 2015, public health agencies in the Brazilian city of Natal observed a cluster of patients who had not travelled presenting a 'dengue-like syndrome' that was soon confirmed as Zika virus: 'the first autochthonous transmission of ZIKV in the country' (Zanluca et al. 2015). No deaths or complications were observed in the Natal outbreak, and the authors postulated that its arrival in Brazil may have been during the football World Cup in 2014 (won by Germany). The authors warned: 'Although most of the patients had mild illness, clinicians

and public health officials should be aware of the risk of expansion of this new emerging virus, especially given the naïve immunological status of the Brazilian population. Spreading of the disease in the country might occur by virtue of the large population mobility and the widespread occurrence of the transmitting vectors' (Zanluca et al. 2015).

Then, in July 2015, Brazilian health authorities began to observe neurological complications, including Guillain-Barre syndrome, in patients with a recent history of Zika virus infection. As the Zika virus outbreak escalated, health authorities began to observe Zika-associated deaths.

By November 2015, Brazilian clinicians started reporting an unusually high number of babies with **microcephaly**—a birth defect in which the baby's head is smaller than the mean for the baby's age and sex (normally two or three standard deviations below the mean, but definitions vary). Microcephaly often occurs because the brain fails to fully develop (PAHO 2015, Kleber de Oliveira et al. 2016). Though normally rare (roughly six cases per 100,000 births), microcephaly rates were being observed at a frightening frequency: 99.7 cases per 100,000 births (PAHO 2015, Kleber de Oliveira et al. 2016, Vogel 2016). Tests revealed Zika virus in amniotic fluid samples of two women whose foetuses had been diagnosed with microcephaly by ultrasound exams. Zika virus was also found in the blood and tissue samples of a newborn baby who presented with microcephaly and other congenital anomalies and tragically died shortly after birth (PAHO 2015). By January 2016, a total of 3530 suspected microcephaly cases had been reported in infants whose mothers lived in or had visited areas where Zika virus was active (Kleber de Oliveira et al. 2016). Grossly abnormal clinical or brain imaging findings, or both, were observed in 42% of infants born to women who were infected with Zika virus during pregnancy, and these outcomes occurred regardless of the stage of pregnancy when the mothers were infected (Brasil et al. 2016). Retrospective studies revealed that the rates of microcephaly had also increased during the Zika virus outbreak in French Polynesia in 2013–2014 (Cauchemez et al. 2016).

By the end of January 2016, autochthonous transmission of Zika virus had been recognized in more than 20 countries or territories in the Americas (Musso and Gubler 2016). Successful colonization

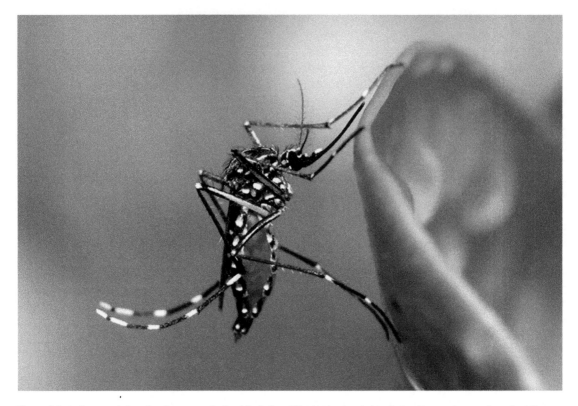

**Figure 9.4** *Aedes aegypti*, the yellow fever mosquito, is originally from Africa but has invaded tropical and temperate areas throughout the world. It is capable of transmitting manifold zoonotic viruses, including yellow fever virus (obviously), Zika virus, Rift Valley fever virus, dengue virus, Chikungunya virus, and others.
Excellent photo by Muhammad Mahdi Karim. Permission to use the document is under the terms of the GNU Free Documentation License, Version 1.2. Link: https://commons.wikimedia.org/wiki/File:Aedes_aegypti.jpg

of the Americas occurred because competent (invasive) mosquito vectors—*Aedes aegypti*, the yellow fever mosquito (Fig. 9.4), and *Aedes albopictus*, the Asian tiger mosquito—were already abundant in these locations. Additionally, sexual transmission fuelled the outbreaks (Deckard et al. 2016, D'Ortenzio et al. 2016). All it took was one or a few imported infections to ignite local epidemics.

What have we learned from this episode? Pathogens can travel the world (Fig. 9.5), whether in infected people or vectors. And pathology associated with a given disease may not always obey the prevailing dogma; neurological impacts, microcephaly, and Zika-associated deaths were not expected from the original consensus of Zika virus epidemiology, but they were relatively common symptoms in the pandemic. It remains unclear whether these unexpected impacts occurred

because Zika virus mutated, infections occurred in naive host populations, or the sample size of infected cases was so much larger than previously observed. We also learned that pathogens with multiple unknown transmission routes (Chapter 2) may be particularly difficult to predict and control during emerging infectious disease events (e.g. sexual transmission and vertical transmission came as surprises during the Zika pandemic).

## 9.4 Wildlife trade drives the spread of infectious diseases

Disease transmission can also be associated with the import and trade of live animals, both domestic and exotic, that can potentially introduce pathogens into naive host populations. **Wildlife trade** involves exchange of wildlife for food, pets, medicines, or

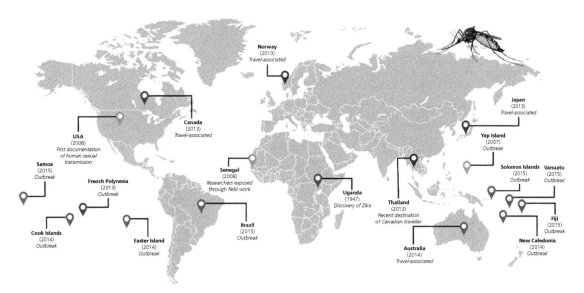

**Figure 9.5** Map of Zika virus outbreaks and travel-associated cases. Created in BioRender.com by Hanna D. Kiryluk.

other goods (e.g. clothing); it occurs both legally and illegally; and it occurs from small to massive ecological scales: from local networks of hunters selling to neighbours, to wet markets in cities selling wildlife that has been harvested or hunted from rural areas, to farmed wildlife (Chapter 11), to internationally smuggled contraband. A recent assessment reported that about 24% of all wild terrestrial vertebrate species are traded globally, that legal wildlife international trade had increased more than five-fold in value between 2005 and 2019, and was estimated to be worth US$107 billion in 2019, with the illegal wildlife trade worth roughly US$7–23 billion annually (IPBES 2020). At the global scale, previous analyses estimated 40,000 live primates, 4 million live birds, 640,000 live reptiles, and 350 million live tropical fish were traded each year (Karesh et al. 2005). In the United States alone, it was estimated that roughly 200 million individual live wildlife specimens were imported every year (2000–2006), predominantly for the pet trade (Smith et al. 2009). This theoretically represents millions of opportunities for exotic pathogens to be imported as well.

On 18 May 2003, a 30-year-old woman . . ., her 33-year-old husband . . ., and the couple's six-year-old daughter purchased two prairie dogs from a pet tradeshow; both animals appeared to be healthy at the time of purchase.

All three family members. . . had extensive contact with the animals. On 21 May, one of the animals was noted to appear to be ill, displaying lethargy, wasting, and anorexia; on 24 May, the animal died. The following day, the second animal developed similar signs, and it died two days later. On 29 May, all three family members developed illness. . .          —Sejvar et al. (2004)

Monkeypox is a disease caused by an orthopoxvirus, a group that includes the smallpox virus that used to infect humans before it was globally eradicated. When we started writing this book, monkeypox was not well known. Now, you probably have heard of it, because monkeypox caused a pandemic in 2022 (>65,000 cases, of which >64,800 occurred in places not historically associated with monkeypox). Unfortunately, outstanding emerging infectious disease events happen so often that we'd never finish writing if we updated our writing each time they occurred! So, this next section only covers the state of monkeypox knowledge in 2021.

Monkeypox virus (Figs. 9.6, 9.7) was first isolated in 1958 from captive monkeys, but in the wild, hosts include rodents and sporadic infections in primates, e.g. a sooty mangabey (*Cercocebus atys*) and a western chimpanzee (*Pan troglodytes verus*). Human cases (until recently) were predominantly observed in the Democratic Republic of the Congo

**Monkeypox virus**

Zoonotic spillover from rodent reservoirs
e.g. rope squirrels, Gambian pouched rats.
Sporadic primate infections.
Stuttering human–to–human transmission.

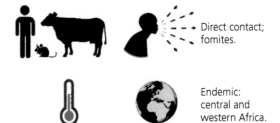

Direct contact;
fomites.

Incubation period = 5–21 days. Illness = 2–4 weeks.
Flu-like symptoms, swollen lymph nodes,
exhaustion, rash and skin lesions. Severe cases can
result in encephalitis and death.

Endemic:
central and
western Africa.

**Globalization and zoonotic spillover**

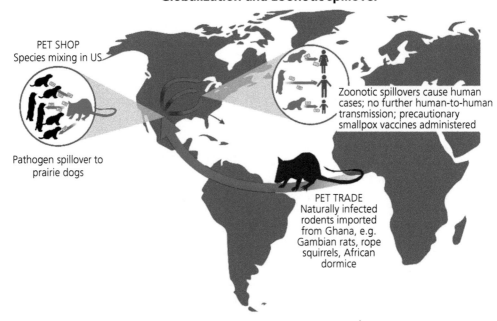

PET SHOP
Species mixing in US

Zoonotic spillovers cause human
cases; no further human-to-human
transmission; precautionary
smallpox vaccines administered

Pathogen spillover to
prairie dogs

PET TRADE
Naturally infected
rodents imported
from Ghana, e.g.
Gambian rats, rope
squirrels, African
dormice

**Figure 9.6** Disease anatomy of monkeypox (aka Mpox) and spread of the virus from Accra, Ghana, to the US in 2003. The shipment from Ghana numbered approximately 800 animals, and included rope squirrels (*Funiscuirus* sp.), tree squirrels (*Heliosciurus* sp.), Gambian pouched rats (*Cricetomys* sp.), brushtail porcupines (*Atherurus* sp.), African dormice (*Graphiurus* sp.), and striped mice (*Hybomys* sp.). Laboratory testing confirmed the presence of monkeypox virus by PCR and virus isolation in one Gambian pouched rat, three African dormice, and two rope squirrels. No human-to-human transmission was evident in the United States after monkeypox spilled over from rodents to humans (Nolen et al. 2016, Hutin et al. 2001, Melski et al. 2003). Note that these circumstances were very different from the 2022 pandemic, where the first zoonotic case or cases were not observed, transmission occurred predominantly among men who have sex with men, and monkeypox spread person-to-person via close contacts (CDC, see 'Notes on sources'; Vusirikala et al. 2022. (In another example of a pet shop disease origin, in 1975, 181 human cases of lymphocytic choriomeningitis virus (an Old-World arenavirus normally found in rodents, as you'll remember from Chapter 2) occurred in 12 states, associated with hamsters from a single pet distributor (CDC 2005).)
Made using BioRender.com; prairie dog icon by Steven Traver from Phylopic.org (public domain).

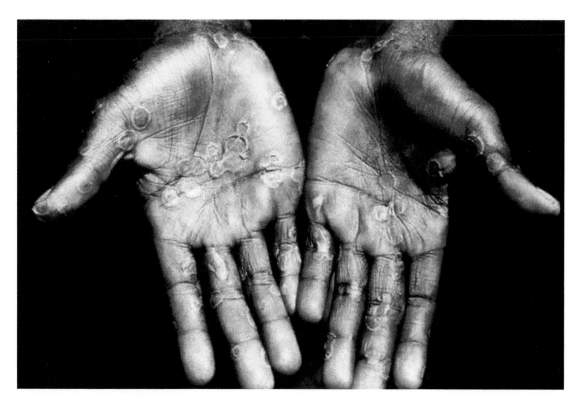

**Figure 9.7** The palms of a monkeypox case-patient from the Democratic Republic of the Congo (DRC) showing maculopapular rash. Monkeypox has been a neglected and emerging disease issue in Africa for decades before it emerged among people in the United States and Europe. Photo credit: CDC/Brian W.J. Mahy; https://phil.cdc.gov/Details.aspx?pid=12761; public domain.

(Rimoin et al. 2010). Infections have varied impacts, ranging from mild to severe to fatal; case fatality ratios (CFR) range from 0–11% in unvaccinated individuals, and infection can cause miscarriage in pregnant women (Beer and Rao 2019). Epidemiologic and sequence data suggest two strains of monkeypox virus (MPXV): a West African strain and a Congo Basin clade, which appears to cause more severe disease, death, and human-to-human transmission (Khodakevich et al. 1986, Hutin et al. 2001, Melski et al. 2003, Radonić et al. 2014, Nolen et al. 2016, Patrono et al. 2020). Most monkeypox virus outbreaks seem to be linked to exposure to wild meat (Beer and Rao 2019).

*Mid-western US, 2003*—The first documented incidents of monkeypox outside Africa were in the mid-western US—Illinois, Indiana, Kansas, Missouri, Ohio, and Wisconsin—in May to June 2003. Most cases (51/53) reported direct or close contact with pet prairie dogs (*Cynomys* sp.); one patient reported contact with a Gambian giant rat (*Cricetomys* spp.) and one patient had contact with a rabbit that became ill after exposure to a sick prairie dog at a veterinary clinic. Monkeypox virus was isolated from the human cases as well as from the lymph nodes of a patient's unwell prairie dog (Melski et al. 2003).

But where did pet prairie dogs get exposed to monkeypox? The infections could not have spilled over from wild prairie dogs, which live further west in the Great Plains (though Kansas does have wild prairie dogs). It is also unlikely that the pet prairie dogs came to the United States from central Africa, where monkeypox is endemic. Instead, pet prairie dogs probably came into close contact with other captive-bred or wild-caught species that had moved from central Africa to facilities in the United States where many animal species are bred and sold.

In this case, an Illinois pet distributor acquired a Gambian pouched rat (*Cricetomys gambianus*), which was unknowingly infected with monkeypox virus and had been shipped to Texas from Ghana. The infected pouched rat then exposed several prairie dogs that were also maintained and sold by the distributor (Fig. 9.6). This unfortunate chain of transmission events led the United States to ban imports of all African rodents to prevent a repeat occurrence of introduced rodent-associated zoonoses (but see Box 9.2).

*Vietnam and Cambodia*—In parts of south-east Asia, greater bandicoot rats (*Bandicota indica*) and rat species (*Rattus* spp.) are a common part of the human diet. Rats are caught, sold, and eaten, including sales by rat traders in large markets and butchering and cooking for prepared dishes in restaurants (Huong et al. 2020). Consumers regard rats as flavoursome and cheap, and they may feature once a week on family menu plans. It's not a small niche market: over 3300 tonnes of live rats are thought to be processed annually, comprising a US$2 million market value. (How many rats is that!?

Roughly 6.6 million, given that a greater bandicoot rat weighs about 0.5 kilograms.) As with consumption of any animal, these rats are potential sources of pathogen spillover to human populations.

Rat parts (heads, tails, and internal organs discarded at slaughter) can also potentially cause pathogen spillover when they are fed to domestic livestock or farmed wildlife (e.g. frogs, snakes, and crocodiles being raised in captivity) (Huong et al. 2020). Wildlife farms in southern Vietnam involve large numbers (roughly 1 million individuals) of diverse wildlife species: rodents, primates, civets, wild boar, deer, crocodiles, softshell turtles, monitor lizards, and pythons. Seventy per cent of the wildlife farms concurrently raise domestic animals, including dogs, cattle, pigs, chickens, ducks, pigeons, geese, and common pheasant. These farms sell wild meat to restaurants that cater to progressively affluent populations domestically and internationally. Clearly, there are many routes by which viruses may follow a pathway from rats to people.

In a 2013–2014 study of rats from various points in the supply chain from farm to table, several

---

**Box 9.2  Rats and health**

It is tempting to simply lump all rodents together as disease-carrying vermin. But that would be unfair. For example, whilst the Gambian pouched rat, *Cricetomys gambianus*, can transmit monkeypox virus, its relative, the Southern pouched rat, *Cricetomys ansorgei*, has saved many lives and made huge contributions to public health (Fig. 9.8).

Southern pouched rats are not your typical wee rodent: they are long-lived enough to be trained and work productively (up to eight years in captivity), and they weigh 1–2 kg (on the order of a small chihuahua), with body lengths of 25–45 cm (half of that is tail) (Poling et al. 2010). Big enough to wear a little backpack! Southern pouched rats are trained by APOPO, a Belgian non-profit organization in Tanzania, to smell TNT (2,4,6-trinitrotoluene), the main explosive charge in most landmines. The rats, which are not heavy enough to set off landmines, can then alert human personnel to the presence of landmines and other explosive remnants of war (ERW). Careful removal and destruction of ERW prevents people from accidentally encountering them (resulting in severe injury and death) and reinstates

people's post-war homes and land. During training, the rats are rewarded with bananas.

The rats' personalities and sense of smell also make them ideal for detecting tuberculosis (TB) (Cengel 2014). A trained rat will walk up a line with ten holes, underneath which is a removeable tray with heat-inactivated human sputum samples. The rat sniffs the samples in order and keeps its nose in the hole for three seconds when it detects TB-positive samples. A single rat can screen 100 samples in under 20 minutes! This is faster than the daily rate of screening by a human lab technician and the rats are more accurate than microscopy techniques, which might miss 50% of cases. These speedy rodent heroes now work in 78 medical centres in four regions, including Dar es Salaam, Tanzania's capital. They also double-check 75% of potential TB samples from medical centres in the Mozambique capital of Maputo. Each rat costs around $6000 to train, but relatively little to maintain over their six-to-eight-year life span. In contrast, rapid diagnostic test equipment can cost $20,000, plus an additional $10–17 per test.

**Box 9.2** *Continued*

**Figure 9.8** Pouched rats (*Cricetomys ansorgei*) have been trained to detect unexploded mines so that they can be removed from minefields, and also to detect tuberculosis in human sputum samples.
Photos provided by APOPO.

different coronaviruses were detected (Huong et al. 2020). (SARS-CoV-2, which emerged five years later, was not detected.) Coronavirus prevalence in the rats increased as they approached the table end of the supply chain: from traders involved in the private sale and butchering of rats for consumption (20.7% or 39/188), to large markets (32.0% or 116/363), to restaurants (55.6% or 84/151). And with multiple coronaviruses circulating, there is a good possibility that a single animal could be co-infected with multiple coronaviruses, which could lead to recombination and the emergence of new coronaviruses, perhaps like SARS-CoV-2. The aggregation of animals along supply chains also increases potential exposure to pathogens for animal traders, transporters, butchers, and cooks (Guan et al. 2003, Kan et al. 2005, Huong et al. 2020).

The examples above demonstrate that mixing species through wildlife trade provides opportunities for pathogen exchange within and between species and subsequent pathogen recombination. This includes opportunities for pathogens to jump to new host species, because species from different geographical regions are often mixed together. The outcomes of species mixtures can be complex and difficult to predict. For example, experimental studies simulating live animal market conditions have demonstrated that H7N9 influenza can be transmitted from virus-shedding chickens (*Gallus gallus domesticus*) to quail (*Coturnix* spp.) that are housed in cages below, whereas other species appeared more resistant to infection (Bosco-Lauth et al. 2016). Susceptibility can also be difficult to predict in these novel environments, because stress, dehydration, and poor nutrition might reduce animals' immune function, unexpectedly increasing pathogen replication and host susceptibility in co-housed animals (Huong et al. 2020). These problems are widespread in wildlife and domestic farming, butchering, transportation, and consumption.

Beyond risks to domestic species and humans, wildlife trade also risks exposing naive wildlife populations to new pathogens. For example, bullfrogs (*Lithobates catesbeianus*) are native to eastern North America, but they are farmed in Asia, Central America, and South America for the international trade in frog legs. To some extent, farming bullfrogs might benefit other native amphibian species

by alleviating the pressures of overharvesting wild populations. However, escaped bullfrogs can be invasive, and they will compete with, and also prey upon, local amphibians. Furthermore, bullfrogs are tolerant of chytrid infections, i.e. the fungal disease agent *Batrachochytrium dendrobatidis* (Chapter 12). In south-eastern Brazil, the fungus can spread rapidly in bullfrog farms (one infected animal is in contact with a high density of fellow bullfrogs), and farms also serve as sources of *Batrachochytrium dendrobatidis* in the outflow water that is released from bullfrog farms into the surrounding environment (Ribeiro et al. 2019). This is just one such example of pathogens spilling over to wildlife from farms or the pet trade.

Though drivers of emerging infectious diseases are complex and numerous, it has been argued that wildlife trade is one of the more manageable in terms of reducing pandemic threats. Increased regulation of the wildlife trade to ensure that traded animals are pathogen-free (e.g. health certificates, quarantines), promote healthy captivity conditions, reduce mixing and aggregations of species, and reduce the dependence on and interactions with wild populations could reduce the potential for disease spillover (Karesh et al. 2005). Importantly, disease mitigation efforts need to include partnerships with the people who rely on wildlife trade for protein, micronutrients, or livelihoods to ensure that their well-being is not negatively affected and to avoid exacerbating illegal trade of wildlife.

## 9.5 Invasive species drive disease emergence

**Invasive alien species** (IAS) are organisms (animals, plants, fungi, or microorganisms) translocated anthropogenically into environments outside their natural species distribution or range, which then establish, spread, and negatively affect the dynamics of local ecosystems (Chinchio et al. 2020). (Note that some species that spread outside of their native ranges do not cause noticeable harm and may even provide important ecosystem function in a world of declining biodiversity; these species are considered **non-native** but not **invasive**.) Some invasive species are pathogens, like the strains of the chytrid fungus that have spread around the world, decimating endemic amphibian populations (Chapter 12). Other invasive species are vectors or reservoir hosts that can bring pathogens with them to new regions or otherwise facilitate local pathogen transmission in their invaded range, such as by becoming competent hosts for pathogens in the invaded range.

Each time individuals of a species are moved to a new region, there are a few possible outcomes for the parasites and pathogens that infect that species in the native range (Macleod et al. 2010). Some parasites and pathogens may 'miss the boat'—by chance, they do not infect the sample of host individuals that are moved to the invaded range. When this happens, invasive species may be successful due to **enemy release**: by chance, the initial individuals have few to no parasites or pathogens, and there are fewer natural enemies in the invaded range, giving the invading species a competitive edge over the natives. Even if parasites and pathogens make it to the invaded range, they may be later 'lost overboard'—the invasive host establishes, but the pathogens cannot successfully maintain transmission in the invaded range and are extirpated (another mechanism of enemy release). Finally, the parasites and pathogens that travel with invasive species may successfully establish in the invaded range, potentially spreading beyond the invasion species by spilling over into native species.

## Invasive alien species as reservoir hosts

*New Zealand*—The Australian brushtail possum, *Trichosurus vulpecula*, was introduced into New Zealand and is now famous for maintaining bovine TB (*Mycobacterium bovis*) and spreading the bacterium to cattle (just like native European badgers can transmit bovine TB in the UK, Chapter 14). It has been hypothesized that possums were first exposed to bovine TB via wild deer: in the 1960s, some lucky possums were able to scavenge the heads and alimentary tracts of hunted and killed wild deer that were left in the field, exposing possums to their retropharyngeal and mesenteric lymph nodes, which are common sites for tuberculous lesions (Nugent et al. 2015). The deer were probably first infected by cattle. But is it difficult to apportion blame for disease transmission in this New Zealand system, because deer, cattle, and possums are all invasive species. (The only terrestrial mammals native to the islands are a few bat species.)

Not many people know this, but squirrels are the cleverest of the woodland creatures. In fact, they're fuzzy little geniuses! They can make a house out of a tree, a bed out of a bunch of leaves, and a box kite out of twigs, dirt, and squirrel spit. They are also excellent at math.
—Adam Rubin, *Those Darn Squirrels* (2008)

*United Kingdom*—In the United Kingdom, there are two squirrel species: the cute, self-deprecating red squirrel (*Sciurus vulgaris*) (Fig. 9.9) is native, but populations have declined as the more brash American grey squirrel (*Sciurus carolinensis*) has increased its distribution across the sceptred isle. Grey squirrels were introduced to the UK at the turn of the 20th century and their larger size allows them to outcompete the more shy and retiring red squirrel for food. But a pathogen plays a role, too.

Squirrel parapoxvirus arrived in Blighty along with its grey squirrel host. Grey squirrel populations maintain the virus with minimal impacts to individuals, whereas individual red squirrels are severely impacted by viral infection (Tompkins et al. 2002). Viral impacts on wild red squirrel populations are hard to quantify, because like many other wildlife species, long-term baseline data are not available, and density is not easy to measure. Mathematical models suggest that when infected grey squirrels make incursions into a naive population of red squirrels, the parapoxvirus incidence in red squirrels skyrockets because of the initially large number of susceptible individuals. This causes a red squirrel population collapse and facilitates the grey squirrel invasion. Even in the absence of noticeable epidemics, pathogen spillover from grey to red squirrels exacerbates the competitive interspecific interactions in modelled populations (Tompkins et al. 2003). Subsequent field investigations have confirmed the role of parapoxvirus in dwindling red squirrel populations (Gurnell et al. 2006, Rushton et al. 2006, Chantrey et al. 2014).

To promote and conserve red squirrel populations, culling of grey squirrels has been instigated, but a more popular management method involves creating or promoting forest habitats that benefit red squirrels more than grey squirrels and reduce

**Figure 9.9** Red squirrel (*Sciurus vulgaris*).
Photo by Martin Eriksson on Unsplash.

interspecific contact (Gurnell et al. 2006). Red squirrels are better adapted to mature boreal coniferous forests of Scots pine (*Pinus sylvestris*) and Norway spruce (*Picea abies*), but they can persist in coniferous plantations of Sitka spruce (*Picea sitchensis*), which are avoided by grey squirrels. Therefore, a successful management plan has involved minimizing non-Scots pine and large seeded broadleaves around Sitka plantations and trapping grey squirrels at forest edges and corridors (Lurz et al. 2003, Shuttleworth et al. 2020).

## Invasive alien species as vectors

The yellow fever mosquito, *Aedes aegypti*, and the tiger mosquito, *Aedes albopictus*, exemplify the role of invasive alien species as pathogen vectors. *Aedes aegypti* is ancestrally from forest habitats of sub-Saharan Africa, where it typically breeds in tree holes or cavities that hold water (Powell 2016). The domesticated form is now distributed in temperate and tropical areas across the globe, and is a human-biting (anthropophilic), day-active mosquito that breeds in human-generated containers (e.g. flowerpots, discarded tyres). *Aedes aegypti* is a competent vector of dengue, Chikungunya, yellow fever, and Zika viruses. *Aedes albopictus* is originally from Asia but possesses many of the same traits as *Aedes aegypti*: a competent vector of the aforementioned arboviruses, anthropophilic, and a fan of laying eggs in standing water, particularly in small pools in items like flowerpots or trash that are found near human dwellings.

Although neither mosquito species is a particularly strong flyer, both mosquito species have spread and continue to spread into new territories. This diaspora can be blamed on unintended human dispersal (e.g. on the ships of colonizers), the presence of suitable climate (which is influenced by climate change, Chapter 10), and urbanization (presence of suitable breeding habitat) (Kraemer et al. 2019). Historically, movement rates of the two species have been approximated at a terrifying 250 km/year.

Over the next few years, both *Aedes* species are predicted to expand into distribution ranges that are already suitable but currently uninhabited, as defined by current climates and urban environments. After saturating this available habitat, future distribution ranges may be more influenced by whether or not climate change provides new fertile grounds for invasion (Kraemer et al. 2019). Unfortunately, areas that are swallowed by the expanding *Aedes* ranges also have high projected human population growth and populations that are naive to these arboviruses.

There are species-specific nuances to *Aedes* invasions: when *Aedes albopictus* changes distribution, it tends to spread from adjacent areas where it is already present, whereas *Aedes aegypti* is more frequently imported over longer distances. Modelling

approaches suggest that in the US, *Aedes aegypti* may reach as far north as Chicago, relying on long-distance introductions across large urban areas. In contrast, European populations of *Aedes aegypti* are only expected in southern Italy and Turkey (Fig. 9.10). In Europe, *Aedes albopictus* is forecast to spread across Germany and France (Kraemer et al. 2019).

Of course, human responses can attempt to prevent these potential distributions from being realized. Yellow fever mosquitoes were recently discovered in central and southern California (Porse et al. 2015) and invoked a public health fury involving enhanced human and mosquito surveillance, education, and intensive mosquito control to try and limit further spread! As with all invasive species, preventing invasions is far cheaper than trying to control or eradicate invasive species after they have established.

Despite our best control efforts, some mosquito species will establish in some new places. These mosquitoes might not bring arboviruses with them,

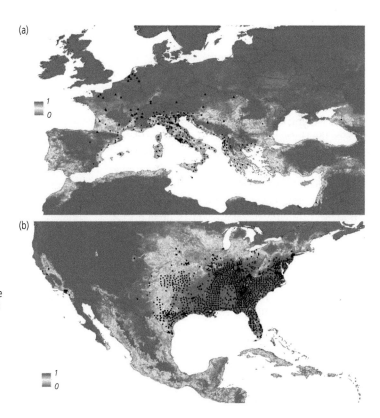

**Figure 9.10** Predicted probability of occurrence of *Aedes albopictus* in Europe (A) and the United States (B), regions in which *Ae. albopictus* is rapidly expanding its range. Points represent known occurrences (transient (triangles) or established (circles)) until the end of 2013.
*Source:* Kraemer et al. 2015, https://elifesciences.org/articles/08347, creative commons CC0 1.0 Universal (CC0 1.0) Public Domain Dedication.

but they may be able to transmit pathogens that arrive in infected travellers, i.e. people returning home or travelling from areas where pathogens like Chikungunya virus or dengue virus are endemic. Infectious humans could infect the mosquitoes, leading to subsequent autochthonous transmission in the new range (Porse et al. 2015). And this isn't just a concern for human health. For example, when the invasive alien malaria pathogen (*Plasmodium relictum*) arrived in Hawaii and established in invasive mosquitoes (*Culex quinquefasciatus*), endemic bird communities were decimated, including several species extinctions (van Riper et al. 1986).

## Invasive alien species as pathogens

We have introduced the concept of an emerging infectious disease being caused by a pathogen spreading to new geographic areas; numerous pathogens have invaded new territories, become well established, and substantially harmed native populations, and they can be viewed like any other invasive alien species. Two classic examples are yellow fever and plague.

Yellow fever virus is currently distributed in the Americas and Africa, but it originally arrived in the Americas from West Africa during the transatlantic slave trade (Bryant et al. 2007). The disease was devastating to naive populations. For example, in 1898, the United States invaded Cuba during the Spanish-American War, and for every soldier killed during fighting, another 13 died from yellow fever (Staples and Monath 2008). Later, Walter Reed was able to prove that the disease was mosquito-borne, a hypothesis originally developed by a Cuban doctor, Carlos Finlay. This led to future mosquito control efforts that saved many lives.

The 'third pandemic' of *Yersinia pestis*, the bacterium that caused bubonic plague, began in roughly the middle 1800s, and it travelled via south China to the then British colonial port of Hong Kong in 1894 (Echenberg 2002). From there, it was transported by British steamships to the Americas, Australia, and Africa. Plague arrived in the US in San Francisco in 1900, and *Yersinia pestis* became established across the western US and persists in native mammalian species with sporadic zoonotic spillovers to this day.

## Invasive alien species affecting local transmission dynamics by indirect mechanisms

As we have seen so far, invasive species can directly alter disease dynamics by being part of the pathogen life cycle (as pathogens, vectors, or hosts). Additionally, even when invasive species play no role in a pathogen's life cycle, they can indirectly alter disease dynamics through impacts on the local pathogen, vector, or host communities. In ecology, **indirect interactions** occur when the effect of one species (e.g. an invasive species) on another (e.g. a native pathogen) is mediated by a third species (e.g. a native host species).

*Florida Everglades, USA*—The Burmese python, *Python molurus bivittatus*, has been politely described as 'a major perturbation to the vertebrate community and ecosystem of southern Florida' (Burkett-Cadena et al. 2021) (Fig. 9.11). Certainly. What else would you expect of a gigantic carnivorous snake that can eat a deer or alligator whole? They average 2.1 to 3 m in length and record specimens reach 84 kg and 5.7 m in length.

Burmese pythons originally hail from south-east Asia. Since before 1985, pythons have been finding their way into the Florida wilds when released pets have outgrown their welcome (Wilson et al. 2011). The authors of this book also heard that python releases could be traced back to exotic dancers tiring of their stage companions, or that massive numbers of pythons escaped when Hurricane Andrew destroyed a python breeding facility in 1995. Unfortunately, no python experts were willing to substantiate these claims in 2023, reducing the dramatic narrative of this textbook. Regardless, pythons invaded, and as the python population grew, it caused dramatic crashes in native wildlife populations (Guzy et al. 2023). For example, marsh rabbits (*Sylvilagus palustris*) radio-tagged and released in python territories overwhelmingly (77%) met their fate in a python's stomach within a year (McCleery et al. 2015). Camera trapping analyses also showed an 85–100% decrease in the frequency of observations of raccoon, opossum, bobcat, and rabbits

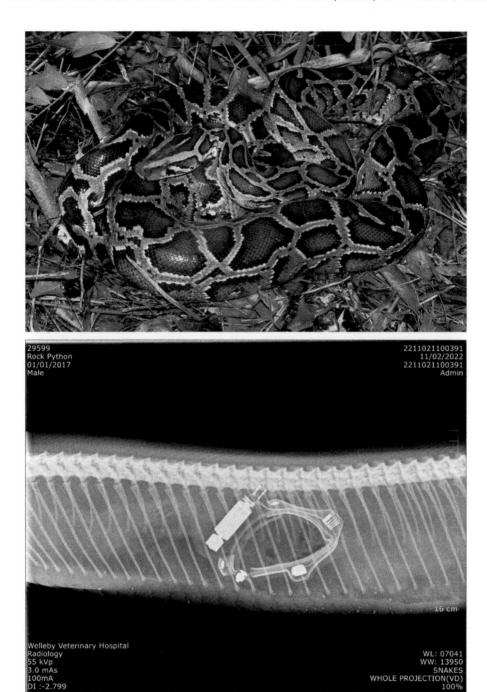

**Figure 9.11** (Top) A Burmese python in Florida, USA. (Bottom) A radio collar from an opossum, which was eventually tracked not to a happy opossum, but to the stomach of an invasive Burmese python in Florida.

(Top) Photograph by J.D. Willson. (Bottom) Photo provided by Mike Cove and his partners from the USFWS and Welleby Veterinary Hospital.

after pythons had become established (Dorcas et al. 2012). And pythons are known to consume endangered species, like the Key Largo woodrat, and substantial numbers of wading birds (e.g. herons, wood storks).

As mammal populations have declined and community structure has shifted, so too has the diet of an endemic mosquito, *Culex cedecei*. *Culex cedecei* will happily feed on a wide variety of Floridian fauna: opossums, armadillos, black bears, Florida panthers (a subspecies of mountain lion), mink, raccoons, white-tailed deer, marsh rabbits, hispid cotton rats, cotton mice, black rats, house mice, and gray squirrels (Burkett-Cadena et al. 2021). In the 1970s, pre-python problem, 50% of *Culex cedecei* blood sources were from white-tailed deer, opossums, and raccoons. But the pythons have obliterated the opossum and raccoon populations, so the mosquito now feeds more frequently on rodents, and especially the hispid cotton rat, *Sigmodon hispidus*. (Hispid means bristly, in case you were wondering.) From 1979 to 2016, mosquito bloodmeals shifted from only 15% being from the hispid cotton rat to 77% being from hispid cotton rat. Medium/large mammals now only constitute 2% of mosquito bloodmeals.

Who cares how much rat blood mosquitoes drink? Potentially people living in Florida. The hispid cotton rat and other rodents are the natural reservoir hosts for Everglades virus, a subtype II variety of the more widespread Venezuelan equine encephalitis virus, which is transmitted by *Culex cedecei*. Mosquito infection rates for Everglades virus were correlated with how often mosquitoes fed on rats; as relative cotton rat host use increased to >50% of bloodmeals, Everglades virus infection rates in mosquitoes increased approximately threefold (Burkett-Cadena et al. 2021). Everglades virus is zoonotic and can cause clinical encephalitis in humans, but the disease is not reportable. Therefore, it is unclear if python-affected ecological communities are increasing human incidence of the disease. But this example illustrates the potential for an invasive predator to modify ecological communities such that they affect patterns of human or wildlife disease.

Considering control interventions: eradicating pythons from Florida is not trivial and may not even be possible. For such a huge snake, they are remarkably difficult to find, because they spend much of their time underground, in the water, or concealed in vegetation. (*Remarkably* difficult. When 19 people were asked to search a 31 × 25 m enclosure containing natural aquatic, terrestrial, and subterranean habitat options for 10 large pythons (2–3 m long) that they were told were inside, only two people saw any of the snakes during their 30-minute search (Dorcas and Willson 2013)! They each saw a single snake, meaning that detectability was only 1%, even when people knew snakes were there.) Programmes have tried paying people bounties for hunted pythons, but the invasive population has continued to spread. Even tricky tactics, like surgically implanting male pythons with radio transmitters and following them to the breeding females, may not be enough to stop these relentless titans.

## 9.6 Summary

'Globalization' is the increasing interconnectedness and interdependence of different countries and regions that exchange goods, services, information, people, and... infectious diseases. When infected people move around the planet, they can spread pathogens to naive populations. People also move pathogens, vectors, and animal hosts around the world, sometimes accidentally (e.g. vector stowaways on ships) and sometimes purposefully (e.g. the pet trade). When a pathogen, vector, or host establishes in a region outside its native range and causes harm to native species, it is considered an invasive alien species (Fig. 9.12). There are also invasive species that do not play roles in the life cycles of native pathogens, but which influence disease dynamics through their indirect effects on the disease system. Of course, sometimes pathogens, vectors, and hosts cannot establish in new places because the ecological conditions do not provide a suitable ecological niche. In the next chapter, we'll discuss how climate change can make some areas more or less suitable for pathogens than they are now, potentially facilitating future pathogen invasions.

## 9.7 Notes on sources

Information on the Marburg virus in travellers: https://www.cidrap.umn.edu/news-

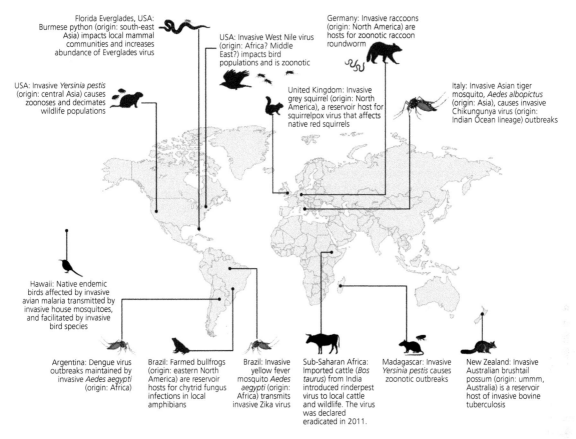

**Figure 9.12** Examples of invasive species and their impacts upon disease ecology in new locations. For example, avian malaria (*Plasmodium relictum*) transmitted by the southern house mosquito, *Culex quinquefasciatus*, impacted native bird communities in Hawaii (Van Riper et al. 1986, McClure et al. 2020); raccoons in Europe infected with raccoon roundworm (*Baylisascaris procyonis*) that are potential zoonoses (Rentería-Solís et al. 2018); and invasive *Aedes aegypti* transmit dengue in Argentina (Estallo et al. 2020). Other examples are discussed in the text. While there are many examples, there are also large knowledge gaps due to inadequate surveillance for both invasive species and their associated pathogens, due to financial, legal, and logistical constraints (Makoni 2020).
Created using BioRender.com with some icons from PhyloPic.org (e.g. cattle icon by DFoidl, modified by T. Michael Keesey, used under the Creative Commons Attribution-Share Alike 3.0 Unported license; bird by Campbell Fleming).

perspective/2008/07/dutch-woman-dies-marburg-fever (accessed 26/3/21); https://www.washingtonpost.com/news/national/wp/2018/12/13/feature/these-bats-carry-the-lethal-marburg-virus-and-scientists-are-tracking-them-to-try-to-stop-its-spread/ (accessed 26/3/21); https://www.joelsartore.com/gallery/the-story-behind/ (accessed 26/3/21).

Information on the ban of imported African rodents to the US: https://www.cdc.gov/poxvirus/monkeypox/african-ban.html (accessed 8/9/2021).

Monkeypox outbreak information: https://www.cdc.gov/poxvirus/monkeypox/response/2022/index.html (accessed 26/9/2022).

Monkeypox global case counts: https://www.cdc.gov/poxvirus/monkeypox/response/2022/world-map.html (accessed 26/9/2022).

Information on status of African swine fever virus: https://www.woah.org/en/disease/african-swine-fever/#ui-id-2 (accessed 30/3/2023).

## 9.8 References

Baker, R.E., Mahmud, A.S., Miller, I.F., et al. (2022). Infectious disease in an era of global change. Nature Reviews Microbiology, 20, 193–205.

Barnes, M. (2014). My own infection is proof that someone can easily carry Ebola into the U.S. Washington Post: https://www.washingtonpost.com/posteverything/wp/2014/08/05/my-own-infection-is-proof-that-someone-can-easily-carry-ebola-into-the-u-s/, accessed 27/3/2021.

Beer, E.M., Rao, V.B. (2019). A systematic review of the epidemiology of human monkeypox outbreaks and implications for outbreak strategy. PLoS Neglected Tropical Diseases, 13, e0007791.

Bosco-Lauth, A.M., Bowen, R.A., Root, J.J. (2016). Limited transmission of emergent H7N9 influenza A virus in a simulated live animal market: do chickens pose the principal transmission threat? Virology, 495, 161–6.

Brasil, P., Pereira Jr., J.P., Moreira, M.E., et al. (2016). Zika virus infection in pregnant women in Rio de Janeiro. New England Journal of Medicine, 375, 2321–34.

Bryant, J.E., Holmes, E.C., Barrett, A.D.T. (2007). Out of Africa: a molecular perspective on the introduction of yellow fever virus into the Americas. PLoS Pathogens, 3, e75.

Burkett-Cadena, N.D., Blosser, E.M., Loggins, A.A., et al. (2021). Invasive Burmese pythons alter host use and virus infection in the vector of a zoonotic virus. Communications Biology, 4, 804.

Cao-Lormeau, V.-M., Blake, A., Mons, S., et al. (2016). Guillain-Barré Syndrome outbreak associated with Zika virus infection in French Polynesia: a case-control study. Lancet, 387, 1531–9.

Cao-Lormeau, V.-M., Roche, C., Teissier, A., et al. (2014). Zika virus, French Polynesia, South Pacific, 2013. Emerging Infectious Diseases, 20, 1084–6.

Cauchemez, S., Besnard, M., Bompard, P., et al. (2016). Association between Zika virus and microcephaly in French Polynesia, 2013–15: a retrospective study. Lancet, 387, 2125–32.

Cengel, K. (2014). Giant rats trained to sniff out tuberculosis in Africa. National Geographic: https://www.nationalgeographic.com/animals/article/140816-rats-tuberculosis-smell-disease-health-animals-world?loggedin=true, accessed 25/3/2022.

Centers for Disease Control and Prevention (CDC). (2005). Lymphocytic choriomeningitis virus infection in organ transplant recipients—Massachusetts, Rhode Island, 2005. Morbidity and Mortality Weekly Report (MMWR), 54, 537–9.

Chantrey, J., Dale, T.D., Read, J.M., et al. (2014). European red squirrel population dynamics driven by squirrelpox at a gray squirrel invasion interface. Ecology and Evolution, 4, 3788–99.

Charrel, R.N., Gallian, P., Navarro-Marí, J.-M., et al. (2005). Emergence of Toscana Virus in Europe. Emerging Infectious Diseases, 11, 1657–63.

Chinchio, E., Crotta, M., Romeo, C., et al. (2020). Invasive alien species and disease risk: an open challenge in public and animal health. PLoS Pathogens, 16, e1008922.

Cohen, J. (2016). Zika's long, strange trip into the limelight. Science (online): https://www.sciencemag.org/news/2016/02/zika-s-long-strange-trip-limelight doi:10.1126/science.aaf4030, accessed 7/6/2021.

Deckard, D.T., Chung, W.M., Brooks, J.T., et al. (2016). Male-to-male sexual transmission of Zika Virus – Texas, January 2016. Morbidity and Mortality Weekly Report (MMWR), 65, 372–4.

Dorcas, M.E., Willson, J.D. (2013). Hidden giants: problems associated with studying secretive invasive pythons. 367–85 in W. Lutterschmidt (Ed.), Reptiles in Research: Investigations of Ecology, Physiology, and Behavior from Desert to Sea. Nova Science Publishers Inc.

Dorcas, M.E., Willson, J.D., Reed, R.N., et al. (2012). Severe mammal declines coincide with proliferation of invasive Burmese pythons in Everglades National Park. Proceedings of the National Academy of Sciences, 109, 2418–22.

D'Ortenzio, E., Matheron, S., Yazdanpanah, Y., et al. (2016). Evidence of sexual transmission of Zika Virus. New England Journal of Medicine, 374, 2195–8.

Duffy, M.R., Chen, T.-H., Hancock, W.T., et al. (2009). Zika virus outbreak on Yap Island, Federated States of Micronesia. New England Journal of Medicine, 360, 2536–43.

Echenberg, M. (2002). Pestis redux: the initial years of the third bubonic plague pandemic, 1894-1901. Journal of World History, 13, 429–49.

Estallo, E.L., Sippy, R., Stewart-Ibarra, A.M., et al. (2020). A decade of arbovirus emergence in the temperate southern cone of South America: dengue, Aedes aegypti and climate dynamics in Córdoba, Argentina. Heliyon, 6, e04858.

Foy, B.D., Kobylinski, K.C., Chilson Foy, J.L., et al. (2011). Probable non-vector-borne transmission of Zika Virus, Colorado, USA. Emerging Infectious Diseases, 17, 880–82.

Fujita, N., Miller, A., Miller, G., et al. (2009). Imported case of Marburg hemorrhagic fever - Colorado, 2008. Morbidity and Mortality Weekly Report (MMWR), 58, 1377–81.

Galindo, I., Alonso, C. (2017). African Swine Fever Virus: a review. Viruses, 9, 103.

Guan, Y., Zheng, B.J., He, Y.Q., et al. (2003). Isolation and characterization of viruses related to the SARS coronavirus from animals in Southern China. Science, 302, 276–8.

Gurnell, J., Rushton, S.P., Lurz, P.W.W., et al. (2006). Squirrel poxvirus: landscape scale strategies for managing disease threat. Biological Conservation, 131, 287–95.

Guzy, J.C., Falk, B.G., Smith, B.J., et al. (2023). Burmese pythons in Florida: a synthesis of biology, impacts, and management tools. NeoBiota, 80, 1–119.

Harley, D., Sleigh, A., Ritchie, S. (2001). Ross River Virus transmission, infection, and disease: a cross-disciplinary review. Clinical Microbiology Reviews, 14, 909–32.

Huong, N.Q., Nga, N.T.T., Long, N.V., et al. (2020). Coronavirus testing indicates transmission risk increases along wildlife supply chains for human consumption in Viet Nam, 2013-2014. PLoS ONE, 15, e0237129.

Hutin, Y., Williams, R., Malfait, P., et al. (2001). Outbreak of human monkeypox, Democratic Republic of Congo, 1996 to 1997. Emerging Infectious Diseases, 7, 434–8.

IPBES. (2020). Workshop Report on Biodiversity and Pandemics of the Intergovernmental Platform on Biodiversity and Ecosystem Services. Daszak, P., das Neves, C., Amuasi, J., et al. Bonn, Germany.

Kan, B., Wang, M., Jing, H., et al. (2005). Molecular evolution analysis and geographic investigation of severe acute respiratory syndrome coronavirus-like virus in palm civets at an animal market and on farms. Journal of Virology, 79, 11,892–900.

Karesh, W.B., Cook, R.A., Bennett, E.L., Newcomb, J. (2005). Wildlife trade and global disease emergence. Emerging Infectious Diseases, 11, 1000–02.

Khodakevich, L., Jezek, Z., Kinzanzka, K. (1986). Isolation of monkeypox virus from wild squirrel infected in nature. Lancet, 327, 98–9.

Kilpatrick, A.M., Randolph, S.E. (2012). Drivers, dynamics, and control of emerging vector-borne zoonotic diseases. The Lancet, 380, 1946–55.

Kleber de Oliveira, W., Cortez-Escalante, J., De Oliveira, W.T., et al. (2016). Increase in reported prevalence of microcephaly in infants born to women living in areas with confirmed Zika virus transmission during the first trimester of pregnancy - Brazil, 2015. Morbidity and Mortality Weekly Report (MMWR), 65, 242–7.

Kraemer, M.U.G., Reiner, R.C. Jr., Brady, O.J., et al. (2019). Past and future spread of the arbovirus vectors *Aedes aegypti* and *Aedes albopictus*. Nature Microbiology, 4, 854–63.

Kraemer, M.U.G., Sinka, M.E., Duda, K.A., et al. (2015). The global distribution of the arbovirus vectors *Aedes aegypti* and *Ae. albopictus*. eLife, 4, e08347.

Lurz, P.W.W., Geddes, N., Lloyd, A.J., et al. (2003). Planning a red squirrel conservation area: using a spatially explicit population dynamics model to predict the impact of felling and forest design plans. Forestry, 76, 95–108.

MacLeod, C.J., Paterson, A.M., Tompkins, D.M., Duncan, R.P. (2010). Parasites lost – do invaders miss the boat or drown on arrival? Ecology Letters, 13, 516–27.

Makoni, M. (2020). Africa's invasive species problem. Lancet Planetary Health, 4, e317–19.

McCleery, R.A., Sovie, A., Reed, R.N., et al. (2015). Marsh rabbit mortalities tie pythons to the precipitous decline of mammals in the Everglades. Proceedings of the Royal Society of London B, 282, 20150120.

McClure, K.M., Fleischer, R.C., Kilpatrick, A.M. (2020). The role of native and introduced birds in transmission of avian malaria in Hawaii. Ecology, 101, e03038.

Melski, J., Reed, K., Stratman, E., et al. (2003). Multistate outbreak of monkeypox—Illinois, Indiana, and Wisconsin, 2003. Morbidity and Mortality Weekly Report (MMWR), 52, 537–40.

Musso, D., Gubler, D.J. (2016). Zika Virus. Clinical Microbiology Reviews, 29, 487–524.

Musso, D., Roche, C., Robin, E., et al. (2015). Potential sexual transmission of Zika Virus. Emerging Infectious Diseases, 21, 359–61.

Nolen, L.D., Osadebe, L., Katomba, J., et al. (2016). Extended human-to-human transmission during a monkeypox outbreak in the Democratic Republic of the Congo. Emerging Infectious Diseases, 22, 1014–21.

Nugent, G., Buddle, B.M., Knowles, G. (2015). Epidemiology and control of *Mycobacterium bovis* infection in brushtail possums (*Trichosurus vulpecula*), the primary wildlife host of bovine tuberculosis in New Zealand. New Zealand Veterinary Journal, 63(sup1), 28–41.

PAHO. (2015). Neurological syndrome, congenital malformations, and Zika virus infection. Implications for public health in the Americas. Pan American Health Organization Epidemiological Alert. 1 December 2015: https://www.paho.org/hq/dmdocuments/2015/2015-dec-1-cha-epi-alert-zika-neuro-syndrome.pdf, accessed 1/7/2021.

Patrono, L.V., Pléh, K., Samuni, L., et al. (2020). Monkeypox virus emergence in wild chimpanzees reveals distinct clinical outcomes and viral diversity. Nature Microbiology, 5, 955–65.

Poling, A., Weetjens, B.J., Cox, C., et al. (2010). Teaching giant African pouched rats to find landmines: operant conditioning with real consequences. Behavior Analysis in Practice, 3, 19–25.

Porse, C.C., Kramer, V., Yoshimizu, M.H., et al. (2015). Public health response to *Aedes aegypti* and *Ae. albopictus* mosquitoes invading California, USA. Emerging Infectious Diseases, 21, 1827.

Powell, J.R. (2016). Mosquitoes on the move. Science, 354, 971–2.

Radonić, A., Metzger, S., Dabrowski, P.W., et al. (2014). Fatal monkeypox in wild-living sooty mangabey, Côte d'Ivoire, 2012. Emerging Infectious Diseases, 20, 1009–11.

Rentería-Solís, Z., Birka, S., Schmäschke, R., et al. (2018). First detection of *Baylisascaris procyonis* in wild raccoons (*Procyon lotor*) from Leipzig, Saxony, Eastern Germany. Parasitology Research, 117, 3289–92.

Rezza, G., Chen, R., Weaver, S.C. (2017). O'nyong-nyong fever: a neglected mosquito-borne viral disease. Pathogens and Global Health, 111, 271–5.

Ribeiro, L.P., Carvalho, T., Becker, C.G., et al. (2019). Bull-frog farms release virulent zoospores of the frog-killing fungus into the natural environment. Scientific Reports, 9, 13422.

Rimoin, A.W., Mulembakani, P.M., Johnston, S.C., et al. (2010). Major increase in human monkeypox incidence 30 years after smallpox vaccination campaigns cease in the Democratic Republic of Congo. Proceedings of the National Academy of Sciences, USA, 107, 16,262–7.

Rubin, A. (2008). Those Darn Squirrels! Clarion Books.

Rushton, S.P., Lurz, P.W., Gurnell, J., et al. (2006). Disease threats posed by alien species: the role of a poxvirus in the decline of the native red squirrel in Britain. Epidemiology & Infection, 134, 521–33.

Sakkas, H., Bozidis, P., Franks, A., Papadopoulou, C. (2018). Oropouche Fever: a review. Viruses, 10, 175.

Sánchez-Seco, M.P., Sierra, M.J., Estrada-Peña, A., et al. (2022). Widespread detection of multiple strains of Crimean-Congo Hemorrhagic Fever Virus in ticks, Spain. Emerging Infectious Diseases, 28, 394–402.

Sejvar, J.J., Chowdary, Y., Schomogyi, M., et al. (2004). Human monkeypox infection: a family cluster in the Midwestern United States. The Journal of Infectious Diseases, 190, 1833–40.

Shuttleworth, C.M., Robinson, N., Halliwell, E.C., et al. (2020). Evolving grey squirrel management techniques in Europe. Management of Biological Invasions, 11, 747–61.

Smith, K.F., Behrens, M., Schloegel, L.M., et al. (2009). Reducing the risks of the wildlife trade. Science, 324, 594–5.

Staples, J.E., Monath, T.P. (2008). Yellow fever: 100 years of discovery. JAMA, 300, 960–62.

Tompkins, D.M., Sainsbury, A.W., Nettleton, P., et al. (2002). Parapoxvirus causes a deleterious disease in red squirrels associated with UK population declines. Proceedings of the Royal Society of London B, 269, 529–33.

Tompkins, D.M., White, A.R., Boots, M. (2003). Ecological replacement of native red squirrels by invasive greys driven by disease. Ecology Letters, 6, 189–96.

van Riper, C. III, van Riper, S.G., Goff, M.L., Laird, M. (1986). The epizootiology and ecological significance of malaria in Hawaiian land birds. Ecological Monographs, 56, 327–44.

Vogel, G. (2016). A race to explain Brazil's spike in birth defects. Science, 351, 110–11.

Vusirikala, A., Charles, H., Balasegaram, S., et al. (2022). Epidemiology of early monkeypox virus transmission in sexual networks of gay and bisexual men, England, 2022. Emerging Infectious Diseases, 28, 2082–6.

Willson, J.D., Dorcas, M.E., Snow, R.W. (2011). Identifying plausible scenarios for the establishment of invasive Burmese pythons (*Python molurus*) in Southern Florida. Biological Invasions, 13, 1493–504.

Zanluca, C., Melo, V.C., Mosimann, A.L., et al. (2015). First report of autochthonous transmission of Zika virus in Brazil. Memórias do Instituto Oswaldo Cruz, 110, 569–72.

# CHAPTER 10

# Climate change and emerging infectious diseases

## 10.1 Climate affects host–parasite interactions

*New England, US*—In the north-eastern USA, formerly regal moose (*Alces alces*; in Europe they are referred to as Eurasian elk) are transforming into 'ghost moose'—their normal dark brown coats are turning pale and patchy, and the moose are becoming emaciated and dying (Fig. 10.1). Indeed, moose populations in New England have been declining, and one mooted cause is the impact of climate change upon ticks. Shorter, less intensely cold winters are enabling 'winter ticks'—*Dermacentor albipictus*—to boom in numbers.

The winter tick seeks a host in late autumn (aka fall; traditionally September–November), when it is in the larval life stage (Fig. 10.2). Winter tick larvae 'quest' for hosts by waiting on vegetation in clumps, 'their tiny limbs [linked] to form long, almost invisible chains. . . and when a big, tall, blood-filled mammal walks by and brushes the branch, one or more of the ticks grasps the animal's fur and holds tight while the rest of the gang swings as a gossamer-thin thread onto the animal' (Dobbs 2019).

People come to Maine, they want to eat lobster, but they want to see moose.

—Lee Kantar

These moose look like the walking dead. . .

—Christine Dell'Amore

The winter ticks feed and remain on the moose throughout their subsequent two developmental moults: from larva to nymph, and from nymph to adult, with each larger stage taking a larger bloodmeal (Fig. 10.2). Consequently, as new larvae are recruited and remain on the moose until they drop off as adults to give birth to the next generation, infestations can accumulate. As you might imagine, having more ticks is bad for moose, because morbidity and mortality for ticks and other parasites is usually **intensity dependent** (Box 10.1).

In days of yore, adult ticks may well have dropped off onto spring snow—even well into April—and this would be detrimental to tick survival (Fig. 10.3). Snowfall in late fall or early winter would also kill the questing clumps of larvae. But climate change is causing shorter, less harsh winters and earlier springs in New England; Maine's warm season—defined as when the daily temperature is above freezing—has increased by two weeks from the early 1900s to the 2000s. This appears to have had a beneficial impact upon tick recruitment, where mild winters allow greater survival of larvae and a longer window for larvae to await wandering moose, leading to mounting tick numbers.

These days, mean tick abundance on calves can be 47,371 with a range of 18,664–95,496—that's the infestation intensity on a single moose, at a single time (Jones et al. 2019). An adult female winter tick's bloodmeal consumes about 1.7–2.6 g of blood (DeBow et al. 2021). Reach for your calculator. At those rates, adult ticks alone could cause blood loss between 32 kg (or 71 lbs) and 248 kg (547 lbs), which is more than the combined weight of this book's three (svelte, nymphish) authors. That constitutes a substantial resource depletion in moose already pressed by winter conditions!

*Emerging Zoonotic and Wildlife Pathogens.* Dan Salkeld, Skylar Hopkins, and David Hayman, Oxford University Press.
© Oxford University Press (2023). DOI: 10.1093/oso/9780198825920.003.0010

**Figure 10.1** Ghost moose in New Hampshire, USA. 'In April, when the gravid females start taking their blood meals, the blood loss over their last two to four weeks aboard the moose "can equate to a calf's total blood volume", according to one recent paper—some three and a half gallons' (Dobbs 2019): https://www.theatlantic.com/science/archive/2019/02/ticks-can-take-down-800-pound-moose/583189/
Photograph by Dan Bergeron, New Hampshire Fish and Game Department.

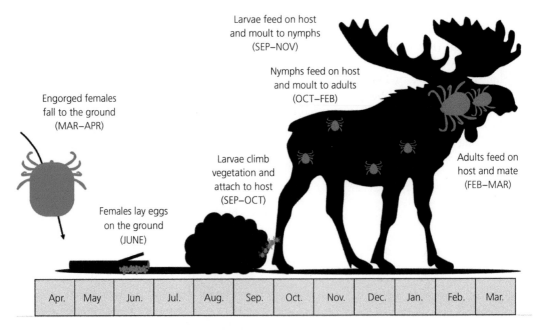

**Figure 10.2** Life cycle of the winter tick, *Dermacentor albipictus*, on moose.
This figure was modified from one published by the government of Quebec (https://www.quebec.ca/en/agriculture-environment-and-natural-resources/animal-health/animal-diseases/list-animal-diseases/moose-winter-tick).

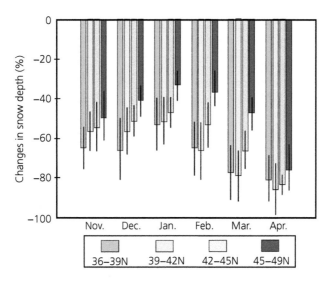

**Figure 10.3** Predicted percentage decreases in future snow depth (2041–2095) compared to historical snow depth (1951–2005) in the north-eastern and upper mid-west United States based on a meteorological model using a single future climate scenario (RCP 4.5). The predicted changes are largest for the spring. Coloured bars show changes for different latitudinal bands.
Modified from Demaria et al. (2016).

Infested moose become anaemic, depleted of fat and protein reserves, immunologically stressed, and prone to mortality. Their haggard condition is caused by more than just blood loss. Irritation caused by the masses of ticks also stimulates excessive grooming (rubbing against trees) that compromises the moose's fur coat. Moose with broken hairs and patchy coats are called 'ghost moose',

---

## Box 10.1  Intensity dependent disease outcomes

We often talk about parasites/pathogens in a present/absent way, as if all infections are equivalent. But **infection intensity**—the number of individual parasites on or in the host—often varies among hosts. For example, macroparasites like ticks, bot flies, and helminths universally have **aggregated distributions** among hosts, where most hosts have few to no parasites and a few hosts have many parasites (Shaw and Dobson 1995, Poulin 2007) (Fig. 10.4). Infection intensity is often difficult to quantify—try counting how many fleas are on your cat—and difficult to include in statistical or mathematical models of disease dynamics. But infection intensity is important to consider, because hosts with more individual parasites usually have higher potential for spreading their infections to other hosts and more severe disease outcomes.

**Intensity dependent disease outcomes** include many possible impacts of parasites on their hosts. In the most obvious examples, hosts with more parasites have the most severe pathology (e.g. too sick to move) or the highest mortality rates. But parasites can also have more subtle impacts on hosts. For example, in the wild, 97–100% of Allenby's gerbils (*Gerbillus andersoni allenbyi*) have fleas, with a subset of the gerbil population having high intensity infestations. Gerbils infested with high densities of fleas are so distracted and irritated by their fleas that they spend less time foraging for food and leave more food behind, especially when predatory foxes are nearby (Raveh et al. 2011). The most distracted gerbils may also be more likely to get eaten by foxes. Whatever the impacts, disease outcomes are usually more severe as infection intensity increases (Fig. 10.5), but the relationship is not necessarily linear; for example, there may be some threshold infection intensity above which hosts die.

**Box 10.1** *Continued*

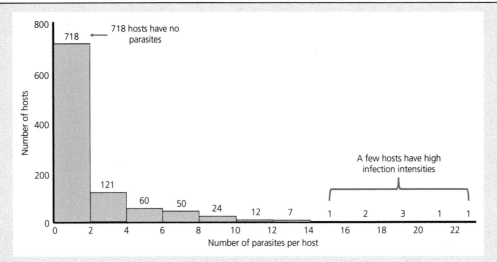

**Figure 10.4** For any given parasite species, most hosts (~80%) have few to no parasites, while a few hosts (~20%) harbour most of the individual parasites, with high infection intensities. This is called the **80/20 rule**, and these aggregated parasite distributions are often well approximated by a negative binomial distribution, as shown in this histogram.

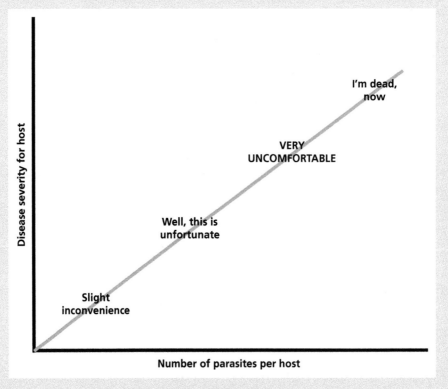

**Figure 10.5** Graphical representation of host disease severity increasing with infection intensity (the number of parasites per host). This graph depicts a linear relationship but there could be thresholds or other relationships.

and they may have reduced survival through the harsh New England winters. These impacts are especially noticeable for moose calves; more than 50% of calves may die in a given winter, and this moose calf mortality is suspected to be driving regional population declines (Dell'Amore 2015, Jones et al. 2019).

However, it is unlikely that climate change's impact on tick demography is the only factor contributing to moose declines (Box 10.2). For example, moose can also die from infections with deer meningeal worm or brain-worm (*Parelaphostrongylus tenuis*) (DeBow et al. 2021), which has also moved northwards in recent decades in concert with white-tailed deer (*Odocoileus virginianus*). And moose can be infected with lungworm (*Dictyocaulus* spp.) that can interfere with breathing; a moose biologist has observed 80% of moose calves have abnormal lung tissue, and occasionally 'discovers big masses of lungworm in the animal's trachea that were trying to leave the dead body "like rats fleeing a ship"' (Dell'Amore 2015). Lungworm infections (prevalence = 17–73%) compound the impacts of winter tick infestations and reduce moose calf survival (DeBow et al. 2021). As if that weren't enough, bot fly infestations in the trachea (Fig. 10.6) and rare protozoal parasitic encephalomyelitis have also contributed to moose mortality events. Any or all of these parasitic infections could be impacted by climate change. For example, moose may be physiologically stressed (DeBow et al. 2021) and more susceptible to parasites in a warmer New England, because their huge bodies and thick dark hair evolved for maximum heat retention. Overall, climate change is undoubtedly playing a role in moose disease and moose declines, but it is difficult to pinpoint exactly which mechanisms are the biggest problems for moose populations.

Short limerick interlude:

> There once was a New England moose
> With ticks that were sucking its juice.
> Plus meningeal worms,
> And protozoal germs...
> Those parasites sure cooked its goose.

End of limerick interlude.

---

**Box 10.2 'Schneiderisms'—The state of the science, preponderance of evidence, and a dose of scepticism**

A while ago, one of the book's authors (the elegant, witty, move-like-a-panther one) had the opportunity to chat briefly with a luminary in the field of climate change and systems science: Dr Stephen Schneider. A 20-minute chat and a guest lecture later, and that same author (now humbled, inspired, and a wee bit insecure) had had their perspective of science, advocacy, and thinking challenged and upended. This is an attempt to convey some of that wisdom, appended with quotations from a videoed presentation by Dr Schneider. Also recommended is his book on cancer treatment: *The Patient from Hell* (Schneider 2005). And tangentially, a lesson: talk to as many people as you can about life and science, no matter the perceived hierarchy. Many of those conversations may be stuck in a rut of grumbling and cliché and blandness, but every now and then it can be magical and life changing. Without further ado, on the topic of climate change...

Predictions from the 1970s are bearing themselves out; nature is cooperating with theory. Scientific consensus that $CO_2$ would increase and be a potential problem has been overwhelming for a long time. The exact impacts—how much will temperatures increase, where will the effects be largest or mitigated—are more debatable. And yet in the 2020s, a half-century after the early warnings about climate change, actions to reduce global change are still woefully inadequate.

Climate science relies on systems science: multiple components interacting, often in complex, non-linear ways. Other examples of systems science are attempts to understand the human body's response to cancer and therapy, the root causes of terrorism or suicide or drug addiction, or emerging infectious diseases. When looking at the system, or breaking it down into component parts, 'rarely do we know everything and rarely do we know nothing'. Consequently, you must determine whether the evidence for different mechanisms and processes are well understood, still debated, or speculative. This can be a difficult path to walk.

It also means that different agendas can be advocated for because evidence can be cherry-picked, or doubted, or adopted out of context. And the more extreme views—at both ends of the spectrum—can start to sound credible. For example, opponents of measures to mitigate climate change can cite the facts that 'carbon dioxide is natural; it's what plants eat; it's a part of the regular life cycle of

Earth...'. At the opposite end, fearmongering might use the arguments that 'insect-borne diseases such as malaria will explode... we're likely to see hundreds of millions of what we'll call environmental refugees... basically, none of the crops will grow... the rest of us would be cannibals...'.

Stephen Schneider would 'confess [his] prejudice: the "end of the world" and "good for you" are the two lowest probability outcomes of global change, and instead we are facing a range of potential outcomes. System scientists try to winnow out the relative likelihood of these multiple outcomes.' Parsing the complex system down into a binary viewpoint is simplistic, but unfair and unmerited, whereas appreciating the nuance can be challenging.

Additionally, you must make judgements on who is telling you what, and whether they have an agenda. 'Experts' wearing scrubs or armed with PhDs might portray expert opinions without confessing whether they're being paid by special interests. People can frame the problem by looking for exceptions to the conventional wisdom and claim that until the exceptions are resolved it isn't proved and it's premature to act. But, as Dr Schneider commented: '[That's] not balance, it is utter distortion...'.

An analogy: supposing a chest x-ray observes a spot or shadow on your lung. It could be a new tumour, or a healed lesion, or a histoplasmoma. Your doctor might suggest a biopsy to investigate, though this would involve surgery because it's a difficult place to access. Surgery is expensive, painful, and risky. In other words, there's a cost associated with assuming that the lung spot is dangerous. Alternatively, rather than do a biopsy, you could wait and monitor whether the spot grows, and if it does then you can perform the biopsy at that point. But there's also a potential cost for waiting: if the spot is cancerous, it could grow and become metastatic, and then the biopsy investigation would be too late, and you could die. In a precautionary mode, you would pay the price now for the surgery that you might not need, but you'll concurrently have a lower likelihood of being dead. What's the right course of action? 'There is no right answer. That's your value judgment.' But whether you choose to act or do nothing, your choice will have consequences (positive or negative).

## 10.2 Thermal performance curves

Organisms live in variable environments, and their performance (e.g. survival, growth, fecundity) can vary drastically with environmental conditions. For example, physiological processes often require a

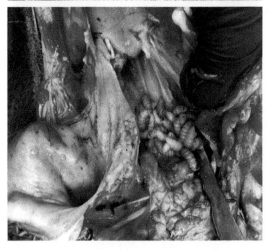

**Figure 10.6** (Top) Winter tick larvae amassing and questing on vegetation, awaiting a passing host. (Middle) Extreme case of engorged winter ticks on a moose hide, New Hampshire, 2015. (Bottom) The trachea (windpipe) of a moose calf revealing heavy infestation of bot fly larvae, an example of the other factors contributing to moose mortality.

Top photo by Cheryl Frank Sullivan, University of Vermont. Middle and bottom photos by Jacob DeBow, New Hampshire Fish and Game.

minimum temperature, peak at the optimal temperature, and can't happen at all when temperatures are just too hot. Like Goldilocks stealing porridge from the three bears, conditions need to be just right, not too hot and not too cold. These non-linear relationships (or 'hump-shaped' or 'unimodal' relationships) are called **thermal performance curves**, and each species (e.g. each pathogen, vector, or host) has species-specific thermal performance curves. As you might imagine, non-linear relationships between pathogen/vector/host performance and temperature complicate how we understand and predict disease dynamics in a changing climate (Lafferty and Mordecai 2016; Fig. 10.7)!

A case in point is *Plasmodium falciparum*, a protist that is transmitted by female *Anopheles* mosquitoes and causes malaria in humans. Back in the day, models that described malaria transmission often predicted optimal transmission rates at 30–33°C (Mordecai et al. 2013). But when people actually measured transmission rates in locations with mean transmission temperatures >28°C during the transmission season, they found that transmission rates were low. So, were the models wrong or were the field observations wrong?

The older models were oversimplified. In those older models, researchers assumed that linear associations could describe the relationships between environmental temperature and the life history traits of mosquitoes and *Plasmodium* (e.g. mosquito biting rates, mosquito mortality rates, *Plasmodium* growth rates). However, the malaria parasite cannot develop if temperatures are too cold (<18°C), and at higher temperatures (>32°C), the mosquitoes tend to die before the pathogen is transmissible (Lafferty 2009, though see below). In fact, even as warm temperatures accelerate some rates, such as a mosquito's activity, growth, development, or reproduction, there is often an associated cost, such as a decrease in survivorship. Re-analysing data from laboratory studies demonstrated that humped curves pervade malaria transmission biology. Incorporating these functions into process-based models suggests that the optimum transmission temperature for *Plasmodium falciparum* is more convincingly 25°C (and declines around 28°C). The newer predictions for optimal temperatures are about 6°C lower than previous models and provide a whole different outlook on understanding the geography of malaria endemicity (Fig. 10.7).

As an added complication, temperatures fluctuate daily, seasonally, annually, and with multi-year climatic phenomena like El Niño or the Pacific Decadal Oscillation. For example, diurnal temperature ranges—the difference between the daily minimum and maximum temperatures—can be 5 to >20°C in areas of Africa where malaria is endemic (Paajimans et al. 2010)! Are *Plasmodium* and mosquitoes most impacted by the minimum, mean, or maximum temperature? Or is it the variability in temperature that is most important? To understand vector-borne disease transmission, careful scientists often strictly control the temperatures of laboratory transmission experiments, e.g. a constant temperature representative of a monthly mean. But experimental manipulations of an animal-model system (rodent malaria, *Plasmodium chabaudi*, and *Anopheles stephensi* mosquitoes) revealed that diurnal temperature fluctuations affect the mosquito's life history traits and the malaria parasite's incubation period (the time it takes from when the mosquito is infected until it is infectious). For example, at a mean of 16°C, temperature fluctuations caused *Plasmodium* to grow faster, but at a higher mean temperature (26°C), temperature fluctuations had a negative impact on the parasite (Paajimans et al. 2010). Therefore, models that ignore diurnal variation would *overestimate* malaria risk in warmer environments and *underestimate* risk in cooler environments. Such models would fail to predict that even small changes in temperature regime could have large biological significance.

To make matters even more complicated, life history traits are influenced by many environmental conditions besides temperature. They can also have 'Goldilocks' relationships with precipitation, humidity, soil pH, and other factors. And of course, many animals can move from habitats with unfavourable conditions to habitats conducive to higher fitness. For example, many ectotherms like ticks and mosquitoes will exhibit behavioural thermoregulation, where they seek micro-habitats with optimal temperatures. Keep these complications in mind as we delve further into the links between climate change and infectious disease,

**Temperature-dependent rate parameters**

**Temperature-dependent net malaria transmission**

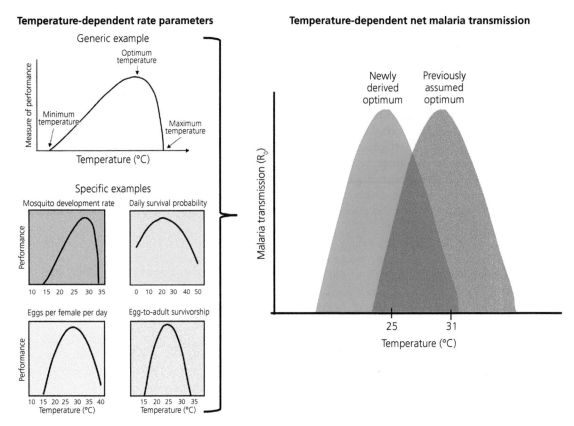

**Figure 10.7** Organismal 'performance' (growth, survival, fecundity) is often temperature dependent, with a minimum, optimal, and maximum temperature (generic example in top left). We show some stylized results from Mordecai et al. (2013), illustrating how mosquito life history traits have non-linear or 'hump-shaped' relationships with temperature. When these temperature-dependent rates are included in process-based malaria transmission models, the peak temperature for malaria transmission ($R_0$) is lower than previous estimates that used linear relationships between temperature and life history traits.

because we must remember that these relationships are nuanced and complicated and can have a big impact on understanding and predicting disease dynamics.

Implicit in almost all the literature on this subject is an assumption that environmental change is more likely to strengthen the transmission potential and expand the range, rather than to disrupt the delicate balance between pathogen, vector and host upon which these systems depend. VBDs, however, are no different from other components of global biodiversity that is predicted to be on the verge of catastrophic collapse, the sixth mass extinction. The arthropods that serve pathogens are as vulnerable as those that deliver the anthropocentrically designated 'ecosystem services.'

—Sarah Randolph (2009)

## 10.3 Climate change and shifting distributions

*Canadian Arctic*—Musk oxen (Fig. 10.8) can be susceptible to a pernicious lungworm, *Umingmakstrongylus pallikuukensis*. The name of this parasite is generated from the Inuinnaqtun word *umingmak*, which means musk ox (*Ovibos moschatus*); 'strongylus' for the strongylate nematodes; *pallik*, which is Inuinnaqtun for the region surrounding the watershed of the Rae and Richardson valleys, west of Kugluktuk; and *kuuk*, which means river (Kutz et al. 2001). It is a relatively newly described Arctic parasite (Fig. 10.9).

Lungworms reach 60–70 cm in length. Imagine trying to go about your day with a piece of spaghetti

**Figure 10.8**  Musk ox grazing on tundra and risking ingestion of nematode-infected slugs.
Photo credit: **Кирилл Уютнов**, https://en.wikipedia.org/wiki/Muskox#/media/File; Creative Commons Attribution-Share Alike 4.0 International license.

**Figure 10.9**  Pratap Kafle bravely collecting musk ox faeces on Victoria Island, Nunavut, Canada. Which brings to mind Apsley Cherry-Garrard, who journeyed to the Antarctic to seek Emperor penguin eggs for scientific research, and who opined, 'If you have the desire for knowledge and the power to give it physical expression, go out and explore. . . . Some will tell you that you are mad, and nearly all will say, "What is the use?" For we are a nation of shopkeepers, and no shopkeeper will look at research which does not promise him a financial return within a year. And so you will sledge nearly alone, but those with whom you sledge will not be shopkeepers: that is worth a good deal. If you march your Winter Journeys you will have your reward, so long as all you want is a penguin's egg' (Cherry-Garrard 1994).
Photo credit: Pratap Kafle.

as long as your arm inside your lungs! Lungworm infections can occupy 5–20% of a musk ox's lung volume (in extreme cases, nearly 50%). These vast infections make their shaggy-haired hosts more susceptible to starvation in harsh winters or more vulnerable to predation, because deteriorating body condition and respiratory compromise make the musk ox less likely to escape wolves.

*Umingmakstrongylus pallikuukensis* has expanded its range northward by 480–560 km in the last two decades (Fig. 10.10). This newly exposed Arctic region has experienced substantial environmental changes in the last few decades, with annual temperature increasing by roughly 2.3°C between 1948 and 2016, compared to a global mean increase of

0.8°C in the same period. As lungworm moves northward, it is infecting musk ox herds that were never exposed to it previously (Kafle et al. 2020). At the same time, worm loads have also increased in the more southerly areas where lungworm was historically established. This suggests that as temperatures continue to warm, the parasite's range may continue to expand and worm loads will increase.

However, changes in species distributions are also dynamic and potentially difficult to predict. A climate that wobbles over time can drive episodes of **environmental sloshing**, i.e. recurrent geographical advances and retreats of organisms in response to environmental change that can involve alternating regimes of warming and cooling (Kafle

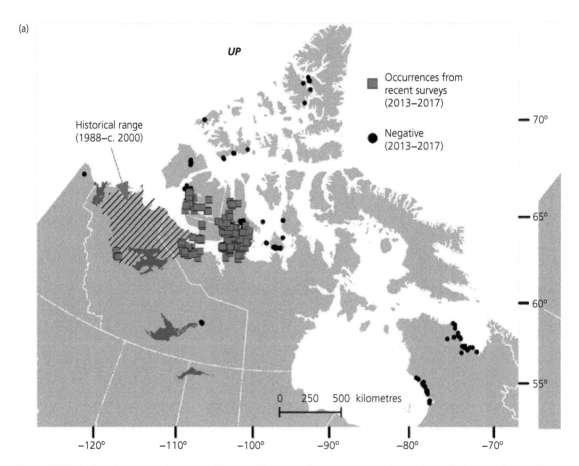

**Figure 10.10** Northward expansion of the geographic range of the musk ox lungworm, *Umingmakstrongylus pallikuukensis*, in the Canadian Arctic. The historical range (shaded) for lungworm was restricted to the mainland but the red squares show sites where lungworm have recently been found in musk ox outside the historical range. Prevalence in musk oxen in the historical range has approached 100%.
From Kafle et al. (2020), used with permission.

et al. 2020). These processes often occur over scales too large for most observational studies. Therefore, to understand these complex and dynamic changes, we often need to turn to process-based models that account for the various non-linear relationships between temperature and biological processes (Box 10.3). Building and validating process-based models works especially well when historical records of climate and disease incidence exist for model validation.

---

### Box 10.3  Process-based models

**Process-based models**, also referred to as mechanistic or biological models (Chapter 5), describe quantifiable relationships in disease systems using mathematical equations, which can then be used to generate predictions or describe biological phenomena. In the case of climate change, the models that incorporate well-described relationships between temperature and vital rates of pathogens, vectors, and hosts (e.g. extrinsic incubation rates, longevity, growth) can be used to simulate disease dynamics under past, present, and future climate scenarios. If the mathematical model generates predictions that match observed field data from the past and present, then the underlying assumptions or 'mechanisms' are likely appropriate. If the model's predictions are different from observed data, re-evaluating assumptions to improve the model can lead to important insights about the disease system. Once a reasonable model has been created, it can be used to explore what could happen in hypothetical scenarios, e.g. temperatures increase.

Take, for example, the musk ox nematode lungworm (Fig. 10.11). Adult lungworms reproduce in the musk ox host and first stage larvae (L1) exit in the burly bovid's faeces. In the Arctic wilds, these free-living larvae penetrate the foot of passing terrestrial slugs and moult through second (L2) and third larval (L3) stages. Some doubly unlucky slugs are eaten by grazing musk oxen (eaten from the inside by worm larvae and the outside by musk oxen! Out of the frying pan and into the fire. . .), whereupon the larvae migrate to the musk ox's lungs and become adults. The sooner a larva develops from an L1 to an L3, the more likely it is to be infectious when the slug is consumed by a musk ox. And if development takes a long time due to environmental conditions, an L1 larva that enters a slug one year might not be able to develop to an L3 before the winter, when slugs aestivate and transmission to musk oxen does not occur, or the slug host shuffles off its mortal coils (Kutz et al. 2001).

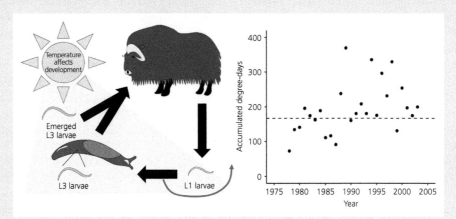

**Figure 10.11** The life cycle of *Umingmakstrongylus pallikuukensis*, the musk ox lungworm, is influenced by environmental conditions. For example, to develop from L1 to L3 larvae in a single year, the lungworm needs conditions to be above 8.5°C for at least 167 'degree-days', where warmer temperatures mean more accumulated degree-days. If there are not 167 degree-days (below dashed line on graph), larval development requires two years and requires the larvae to overwinter in the slug host. As temperatures have increased due to climate change, the number of degree-days per year has also increased, such that the lungworm larvae can more often complete their L1 to L3 development in a single year now than a few decades ago.
The degree-day data are from Kuntz et al. (2005).

---

**Box 10.3** *Continued*

A process-based model was used to explore how climate change might influence *Umingmakstrongylus pallikuukensis* development from L1 to L3 larvae. The model includes a minimum temperature for development (8.5°C) and an hourly temperature-dependent growth rate that described how larval development rates depend on soil-surface temperatures (based on laboratory data). The model was especially focused on 'degree-days'—the time where conditions were above the threshold for larval development. For past and current climate conditions, weather data could be used to calculate accumulated degree-days each year, and for future climate scenarios, projected increases in hourly temperatures were used to predict accumulated degree-days. This model allowed investigators to predict how rates of larval development might increase in a climate future where the region increases in mean temperature by $\geq 2°C$ by the 2020s, $\geq 4°C$ by the 2050s, and $\geq 6°C$ (to a maximum of 8°C) by the 2080s (Kutz et al. 2005).

Historic data (i.e. 1978–2003) showed considerable yearly variability in the degree-days that generated lungworm larval growth. Only in some years were there sufficient degree-days (167) for development to L3 within a one-year development cycle; such years were more common from 1991 to 2003 (12 of 13) than from 1978 to 1990 (5 of 13). When sufficient degree-days were accumulated in a single year for development to L3, the window for transmission of L3 from slugs ranged from 6–72 days. When there were insufficient degree-days for the larvae to develop through the L3 stage, the larvae would have had to overwinter in gastropods, i.e. a two-year development cycle.

Scenarios of climate warming suggested that even with temperatures increasing by just 1°C above the 1978–2000 average, lungworm larvae development times would shift from a two-year cycle to a one-year cycle. What's more, warming temperatures also expanded the window of opportunity for transmission of L3 in slugs to musk oxen (defined as the time from when the first L3 became available in slugs to the last day before five consecutive days with average temperatures below 0°C, when slugs are assumed to move deeper into the soil and begin hibernation). These results suggest that warming climates are increasing lungworm prevalence by increasing the temporal overlap between musk oxen grazing and the presence of infectious slugs.

---

When historical data do not exist (as is often the case), one may be able to extrapolate the potential impacts of climate change after quantifying the current distribution of hosts/vectors/pathogens as a function of climate variables (e.g. temperature or rainfall patterns). If the temperature or rainfall or seasonality patterns were to change, would the distribution of the host/vector/pathogen increase, decrease, or shift? These models are called **species distribution models** (formerly **ecological niche models**) and they are forms of **statistical models** (Box 10.4): pattern matching that shows the strength of correlation between field observations of the species' presence (or presence–absence or abundance) and environmental factors (e.g. average temperature, average temperature of three warmest months, summer rainfall, etc.). After a correlation has been established, predictions can be made by **interpolation** (making predictions within the range of sampled data) or **extrapolation** (making

predictions outside the range of sampled data). Climate change studies often rely on extrapolation to conditions that have never been observed in a given region, and thus these studies must be conducted and interpreted with caution; we cannot necessarily describe non-linear relationships or predict thresholds using existing distribution data and correlational models.

And of course, studies can be biased due to their geographical foci. For example, early studies of climate change often focused on the temperate north, predicting that species distributions would expand northwards and shift to higher altitudes, e.g. the movement of plague (*Yersinia pestis*) in California, or the movement of sasquatch away from Pacific coastal areas and into south-western Canada and montane habitats of the American south-west (Arizona, Nevada, Utah) (Holt et al. 2009, Lozier et al. 2009). This led to some alarming overgeneralizations, such as that infectious diseases would broadly increase with climate change. While it is true that pathogens' distribution shifts will expose many naive populations to many new diseases, it is also important to remember that climate change will not universally increase pathogens' ranges or disease burdens. In many places, hosts, vectors, and/or pathogens might disappear when climates become unsuitable (Lafferty 2009).

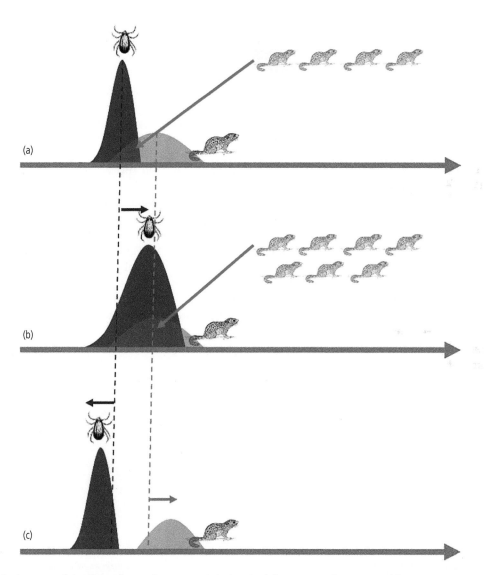

**Figure 10.12** Impacts of altered phenology on disease transmission in a simple host–vector pathogen system. (A) Disease transmission (red squirrels) occurs as a result of vector abundance (blue) overlapping host exposure (orange). Changes in vector phenology, i.e. a longer activity season, increases the window of overlap with host animals, increasing the prevalence or abundance of infection (B). In contrast, increased asynchrony in the phenology of vector and host disrupts pathogen transmission (C).

Climate change may be especially disruptive to parasites and pathogens with complex life cycles that rely on multiple different host species. For example, vector-borne pathogens often have an arthropod vector host and at least one competent vertebrate host. In technical terms, they are **ontogenetic niche specialists**, where each stage in their life cycles requires a different set of habitat conditions (i.e. host species). When multiple host species are required for life cycle completion, the different host species must overlap in space in a single pathogen generation. If the distribution range of either host changes, the pathogen's distribution can also change—more overlap could increase a parasite's range, but less overlap could create an 'ecological mismatch' and a reduction in parasite range (Pickles et al. 2013). Many pathogens also require that their hosts overlap in time. Climate-induced changes in **phenology**—the timing of life history—can create more temporal overlap (e.g. between infectious slugs and musk ox) or reduce the temporal overlap and the window for transmission (Fig. 10.12). Given these complications, models based on a single stage of the parasite life cycle might fail to generate good predictions of how disease dynamics will respond to climate change.

For example, you may recall that deer meningeal worms are now more common in the north as their normal definitive host, the white-tailed deer, has been able to expand its northern range (Section 10.1). Similar to *Umingmakstrongylus pallikuukensis*, *Parelaphostrongylus tenuis* (yes that was a deliberate placing of two indigestible Latin binomials) is shed from the vertebrate host (deer) in faeces, and the free-living nematode larvae are ingested by gastropods, before being trophically transmitted back to deer via gastropod consumption. A species distribution model for *Parelaphostrongylus tenuis* alone predicts a large range expansion by 2080 due to climate change. However, the predicted expansion is much smaller in models that account for the habitat requirements of *Parelaphostrongylus tenuis and* its deer and gastropod hosts. While the definitive and intermediate host species are also expected to increase their ranges, their species-specific expansions will create areas of 'ecological mismatch' for *Parelaphostrongylus tenuis* where only deer or only gastropods are present, and the life cycle cannot be completed (Pickles et al. 2013) (Fig. 10.13). Habitat suitability for *Parelaphostrongylus tenuis* is even predicted to decline in the Great Plains and southeastern United States, reminding us that while parasites may move into new regions due to climate change, they may also be lost from other regions.

**Figure 10.13** One of the textbook co-authors (the courageous, warrior-poet one) enjoys anthropomorphizing parasites. This particular cartoon shows *Parelaphostrongylus tenuis*, the deer meningeal worm, anxiously failing to increase its range with its host species, because the host species are expanding into different, non-overlapping regions. If you listen closely, you can hear the nematodes singing, 'Oh, give me a home where the parasites roam, where the deer and the gastropods play. . .'

## Box 10.4   Statistical models and mapping species distributions

Statistical models are critical tools in science because they allow us to look at complex, noisy data and determine whether there are correlations among the variables (or if the patterns are more likely to occur due to random chance) and to quantify the strength of those correlations.

For example, Gregory et al. (2022) wanted to know why *Ixodes scapularis* ticks can be found in some urban backyards but not others, so they measured tick presence (the response variable) and a suite of environmental factors in backyards. They found that tick presence was strongly correlated with variables like large areas of tree canopy cover near yards, presence of wood or brush piles in yards, and whether the yard was fenced (and thus likely kept out some tick wildlife hosts). They also found that tick abundance was *not* correlated with some variables, like whether the yard had a vegetable or flower garden. This is an example of a **species distribution model (SDM)**—a complex statistical model that uses georeferenced data (all the variables are associated with geographical coordinates) to understand spatial correlations among species' distributions and abiotic and biotic conditions (Fig. 10.14).

Like many statistical models, species distribution models make minimal assumptions about biological mechanisms or processes, which is both good and bad (Lafferty 2009)! While these models may not allow us to understand *why* we see patterns in nature, they are often our best tool for making sense of complex observational data and using those data to generate hypotheses for future studies. Species distribution models have become especially popular in our 'big data' era, because these models can incorporate readily available spatial data sets for environmental factors (e.g. satellite data for forest cover) and georeferenced site local-ity data for species occurrence (e.g. museum records) (Lozier et al. 2009). Using these data, SDMs can then produce maps that describe an organism's distribution and allow one to make predictions about how that distribution will be impacted by environmental change (Fig. 10.15).

When models will be used for prediction, their predictive capability can (and should) be explicitly quantified using **cross validation**. During cross validation, the statistical model is built using a subset of the data (often 80% of the observations). For example, we could fit the tick abundance model described above using tick and environmental data from 80% of backyards. Then the predictive ability is tested on the remaining data set (often 20% of the observations). For example, if we took the SDM fit to 80% of the data and the environmental data from 20% of the backyards, we could use the SDM to predict tick presence in those remaining backyards. Then we could evaluate how closely the predictions match the tick presence that was actually observed. Well-built and validated SDMs can describe *and* predict distributions, which makes them useful for conservation or public health.

Of course, no statistical method is perfect, and models can even be dangerous in the hands of someone who does not recognize the approach's limitations. For example, SDMs describe and predict distributions using correlations, and correlations do not necessarily reflect causation. Correlations can be especially misleading if we assume that relationships are linear (Lafferty 2009), because variables are often only correlated over part of their ranges; for hump-shaped temperature curves, the correlations even switch from being positive to negative on either side of the optimal temperature!

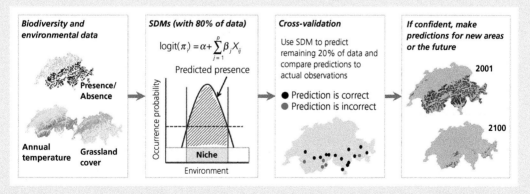

**Figure 10.14** General steps for creating a species distribution model and using it to predict the distribution in new sites (outside the spatial coverage of the original sample) or into the future (outside of the temporal coverage of the original sample). This figure was modified with permission from Damaris Zurell (2020).

**Box 10.4** *Continued*

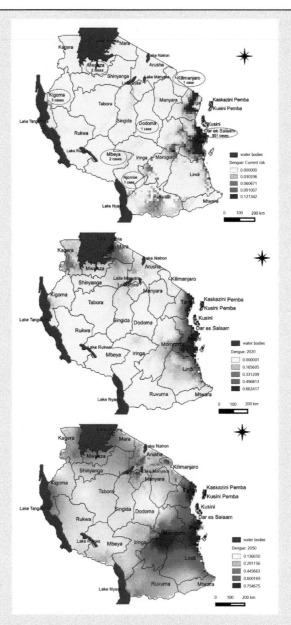

**Figure 10.15** An example of a species distribution model. This model was used to forecast changes in dengue virus epidemic risk areas, using 2014 presence data for *Aedes aegypti* infected with dengue virus and several bioclimatic predictors. Currently dengue is highly localized in coastal areas (left panel). Dengue is likely to remain in coastal areas, but climate change may also cause dengue epidemics in more central and north-eastern Tanzania, with intensification in areas around all major lakes (middle: 2020, right: 2050). From Mweya et al. (2016).

---

**Box 10.4** *Continued*

---

SDMs are also highly sensitive to the choice of underlying data used to build the models. For example, in the backyard tick model, what if there is a variable that is critically important for ticks, but we didn't know to measure it, so it is missing in the data set and models? Other such missing variables could include things like dispersal barriers (e.g. the climate on either side of a large river may be identical, but the river might prevent species dispersal to the other side of the river) and anthropogenic impacts on species (e.g. the climate might be suitable for mosquitoes, but humans remove possible mosquito breeding sites). Even if all the important variables have been included, the scale and extent at which the variables were measured could be problematic.

For example, data may be collected from only part of a species' distribution (e.g. only from a certain distance from the investigator's home or institution) or may only be coarse-grained due to data set availability (e.g. regional data were only available from satellite imagery at 10 × 10 km scale). These questions and complexities likely explain why three recent attempts (Porter et al. 2021, Alkishe et al. 2021, Hahn et al. 2021) to describe the potential changes in future tick distributions (western black-legged tick, *Ixodes pacificus*) in the western US predicted three different outcomes: a decreasing range, an unchanged range, and an increasing range! Like all models, the inputs have a big influence on the outputs.

## 10.4 Extreme weather events

Climate change is not simply an increase in average temperature. Precipitation patterns can also change, as can the frequency of **extreme weather events**, e.g. heat waves, droughts, hurricanes, or floods. These events can interact with a range of other stressors and result in disease outbreaks.

*Southern Norway, 2006*—Musk oxen seem to lead particularly hard lives, with a spectrum of odd mortality causes, e.g. avalanches, poachers, railroad accidents, culling of rogue animals, lightning strikes, falling from cliffs, and phosphorus poisoning. But early autumn was not usually associated with mortality in a musk ox population in Norway, until an unusual event in 2006. Unusually, in August and September, multiple musk oxen were observed with clinical disease, and several were found dead. The index ox (indox?) case was noticed on 14 August, and in the next six weeks, a further 24 animals were found dead (n = 21) or dying (n = 3). A population census the following March (2007) suggested that roughly 20% of the herd had been lost to the disease outbreak (Ytrehus et al. 2008). It appeared that the animals died from pasteurellosis caused by *Pasteurella multocida*.

Bacteria in the genera *Pasteurella* normally reside in the pharynx of healthy animals. However, the bacteria are opportunistic pathogens and can cause pasteurellosis, resulting in both pneumonia and sepsis if the hosts' immunity is suppressed by stress. (Side note: humans also have bacteria in our bodies that can go rogue and negatively impact our health, like *Staphylococcus aureus*.)

The onset of the disease outbreak coincided with an extreme weather event: record-breaking air temperatures and humidity in late summer and autumn 2006. Average air temperature in the study site in southern Norway in August to September 2006 reached 10.1°C, which is 3.2°C above the 1961 to 1990 average, and the warmest temperature since records began in 1923. There were no days with temperatures below freezing in August and September, a situation that had only happened once before (in 1988). Humidity was also the highest since records began (in 1973). An *extreme* weather event.

The unusually high temperatures and high humidity occurred when the musk oxen had already well-developed winter fur—a massive pelage with a thick layer of underwool, the *qiviut*, as well as a thick layer of subcutaneous fat, giving even more insulation. When facing gale force winds and −40°C (−40°F), these protective measures benefit the musk ox. But being well-insulated during abnormally warm temperatures may cause heat stress and immunosuppression and favour colonization of *Pasteurellosis*-causing bacteria.

The musk oxen might also have been stressed because it was rutting season and reindeer hunting season, so human hunters were active. Rutting and reindeer hunting are carried out every fall, but perhaps were more important if the musk oxen were also stressed from the warm autumnal conditions. While other causes of the *Pasteurella* epizootic cannot be definitively ruled out, it is a compelling example of how climate change-induced extreme weather events may facilitate disease outbreaks of opportunistic pathogens, with ramifications for the population health and conservation of wildlife species (Ytrehus et al. 2008).

*Kazakhstan, 2015*—The saiga antelope (*Saiga tatarica*), gifted with a distinctive pendulant proboscis (a snout worthy of an alien lifeform), gathers annually in May in Kazakhstan for calving (Fig. 10.16). Like musk oxen, saiga antelope are a species important to conservation efforts: saiga are a critically endangered species and exist in increasingly fragmented herds.

On 10 May 2015, a die-off of saiga antelope began, culminating in ~200,000 deaths across several calving groups within just three weeks (Fig. 10.17). The number of dead animals constituted more than two-thirds of the global population of saiga antelope at the time, constituting a **mass mortality event**: a demographic catastrophe that can simultaneously affect all life stages and rapidly remove a substantial proportion of a population over a short period relative to the generation time of the organism (Kock et al. 2018, Fereidouni et al. 2019).

The killer in the saiga case was the same bacteria species as the one that decimated the musk oxen: *Pasteurella multocida*, which caused haemorrhagic septicaemia. Observed animals died horrible deaths within hours of clinical disease onset: laboured breathing (dyspnoea) and grunting, frothing at the

**Figure 10.16** Saiga antelope, Stepnoi Sanctuary, Astrakhan Oblast, Russia.
Photo by Andrey Giljov, CC BY-SA 4.0, https://commons.wikimedia.org/w/index.php?curid=73737597

**Figure 10.17** Dead saiga antelope and burial pits following the mass mortality event attributed to extreme climate events and infection with *Pasteurella multocida.*
Photos by Sergei Khomenko/FAO.

mouth, and terminal diarrhoea. Disturbingly, saiga calves were seen suckling from their mothers until, and even after, their mother's death, and then they would die too. Within-herd mortality approached 100%.

These different saiga herds were spread discretely across a vast steppe-landscape of several hundreds of thousands of square kilometres. The nearly simultaneous mortality events suggested that there was a regional driver of the *Pasteurella* outbreaks, rather than rapid transmission between herds. And like southern Norway, the outbreak in Kazakhstan appeared to have been triggered by unusually warm and humid environmental conditions, though the mechanisms by which climatic factors triggered bacterial invasion in the saiga mass mortality event remain unknown (Kock et al. 2018, Fereidouni et al. 2019).

## 10.5 Summary: interpreting complex climate–disease patterns

There is a perception that a warmer world is a sicker world. But this paradigm may be a result of over-simplifying relationships between temperature and disease processes (Randolph 2009, Thomas 2020). Climate change will impact the spatial distributions of hosts, vectors, and pathogens, but may just as often lead to declines or shifts rather than expansions. Some habitats or environments may become too hot or too dry or too wet, whereas others that have previously been unsuitable habitats could perhaps shift to facilitate disease transmission (Lafferty and Mordecai 2016). Some pathogens and vectors may be resilient to climate change, and even thrive, and the incidence of these diseases will increase; but other vectors and pathogens may decline or disappear altogether. Furthermore, even if habitat suitability becomes more favourable in some places, hosts, vectors, and pathogens may never reach those suitable habitats due to factors that impede organisms' movement, e.g. barriers to dispersal, competition, predation, or public health interventions. All of this is to say that some infectious diseases are going to get worse in some places while others will likely become less problematic.

As the ghost moose example illustrates, it is difficult to unambiguously identify the links between infection disease and climate change. Climate change itself is the consequence of human population growth, land-use change, industrialization, globalization, etc., and thus many of these correlated drivers will impact a given area, potentially synergistically. To quantify drivers and impacts, one needs baseline data on climate patterns and pathogen occurrence. Yet we are often left wanting in terms of long-term monitoring programmes that can provide these data (Kovats et al. 2001, Ytrehus et al. 2008).

In the absence of perfect or good data, it becomes difficult for a wise ecologist or epidemiologist to draw scientific conclusions. If investigators are overly careful, they may resort to highlighting the many confounders and steadfastly maintain a reluctance to attribute climate change as a driver of new disease outbreaks. They may even require careful randomized experimental trials before they are willing to concede that climate change is indeed *the* factor that is responsible for such-and-such emergence of infectious diseases. And yet—climate change is complex (messy), we suffer a dearth of appropriate historical biological data that can reliably illustrate historical baselines, and even good data may be subject to other confounding variables (e.g. historically smaller human populations). Perhaps we ought to be more forgiving and entertain the impacts of climate change with a trade-off between rigour and the recognition that complex systems are hard to understand.

There will be a lot of people out there who are not going to like this. They like to do things very slowly and consider every option, have another cup of tea, have another Kipling cake, have another wee chat in four months' time. If we keep on course with this we'll be left living on a planet full of pigeons and dogs on the beaten-down crust of our own excrement.

—Derek Gow

Even if we cannot predict how many diseases or which diseases will emerge or worsen in the coming decades, we can say with certainty that climate change is a public health crisis. Increasing wildfires, hurricanes, floods, droughts, heat waves, and other climate phenomena directly harm people, without factoring in the role of infectious diseases. Therefore, climate change mitigation, universal healthcare, and strong preventative public health programmes are critical investments for a healthier future.

## 10.6 Notes on sources

Information on musk ox and lungworm: https://weberarctic.com/stories/the-muskox-of-somerset-island (accessed 9/9/2021).

Quotes from Stephen Schneider were transcribed from 'The Truth About Global Warming – Science and Distortion', https://www.youtube.com/watch?v=4_eJdX6y4hM

Quote from Derek Gow originally reported in: '"It's going to be our way now": the guerrilla rewilder shaking up British farming', by

Phoebe Weston in The Guardian: https://www.theguardian.com/environment/2020/sep/04/its-going-to-be-our-way-now-the-guerrilla-rewilder-shaking-up-british-farming-aoe (accessed 9/9/2020).

## 10.7 References

Alkishe, A., Raghavan, R.K., Peterson, A.T. (2021). Likely geographic distributional shifts among medically important tick species and tick-associated diseases under climate change in North America: a review. Insects, 12, 225.

Cherry-Garrard, A. (1994). The Worst Journey in the World: Antarctic, 1910–13. Pan Books Ltd.

DeBow, J., Blouin, J., Rosenblatt, E., et al. (2021). Effects of winter ticks and internal parasites on moose survival in Vermont, USA. Journal of Wildlife Management, 85, 1423–39.

Dell'Amore, C. (2015). What's a ghost moose? How ticks are killing an iconic animal. National Geographic online: https://www.nationalgeographic.com/animals/article/150601-ghost-moose-animals-science-new-england-environment, accessed 17/8/2021.

Demaria, E.M.C., Roundy, J.K., Wi, S., Palmer, R.N. (2016). The effects of climate change on seasonal snowpack and the hydrology of the Northeastern and Upper Midwest United States. Journal of Climate, 29, 6527–41.

Dobbs, D. (2019). Climate change enters its blood-sucking phase. The Atlantic, 21 February 2019: https://www.theatlantic.com/science/archive/2019/02/ticks-can-take-down-800-pound-moose/583189/, accessed 14/10/2021.

Fereidouni, S., Freimanis, G.L., Orynbayev, M., et al. (2019). Mass die-off of saiga antelopes, Kazakhstan, 2015. Emerging Infectious Diseases, 25, 1169–76.

Gregory, N., Fernandez, M.P., Diuk-Wasser, M. (2022). Risk of tick-borne pathogen spillover into urban yards in New York City. Parasites & Vectors, 15, 1–14.

Hahn, M.B., Feirer, S., Monaghan, A.J., et al. (2021). Modeling future climate suitability for the western black-legged tick, *Ixodes pacificus*, in California with an emphasis on land access and ownership. Ticks and Tick-borne Diseases, 12, 101789.

Holt, A.C., Salkeld, D.J., Fritz, C.L., et al. (2009). Spatial analysis of plague in California: niche modeling predictions of the current distribution and potential response to climate change. International Journal of Health Geographics, 8, 38.

Jones, H., Pekins, P., Kantar, L., et al. (2019). Mortality assessment of moose (*Alces alces*) calves during successive years of winter tick (*Dermacentor albipictus*) epizootics in New Hampshire and Maine (USA). Canadian Journal of Zoology, 97, 22–30.

Kafle, P., Peller, P., Massolo, A., et al. (2020). Range expansion of muskox lungworms track rapid arctic warming: implications for geographic colonization under climate forcing. Scientific Reports, 10, 17323.

Kock, R.A., Orynbayev, M., Robinson, S., et al. (2018). Saigas on the brink: multidisciplinary analysis of the factors influencing mass mortality events. Science Advances, 4, 2314.

Kovats, R.S., Campbell-Lendrum, D.H., McMichael, A.J., et al. (2001). Early effects of climate change: do they include changes in vector-borne disease? Philosophical Transactions of the Royal Society of London B: Biological Sciences, 356, 1057–68.

Kutz, S.J., Hoberg, E.P., Polley, L. (2001). A new lungworm in muskoxen: an exploration in Arctic parasitology. Trends in Parasitology, 17, 276–80.

Kutz, S.J., Hoberg, E.P., Polley, L., Jenkins, E.J. (2005). Global warming is changing the dynamics of Arctic host–parasite systems. Proceedings of the Royal Society of London B, 272, 2571–6.

Lafferty, K.D. (2009). The ecology of climate change and infectious diseases. Ecology, 90, 888–900.

Lafferty, K.D., Mordecai, E.A. (2016). The rise and fall of infectious disease in a warmer world. F1000 Research, 5, 2040.

Lozier, J.D., Aniello, P., Hickerson, M.J. (2009). Predicting the distribution of Sasquatch in western North America: anything goes with ecological niche modelling. Journal of Biogeography, 36, 1623–7.

Mordecai, E.A., Paaijmans, K.P., Johnson, L.R., et al. (2013). Optimal temperature for malaria transmission is dramatically lower than previously predicted. Ecology Letters, 16, 22–30.

Mweya, C.N., Kimera, S.I., Stanley, G., et al. (2016). Climate change influences potential distribution of infected *Aedes aegypti* co-occurrence with dengue epidemics risk areas in Tanzania. PLoS ONE, 11, e0162649.

Paaijmans, K.P., Blanford, S., Bell, A.S., et al. (2010). Influence of climate on malaria transmission depends on daily temperature variation. Proceedings of the National Academy of Science, USA, 107, 15,135–9.

Pickles, R.S.A., Thornton, D., Feldman, R., et al. (2013). Predicting shifts in parasite distribution with climate change: a multitrophic level approach. Global Change Biology, 19, 2645–54.

Porter, W.T., Barrand, Z.A., Wachara, J., et al. (2021). Predicting the current and future distribution of the western black-legged tick, *Ixodes pacificus*, across the Western US using citizen science collections. PloS One, 16, e0244754.

Poulin, R. (2007). Are there general laws in parasite ecology? Parasitology, 134, 763–76.

Randolph, S.E. (2009). Perspectives on climate change impacts on infectious diseases. Ecology, 90, 927–31.

Raveh, A., Kotler, B.P., Abramsky, Z., Krasnov, B.R. (2011). Driven to distraction: detecting the hidden costs of flea parasitism through foraging behaviour in gerbils. Ecology letters, 14, 47–51.

Schneider, S.H. (2005). The Patient from Hell. Da Capo Press.

Shaw, D.J., Dobson, A.P. (1995). Patterns of macroparasite abundance and aggregation in wildlife populations: a quantitative review. Parasitology, 111, S111–S133.

Thomas, M.B. (2020). Epidemics on the move: climate change and infectious disease. PLoS Biology, 18, e3001013.

Ytrehus, B., Bretten, T., Bergsjø, B., Isaksen, K. (2008). Fatal pneumonia epizootic in musk ox (*Ovibos moschatus*) in a period of extraordinary weather conditions. EcoHealth, 5, 213–23.

Zurell, D. (2020). Introduction to species distribution modelling (SDM) in R. Available at: https://damariszurell.github.io/SDM-Intro/, accessed 6/2/2023.

# CHAPTER 11

# Land use change and emerging infectious diseases

The road to conservation is paved with good intentions that often prove futile, or even dangerous, due to a lack of understanding of either land [the ecosystem] or economic land use.

—**Aldo Leopold**

## 11.1 Introduction

*Brisbane, Queensland, Australia, 1994*—In the Hendra suburb, racehorses that had been outside in a paddock were brought to a stable facility because one of the horses, a pregnant mare, had symptoms of fever and respiratory illness. The cause of illness was unclear—possibly poisoning?—and the symptoms rapidly worsened, leading to traumatic death. A few days later, other horses housed in the stable developed similar respiratory symptoms and mild neurological symptoms; most deteriorated rapidly and died or needed to be euthanized (Fig. 11.1). Two people, a horse trainer and a stablehand who had cared for the sick horses, also fell seriously ill, and the trainer died despite receiving intensive medical care (Murray et al. 1995).

The source of the 1994 outbreak was a paramyxovirus that was soon named Hendra henipavirus. Dozens of other outbreaks in horses have occurred in eastern Australia since, especially in the winter months. In most cases, a single horse out to paddock becomes acutely ill and rapidly dies. In some cases, infections spread among horses housed together, and people closely interacting with sick horses have also become sick.

Although outbreaks of Hendra henipavirus are new in horses and people, the virus is old. The reservoir hosts for Hendra virus are 'megabats' known as 'flying foxes' (*Pteropus* spp., Fig. 11.2) (Halpin et al.

2011), which have probably co-existed with Hendra virus in Australia for millions of years. So why are Hendra outbreaks in horses suddenly popping up in eastern Australia? More than one ultimate driver of disease has led to these outbreaks (climate change, spread of domestic animals to Australia) (Martin et al. 2018), but one culprit is to be the topic of this chapter: land use change.

Flying foxes eat nectar, pollen, and fruit, and they use their metre-long wingspans to fly long distances among food resources, such as patches of flowering eucalypt trees. When patches of trees are cleared by humans, flying foxes can no longer sustain their nomadic foraging habits (Eby et al. 2023). Instead, increasing numbers of bats have moved into towns and cities in eastern Australia to roost near year-round, suboptimal urban food sources. These human-associated food resources include fruit trees planted in horse paddocks in more rural areas.

Land clearing and paddock creation by humans seems to have led to new interactions among flying foxes and highly susceptible domestic animals: bats with virus-laden urine literally hang out in trees above paddocked horses (which were absent from Australia before the arrival of European colonizers in the late 1700s). In winter in subtropical Australia, food resources are more scarce for bats, especially following years when winter Eucalyptus flowering is reduced by long-term El Niño and La Niña cycles

*Emerging Zoonotic and Wildlife Pathogens.* Dan Salkeld, Skylar Hopkins, and David Hayman, Oxford University Press.
© Oxford University Press (2023). DOI: 10.1093/oso/9780198825920.003.0011

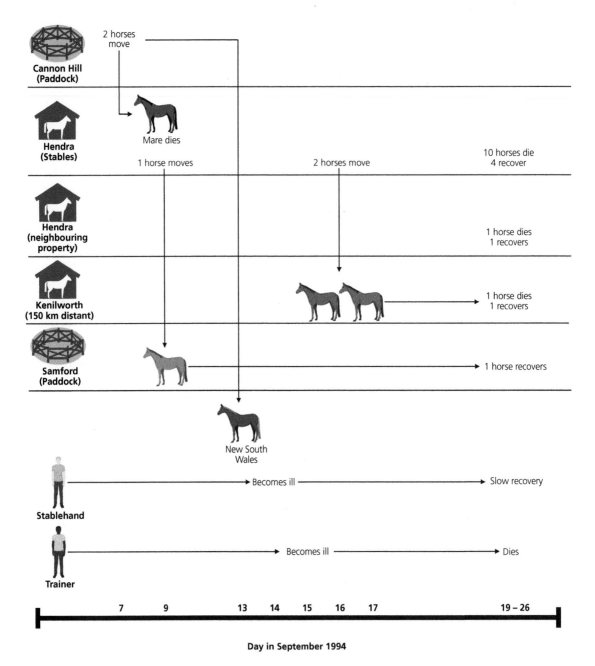

**Figure 11.1** Timeline of the 1994 Hendra henipavirus outbreak in the Hendra suburb.
Figure modified from Murray et al. (1995) by Hanna D. Kiryluk using BioRender.com and additional drawings by Skylar Hopkins.

(Giles et al. 2018, Eby et al. 2023). Together, these conditions send nutritionally stressed bats that are shedding more virus than normal into closer contact with horses (Plowright et al. 2008, Becker et al. 2023), leading to winter outbreaks and bat–human conflicts.

**Figure 11.2** Three black flying foxes (*Pteropus alecto*), the main Hendra virus reservoir, co-roosting with a grey-headed flying fox (*Pteropus poliocephalus*).
Photo by Vivien Jones.

## 11.2  What is land use change?

**Land use change**, also known as **land conversion**, is a catch-all term for ways that humans modify abiotic and biotic environments: paving surfaces, building roads through the middle of forests, turning grasslands into agricultural fields, etc. Taken together, land use change activities have been estimated to drive 31% of emerging infectious disease events (Loh et al. 2015, UNEP 2016). Many of these outbreaks occur in the poorest countries, where healthcare access and disease surveillance capacity are most limited. In these regions, rapid and extensive land use change is often ultimately driven by global demand and global markets for food, timber, and other products desired by people in the wealthiest countries. Therefore, while land use change is superficially easy to understand,

the proximate and ultimate mechanisms by which land use change leads to spillover will likely differ between outbreaks (Fig. 11.3). We cannot cover all these mechanisms in one chapter, but in the following sections, we describe some common drivers of land use change and give specific examples of how changing ecology caused a specific spillover event.

Before digging into these various mechanisms, it is helpful to know more about the geometry of land use change in order to distinguish between **habitat destruction**, where the previous wildlife habitat is completely destroyed/removed; **habitat degradation**, where the previous habitat remains the same size and shape, but the habitat's ability to support the resident wildlife species has decreased due to human activities; and **habitat fragmentation**, where previously contiguous habitat has been transformed into smaller habitat patches separated by altered space that is now unsuitable for wildlife. The distance between habitat fragments will determine how likely species are to move from one fragment to another; a given distance could be passable by long-ranging species (e.g. large mammals) and impassable for short-ranging species (e.g. salamanders). Together, these changes affect the separation among habitats and the shapes of the habitats (Fig. 11.4).

After land use change has occurred, the spaces that are now unsuitable for long-term wildlife

survival or reproduction are called **matrix** and the remaining, but potentially degraded, habitats are generally referred to as **core habitats**. There is also a geometrical distinction for the edges of the remaining habitat fragments, conveniently called **edge habitats**, which are often altered due to proximity to the matrix (e.g. more UV light, more wind).

Edge habitats may also be interfaces where species interactions are especially likely to change or increase, leading to higher risks of disease emergence. For example, the mosquitoes that transmit

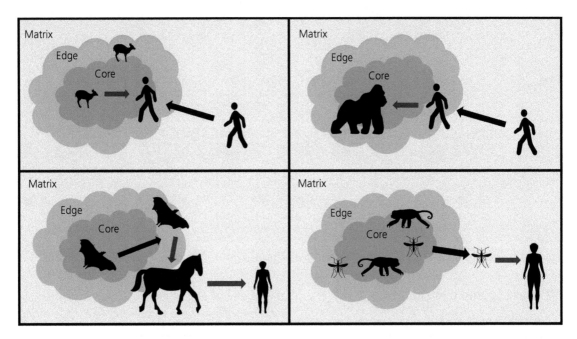

**Figure 11.3** Spillover can occur when humans or domestic species enter core or edge wildlife habitats for hunting, ecotourism, or other activities (top panels). This includes anthroponoses, where humans spread infectious diseases to wildlife (top right panel). Zoonotic spillover (bottom panels) can also occur when wildlife or vectors move from core habitats into edge habitats or human-dominated matrix habitats and come into closer contact with domestic species and humans. Black arrows indicate host movement and purple arrows indicate transmission.

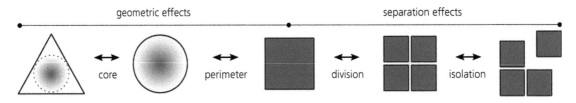

**Figure 11.4** There is more than one kind of habitat fragmentation. Fragmentation can affect a habitat's geometry, such as by reducing the distance from any point in the habitat to the edge of the habitat, or by changing how much of the habitat is core versus perimeter/edge. Fragmentation can also affect how habitats are separated, including the number of patches that the original habitat is divided into and how far apart the remaining patches are.

This figure was originally published in Wilkinson et al. (2018) and has been reproduced with permission from the authors.

yellow fever virus are typically forest dwellers, but they can move quickly along road and matrix corridors adjacent to forests; therefore, the spread of yellow fever virus is fastest along forest edge habitats adjacent to roadways, agricultural areas, and wetlands, whereas the virus is less likely to spread across core forest regions or contiguous matrix habitats (e.g. urban areas, agricultural areas) (Ribeiro Prist et al. 2022). As this example illustrates, emergence depends not just on how habitats are changed (e.g. forests turned into agricultural areas), but also on how species move in and out of core, edge, and matrix habitats (Fig. 11.3).

## 11.3 Deforestation and forest fragmentation

Deforestation and forest fragmentation are some of the most prominent examples of land use change and the quintessential examples of core versus edge versus matrix habitats. For example, in Fig. 11.5, the maps of West and Central Africa show core habitats with a high percentage of forest cover, and many edge habitats between the core habitats and the brown matrix regions with low percentage forest cover. When looking across these landscapes, 11 independent index cases of *Ebolavirus* infection

**Figure 11.5** Top: Maps showing forest cover in Central (a) and West (b) Africa, where contiguous regions of dark green are core habitats and areas mixed with brown are more degraded edge habitats. The yellow triangles show the locations of independent first infection events for *Ebolavirus* outbreaks, which tend to occur in areas that recently experienced rapid fragmentation. Bottom: An example of land use change at the border of Bwindi Impenetrable National Park in south-western Uganda.

Top: This figure was published by Rulli et al. (2017) under a CC BY license. Bottom: Photo by David Hayman.

(*Ebolavirus* is a genus that includes several species) were found to occur in forested areas where rates of forest fragmentation had been particularly high the year of the outbreak (Rulli et al. 2017). This suggests that spillover of ebolaviruses is more likely to occur near edge habitats, where human–wildlife interactions are most rapidly changing, than in matrix habitats, where humans and wildlife are less likely to interact.

However, it is unclear how wildlife movements and behaviour contributed to the virus spillover events. It might be that frugivorous bats (Fig. 11.5), which are probably the reservoir for Ebola virus (Box 11.1), are more abundant in fragmented habitats. Or it might be that other animals that become periodically infected and spread infections to people—like gorillas, chimpanzees, and duikers—increase in abundance in recently fragmented areas and thus have more interactions with humans. Or it might be some combination of

these and other effects. If we knew exactly how fragmentation leads to spillover of ebolaviruses and other zoonotic diseases, we might be able to intervene to reduce spillover risks, so this remains an active research focus.

Deforestation and forest fragmentation do not always increase infectious disease transmission to humans. For example, logging is sometimes associated with increased malaria and sometimes with decreased malaria (Tucker Lima et al. 2017). The outcomes of land use change depend on the specific ecologies of the abiotic and biotic reservoirs, vectors, and people.

## 11.4 Wild meat

Terrestrial wildlife is the primary source of meat for hundreds of millions of people throughout the developing world.

—Golden et al. (2011)

---

### Box 11.1 Bats as reservoir hosts

Several deadly zoonotic diseases are known to have spilled over from bats, including Ebola (Fig. 11.6) and Marburg viruses, Hendra and Nipah viruses, the ancestors of the SARS and MERS coronaviruses, and rabies virus. Does this mean that bats are especially important reservoir hosts, and if so, why?

This is still an active topic of research and debate, but there are several proposed hypotheses (Hayman et al. 2013, Luis et al. 2013, Olival et al. 2017, Guth et al. 2022). For example, bats can be gregarious, roosting at high densities in a mix of bat species. This might allow for frequent cross-species transmission (host switches) and opportunities for viruses to evolve to be less host-specific—a potential recipe for viral spillover to non-bat species. Adaptations to flight, such as high metabolic rates, have led to specific cellular repair mechanisms and inflammatory responses that are unique to bats, and these physiological conditions might select for viral adaptations that cause bat viruses to be especially virulent in people if they spill over (O'Shea et al. 2014, Hayman 2019). Flight also allows long distance movements, potentially spreading viruses broadly throughout the landscape (O'Shea et al. 2014, McKee et al. 2021). All these conditions are ideal for viruses moving among individuals and species and mutating as they go.

Given all the reasons that bats might be especially good reservoir hosts, are they more likely than other vertebrate groups to be reservoirs of viruses that could spill over into human populations? Some large comparative reviews have found that bats hosted proportionally more zoonotic viruses than rodents or primates (Luis et al. 2013, Olival et al. 2017)—two other taxonomic groups that are known to be reservoirs for zoonotic diseases—and that bat viruses are more likely to be highly virulent (Guth et al. 2022). Other studies have suggested that bats aren't particularly special; they *do* host many zoonotic viruses, but that's just because there are so many bat species, and viral diversity increases with host diversity (Mollentze and Streicker 2020). Instead, bats might have equal reservoir potential to other vertebrate hosts, but perhaps some bat species are strongly affected by the drivers of disease discussed in this and previous chapters; as people are increasingly encroaching on bat habitats in larger numbers with their livestock, contacts with bats and spillover risk are also increasing.

Whether bats are special reservoirs or not, contacts between humans and bats can be very risky, so education efforts to improve human hygiene during interactions with bats are important (e.g. making houses and food storage 'bat proof', providing rabies vaccinations to people working closely with bats).

**Box 11.1** *Continued*

**Figure 11.6** Hammer-headed fruit bat (*Hypsignathus monstrosus*), a putative Ebola virus reservoir. This fascinating bat has a spectacular nose that helps males attract females during a process called lekking.
Photo by Sarah Olson (Olson et al. 2019).

When forests are degraded or fragmented and human populations grow at their edges, human–wildlife contacts can increase, including contacts involved with hunting, transporting, and eating wildlife. These practices are not limited to just a few geographic areas: wildlife consumption is ubiquitous around the world and wildlife are consumed by people from all socioeconomic backgrounds. For people from relatively low socioeconomic back-grounds, wildlife can be a culturally important source of food and an important source of macro and micronutrients (Brashares et al. 2011, Golden et al. 2011). For people from relatively privileged socioeconomic backgrounds, wild meat can also be a culturally important source of food, and it can also be a luxury good or a status symbol (Huong et al. 2020, Brashares et al. 2011, Chapter 9).

Regardless of the motivation, wildlife consumption poses risks for zoonotic spillover, but the specific risks vary with context. For example, each time a hunter kills an animal and shares it locally, perhaps just with their family, there could be low spillover risk. But those people might have limited access to affordable healthcare, so there may be a higher risk of extreme sickness or infections spreading widely before public health agencies can intervene. At the other extreme, there is a massive flux of living or dead wildlife being transported away from natural and rural areas to urban areas to meet demands for luxury wild meat. These transport chains expose many people to wildlife and their potential pathogens, from the hunters, to butcherers, to transporters, to vendors, to restaurant staff, to consumers (see also Chapter 9). While these people might have better access to healthcare, they might also live in higher density populations where an outbreak can spread more rapidly. In both scenarios, the risk associated with consuming wildlife will depend on the animal in question—is it a primate species that already shares many pathogens with humans? Or perhaps a more distantly related species, like a crocodile?

And to be extremely clear, there is a risk of zoonotic spillover for all hunters and wildlife consumers, not just those consuming primates! For example, a hunter in Arizona, USA, contracted bubonic plague from hunting a bobcat (Poland et al. 1973). And in parts of neighbouring Colorado, hunters who successfully kill mule deer (*Odocoileus hemionus*) must have the deer heads tested for chronic wasting disease (CWD) prions, to reduce the possibility of human infections. To date, the prion that causes chronic wasting disease in deer has not been reported in humans, but testing deer samples prevents the prion from even entering the human food chain. These risks are often perceived differently (e.g. some hunters continue hunting despite possible risks; others may desist). The ways that people perceive risks and behave in response can be complex, and thus the study of emerging infectious diseases should rely heavily on expertise in the social sciences.

*South Luangwa River, north-eastern Zambia, 2011—* In Zambia, August and September are so dry and food is so scarce that it's called the lean season. As forage becomes less and less available, hungry hippos (Fig. 11.7) may graze grasses right down to the dirt (Stears et al. 2021). This may be how more than 85 hippos died in 2011 from anthrax (*Bacillus anthracis*), which has resilient spores that survive for decades in the soil (Chapter 6).

The hippo carcasses were a bounty of wild meat for local villagers. After many people shared and consumed the hippos, more than 521 people became sick, mostly with cutaneous anthrax. Fortunately, anthrax was treatable by antibiotics available at a nearby health clinic, but five people still died during the outbreak (Lehman et al. 2017).

The link between hippo meat and anthrax exposure was recognized, and this public health information was shared throughout local communities. But in a subsequent survey, 23% of residents said that they would continue to butcher dead hippos, because their families were hungry and they had no other source of protein to complement nshima, a maize-based stable food (Beaubien 2017). People perceived the risks from anthrax to be smaller than the risks from starvation and malnutrition. And besides, hippo scavenging was not illegal.

What if local, national, or international governments were to declare wild meat consumption to be illegal? This has happened during disease outbreaks around the world. Sometimes, these wild meat bans have not been enforced strongly enough to have any impact (Cronin et al. 2015). Bans have even favoured illegal markets, especially if people were not provided with acceptable sources of alternative nutrition or livelihoods and if public health messaging about disease risks did not match people's lived experiences (Bonwitt et al. 2018). Reducing wild meat consumption has also led vulnerable people to consume fewer meals per day, reflecting food shortages. These examples suggest that we cannot simply outlaw people's livelihoods to prevent local outbreaks or pandemics. Better warnings and education can alert people to the risks of wild meat consumption, but the most important

**Figure 11.7** A hippo (*Hippopotamus amphibius*) in Kruger National Park, South Africa; this hippo appears to have abundant forage, unlike the drought-afflicted hippos involved in the 2011 anthrax outbreak in Zambia.
Photograph by Bernard Dupont, published under a CC-BY-SA 2.0 license on Wikipedia.

interventions focus on the ultimate driver of wild meat consumption for subsistence: poverty and food insecurity (Friant et al. 2015, Lehman et al. 2017).

The examples of hunting or consuming wild meat that we have discussed so far aren't examples of land use change, per se, but these human behaviours and the disease risks they involve may increase due to land use change. Deforestation and fragmentation allow more and more humans to access forested regions with relatively undisturbed wildlife and pathogen communities, and thus increase the chances of pathogen spillover during hunting. Urbanization (Section 11.8) creates cities full of people who want exotic wild meat options, so more live animals and wild meat are moved from rural to urban areas to meet that demand, exposing many people to potential zoonotic diseases along the way.

## 11.5 Wildlife farming

In pet stores and zoos, we *hope* that the animals are 'bred in captivity' instead of being collected from the wild; when done ethically, captive breeding can benefit species conservation by reducing how many animals are taken from the wild, and even creating captive 'insurance' populations for threatened species.

Captive breeding is also used in a different setting to create a surplus of wildlife: **wildlife farming**. This oxymoronic term refers to captive breeding of wildlife or semi-domesticated wildlife for human consumption or to create other wildlife products (e.g. leather, feathers). On commercial wildlife farms, farmers breed undomesticated or semi-domesticated wildlife species for legal sale using methods synonymous with traditional livestock. A wide range of wildlife species are farmed, including elk, bamboo rats, crocodiles, snakes, porcupines, civets, ostriches, and more. Many such wildlife farms exist in relatively rural, undeveloped regions where biosecurity practices are often limited. But as the next case study illustrates, many other wildlife farms occur in highly developed and regulated regions, and disease outbreaks still happen there. In both cases, wildlife farms are places on the landscape where land use has changed, and the risk of disease emergence can be high even when farming practices are strongly regulated.

*North Brabant Province, Netherlands, April 2020*—In mid-April 2020, just over a month after the World Health Organization declared the COVID-19 pandemic, mink mortality increased in two mink farms, coinciding with an outbreak of respiratory disease (watery nasal discharge with occasional severe respiratory distress). The two farms had a total of 21,200 mink maintained at high densities (Oreshkova et al. 2020, Oude Munnink et al. 2021). Although the mink were caged individually to keep their fur in good condition, that did not prevent the spread of an airborne virus. And the dying mink did, in fact, have SARS-CoV-2; post-mortem dissections and sequencing revealed that dead mink had signs of pneumonia and tested positive for the virus (Fenollar et al. 2021). Subsequent serosurveys found that 98% of surviving mink had developed antibodies to SARS-CoV-2 (Oreshkova et al. 2020).

The two farms were operated by different families, 14 km apart, and no personnel connected the two of them. According to phylogenetic analysis of viral sequences, the virus was introduced by separate events to each farm, with the most likely culprits being infected farm workers. (Some wandering stray cats also had evidence of infections or exposures, but they could have been infected from the mink.)

Until recently, American mink (*Neovison vison*) were heavily farmed in the Netherlands for their fur—approximately 125 farms, comprising 5000 breeding female mink, produced roughly 4 million mink each year. But following those first two outbreaks, SARS-CoV-2 soon became rife in mink farms in the south-eastern Netherlands. Retrospective analyses always tied the outbreaks to infected farm workers who came to work and infected the mink—**anthroponotic spillover** from people to

animals. And the pathogen didn't just move from people to mink; genetic analyses show that as the virus circulated in mink populations, new variants appeared, and these variants subsequently spread back to farm workers (**spillback**), infecting dozens of people (Oude Munnink et al. 2021). Therefore, on 3 June 2020, the Dutch Ministry of Agriculture decided to cull all mink on SARS-CoV-2 infected farms (Oreshkova et al. 2020). Soon after, the government banned all mink farming (starting January 2021), leading all uninfected farms to close.

Outside of the Netherlands, something was also rotten in the state of Denmark, the world's largest mink pelt producer. Here too, SARS-CoV-2 was infiltrating mink fur farms and then spilling back into the human community. During this time, SARS-CoV-2 accumulated mutations in the spike protein gene, raising the worrying prospect that evolution could make vaccines and therapeutic antibodies ineffective against the new variants (Koopmans 2021, Fenollar et al. 2021, Hammer et al. 2021). Consequently, on 5 November, the Danish Ministry of Environment and Food announced plans to cull the country's mink—approximately 17 million animals. The horror of the situation was exacerbated by the mass graves; some slaughtered mink were not buried deep enough, and when they began to rot, the bodies expanded with gas and pushed themselves out of their shallow graves like the living dead (Henley 2020). The entire situation could have been prevented by better biosecurity practices, such as stricter measures to ensure that COVID-infected people did not come to work. However, since it did happen, the rapid response—only possible due to surveillance at the human–animal interface and communication between virologists, epidemiologists, and policymakers—may have saved many human lives.

Though much recent attention has focused on the potential role of wildlife farming in the emergence and subsequent spread of SARS-CoV-2, many other pathogens are also known to circulate in and emerge from wildlife farms. For example, multiple avian influenza strains have been introduced to ostrich farms in South Africa (Abolnik et al. 2016) and rabies was maintained independently from wild canids in farmed kudu (a species of ungulate) in Namibia (Scott et al. 2013). And of course, it isn't just

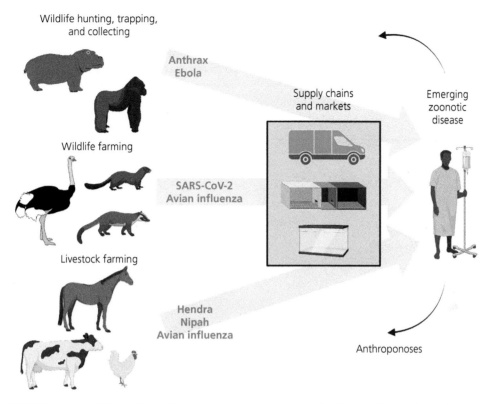

**Figure 11.8** Wildlife hunting, wildlife farming, and livestock farming can all lead to zoonotic spillover of infectious diseases from animals to people. In turn, the same activities can lead to anthroponotic spillover from people to animals. The red text gives just a few examples of relevant zoonotic diseases.

Figure made with BioRender.com and additional animals drawn by Skylar Hopkins.

zoonotic diseases that can be amplified and spread by wildlife farms; the chytrid fungus spills over into wild frog populations from bullfrog farms in Brazil (Ribeiro et al. 2019, Chapter 9).

There are several reasons why farms for wildlife and domestic livestock are primed for disease emergence and outbreaks. Farms usually house animals at densities much higher than would be found in the wild, and farm workers are more likely to interact at higher intensity with these farmed animals than an average wild animal. Farmed animals also tend to have low genetic diversity and thus limited variability in disease resistance. For example, farmed mink may be bred from just a few males with the best fur. Farmers often work in unsanitary conditions with limited protections (e.g. no respirators or medical gloves), and both the animals and people may have compromised immune systems due to the stressful

conditions in which they exist. Farmed animals and farmers may also travel more widely than an average wild animal, spreading infections among farms and other locations.

## 11.6  Livestock farming

'The Vampire Bat,' says Mr. Darwin, . . . 'is often the cause of much trouble, by biting the horses on their withers. The injury is generally not so much owing to the loss of blood, as to the inflammation which the pressure of the saddle afterwards produces. . . Before the introduction of the domesticated quadrupeds, this Vampire Bat probably preyed on the guanaco, or vicuña, for these, together with the puma, and man, were the only terrestrial mammalia of large size, which formerly inhabited the northern part of Chile.'

—*The Zoology of the Voyage of H.M.S. Beagle . . . During the Years 1832 to 1836*, Bell et al. (1839)

Common vampire bats, *Desmodus rotundus*, inhabit Mexico, Central America, and parts of South America (Fig. 11.9). The bats don't actually suck blood vampirishly, but rather use their sharp teeth to make small incisions on their victims and then lap at the bleeding wound, which fails to clot because of anticoagulants in the bats' saliva. Historically, the victims of these nocturnal attacks would have been warm-blooded wildlife (even penguins) and the occasional human, but these days, vampire bats commonly feed on livestock, such as horses and cows. Of course, 'these days' includes the

**Figure 11.9** Known locations where vampire bats (*Desmodus rotundus*) have been observed are indicated with black points and the estimated suitable habitat for vampire bats based on environmental conditions is indicated by green shading.

past hundreds of years of presence of domesticated species, including the first description of a vampire bat in the early 1800s by a Western scientist, Charles Darwin, who watched it feeding on a horse (Fig. 11.10, see quote).

In fact, vampire bats don't just haphazardly feed on livestock, they *prefer* them to local wildlife, perhaps because livestock are larger, relatively inactive, and more predictable because herds stay in the same locations (Lee et al. 2012). Vampire bats (Fig. 11.11) have been known to roost near cattle herds so that they can feed repeatedly without a lengthy commute; some cattle can receive 12 bites in a single night and have up to four bats feeding on them simultaneously.

These nightly blood-feeding attacks potentially impact meat and milk production in cattle, but it is the bat's ability to transmit rabies (rabies virus: family Rhabdoviridae, genus *Lyssavirus*) that poses a major health and economic burden. In 1968, over half a million cattle died from bat-transmitted rabies in Latin America. In more recent years, bat control programmes and vaccination have reduced cattle rabies cases to fewer than 10,000 annual cases. However, these reports are probably under-estimates given the scarcity of diagnostic labs available to test dead cattle for rabies (Lee et al. 2012). At a local scale in the Peruvian highlands, surveys that calculated

livestock-owners' propensity to report deaths of their cattle were used to estimate that over 500 cattle succumbed to rabies in southern Peru in 2014, causing an economic loss of roughly $US170,000—the equivalent of 700 months of local income (Benavides et al. 2017).

Human cases of rabies from vampire bat bites are not as common as cattle cases, because cattle provide an easier target. But humans do indeed get infected with the rabies virus after vampire bat bites, especially if their dwellings don't possess bat barriers, like window screens. This includes the temporary structures often used by indigenous peoples, loggers, and miners in the Amazon, which do not prevent entry by vampire bats (Johnson et al. 2014). And note that anyone who has not had pre-exposure immunization against the rabies virus or that does not receive post-exposure prophylaxis soon after infection will develop disease symptoms, and the disease is always fatal.

In a particularly dark example from September 2005, a transformer failure eliminated electrical power to 223 of 256 homes in Antônio Dino—a remote village in north-east Brazil. During the 45 days without power, 57 people were bitten by vampire bats (8.7% of the population!). A total of 16 people developed rabies symptoms and died. Being under 17 years of age and in a house without

VAMPIRE BAT (DESMODUS D'ORBIGNYI). CAUGHT ON BACK OF DARWIN'S HORSE
NEAR COQUIMBO. HEAD, FULL SIZE.

**Figure 11.10** Darwin's drawing of the vampire bat he saw feeding on a horse (Darwin 1890).

**Figure 11.11** Vampire bats, Peru.
Photos by Hollie French (left) and Daniel Streicker (right).

electricity were identified as risk factors for developing rabies in this outbreak, potentially because electric lights usually deter vampire bats and allow people to see them (Mendes et al. 2009).

Since vampire bats prefer to feed on livestock, we might predict that livestock could protect people from rabies, a phenomenon termed **zooprophylaxis**: using non-human animals to attract bites of vector species (or in this case, reservoir species) away from people. However, zooprophylaxis only works if human risk decreases in the presence of the non-human animals, and it is unclear if livestock protect humans or increase their risk. On the one hand, vampire bat colonies are bigger near locations with high livestock densities (Lee et al. 2012), which could translate to higher human risks. On the other hand, rabies seroprevalence in vampire bats is not necessarily higher in colonies that live in areas with high livestock density (Streicker

et al. 2012). All things being equal, when comparing a big colony and a small colony with the same seroprevalence, the bigger one should pose the greatest risk to humans. But given the complexity in this system, there are probably some cases at some ecological scales where livestock increase human risk and some cases where they decrease human risks.

So far in this chapter, we've covered the roles that horses and cattle play in some emerging infectious diseases. There are many more where that came from (Fig. 11.8)! Livestock may create the perfect storm for disease emergence because they represent large populations of susceptible individuals (Fig. 11.12) that provide opportunities for recombination and amplification; they often have close interactions with both wildlife and humans; and we ship them around the world (Chapter 9).

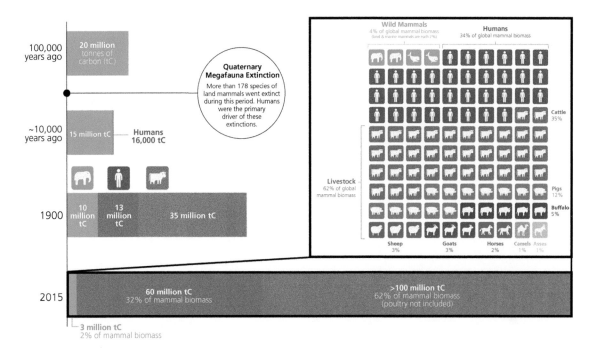

**Figure 11.12** The time series shows how land mammal biomass has changed over geological time, where wild mammal biomass has declined ~85% since the rise of humans. The inset shows the breakdown of total mammal biomass (both terrestrial and marine) in 2015, where cows, humans, and pigs dominate the planet. Pets (e.g. dogs and cats) are not shown, because they are <1% of the biomass. One tonne of carbon is equal to two elephants, 20 cows, or 100 people.

This figure was modified from two figures created by Hannah Ritchie and Klara Auerbach and published under a CC BY license on ourworldindata.org. The data were compiled from Barnosky (2008), Smil (2011), and Bar-on et al. (2018).

## 11.7 Agriculture and water use

Modern society would not exist without agriculture, and agricultural advancement is critical for increasing food security and thus reducing spillover risks due to wild meat consumption. However, it is not without risks and challenges. By one estimate, agriculture has been the primary driver of more than 25% of all emerging infectious diseases in humans, and more than 50% of emerging zoonotic diseases (Rohr et al. 2019). Of course, 'agriculture' involves a suite of environment changes that can vary in different places, and some may be more or less likely to alter disease dynamics. In this section, we describe how agriculture impacts water and infectious disease dynamics.

Hydrological manipulations are often required to irrigate crops. Some crops, like rice, require flooded fields or 'paddies' to grow. Other crops don't need to be underwater, but they require so much water

that rivers are dammed to provide consistent freshwater reservoirs upstream from the dam. Dams are important both for agriculture (livelihoods, food security) and renewable energy (climate change mediation), especially in low- and middle-income countries. But dams have also displaced millions of people who used to live upstream and exposed people who remain near dams to a suite of infectious diseases with water-borne transmission or aquatic intermediate hosts (Lund et al. 2021).

*Diama Dam, Senegal*—In the Senegal River Basin in West Africa, cases of schistosomiasis started to rapidly increase in the late 1980s (Fig. 11.13). Before 1986, schistosomiasis had been relatively rare and associated with limited freshwater habitats used by *Bulinus* species of snails infected with the trematode *Schistosoma haematobium*. But *Biomphalaria* snails infected with *Schistosoma mansoni* soon appeared in the region and together the two parasites infected more than 50% of people in some

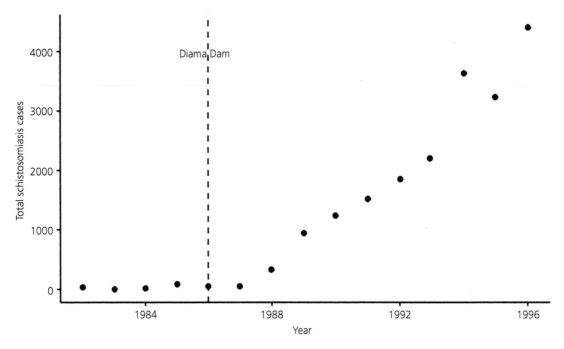

**Figure 11.13** After the Diama Dam was built, the number of schistosomiasis cases in the Senegal River Basin rapidly increased. This figure was modified from Jones et al. (2018) and created using data from Sow et al. (2002).

villages (Jones et al. 2018). The impact on local communities was severe because chronic schistosomiasis is a debilitating disease, especially for children; it causes inflammation and organ damage, which can lead to anaemia, malnutrition, and developmental issues. Before we can understand why this devastating change occurred, we need to understand the life cycle of the parasites that cause schistosomiasis (Fig. 11.14).

Schistosomes are a type of trematode parasite or 'fluke' with a complex life cycle involving multiple host species. As an adult worm, they live in blood vessels of a **definitive host** (the host in which the adult parasites reproduce). Definitive hosts vary among schistosome species and can include humans, cattle, and many bird and mammal wildlife species (Rudge et al. 2013). Inside the definitive host, male and female trematodes mate and produce eggs that migrate to the bladder or intestines (depending on the parasite species), where they often cause inflammation and damage to the host. The eggs are released into the environment in the host's faeces (*Schistosoma mansoni* and *Schistosoma japonicum*) or urine (*Schistosoma haematobium*). If the

urine or faeces enter a toilet or a latrine far from water, the life cycle ends there. But if the eggs can make it to a nearby water body containing snails, the life cycle continues (Fig. 11.14).

When they reach water, schistosome eggs hatch into a free-swimming larval stage called a **miracidium** (plural miracidia), which swims around looking for an intermediate aquatic snail host: *Biomphalaria* snails for *S. mansoni*, *Bulinus* snails for *S. haematobium*, and *Oncomelania* snails for *S. japonicum*. When schistosomes infect these snails, they specifically infect the snail gonads, replicating so extensively in the reproductive tissues that the snails are castrated. (This is a good reason to avoid being reincarnated as a snail in a future life.) The infective stage inside the snail produces another free-living, fork-tailed larval stage, called a **cercaria** (plural cercariae); thousands of cercariae are created from a single snail's gonads during the snail's lifetime. Schistosome cercariae leave the snail host and use chemical and physical cues in the water to find a definitive host, such as humans who are swimming, bathing, washing their clothes, or collecting water. When they find the definitive host, cercariae lose

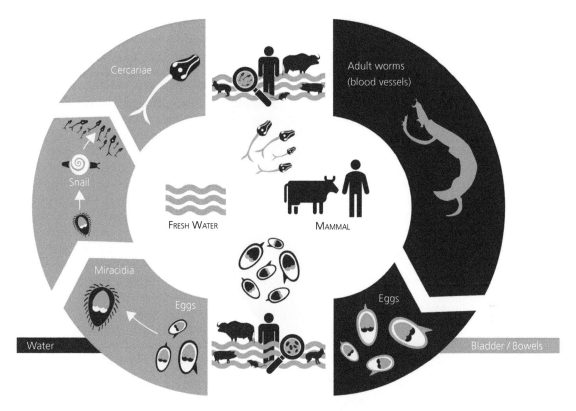

**Figure 11.14** General life cycle of *Schistosoma* parasites that cause schistosomiasis in humans and livestock. Note that the life cycle details differ between the three species and those details are not all shown here (but are given in the text).
Adapted from previous versions used by Janoušková et al. (2022), SCI foundation, and Genome Research Limited.

their forked tail, penetrate through the host's skin, and make their way to the blood vessels. Inside the host, cercariae that evade host defences will eventually become adult worms, and the life cycle will start again.

Humans can be treated to clear their schistosome infections using a fairly cheap and effective drug called praziquantel. Though the drug is cheap, tens of millions of people need treatment for their infections globally (WHO 2019), and those costs add up! Furthermore, praziquantel only kills existing adult worms; it does not prevent future infections, so people can become re-infected after treatment. In Senegal, mass drug administration of praziquantel is used to treat infected people, including school children, but schistosomiasis prevalence remains high due to rapid re-infection in high transmission sites. Therefore, in areas of high transmission, people must be treated repeatedly, as often as every six months or less. In addition to mass drug

administration, could human health be improved by other measures to reduce exposure and infection (Sokolow et al. 2018; Janoušková et al. 2022)? This returns us to our original question: how did the Diama Dam increase human exposure to infected snails and cercariae in the Senegal River Basin in 1986?

The Diama Dam (Fig. 11.15) was built near the mouth of the Senegal River to create a saltwater barrier, leaving the water upstream from the ocean fresh and suitable for agriculture. As intended, the dam drastically changed the environment: salinity *was* reduced, which incidentally increased habitat availability for freshwater snails. There were also changes to water flow and the input of agricultural nutrients (nitrogen, phosphorous) following agricultural intensification, which likely increased the amount of aquatic vegetation that serves as habitat and food for snails (Haggerty et al. 2020, Hoover et al. 2020). The dam also caused population declines for an important snail predator: giant river prawns (*Macrobrachium vollenhovenii*) (Sokolow et al. 2015) (Fig. 11.15). Giant river prawns used to migrate to the ocean to reproduce, before returning upstream, but the dam blocked prawn migration and thus reduced predation on snails. Lastly, by providing stable access to freshwater for subsistence agriculture, the dam also supported an increased local human population. All these mechanisms could have increased snail populations (and led to better-fed snails that can produce more cercariae) and people living and working in cercariae-contaminated water, thus accelerating the transmission cycle of schistosome parasites.

Since many proximate mechanisms could have increased schistosomiasis incidence in Senegal, how do we know how important any given mechanism was? For example, what if the disruption to giant river prawn migration didn't really have an effect compared to increased snail habitat or numbers of people at risk of schistosomiasis? We can start to answer this question by looking across many damming events, because the Diama Dam wasn't the only dam that was associated with increases in schistosomiasis prevalence. Dams have been built across Africa for agriculture and hydropower, and while many of them are in the ranges of native river prawns, others are not. In a comparative study, Sokolow et al. (2017) analysed three kinds of watersheds in Africa: (1) watersheds where prawns previously existed and where dams were built that blocked migrations; (2) watersheds where prawns were absent and dams were built; and (3) watersheds where dams had not been built (Fig. 11.16). Schistosomiasis increased after damming in watersheds with *and* without prawns, whereas

**Figure 11.15** (Top) The Diama Dam on the Senegal River in Senegal. (Middle) River prawn in the hand, Senegal. (Bottom) The tiny snails that serve as intermediate hosts for schistosomes.
(Top) Photograph published in the public domain by the US Geological Survey. (Middle and bottom) Photos by Hilary Duff.

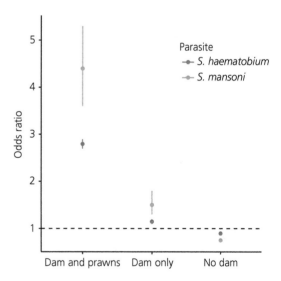

**Figure 11.16** Schistosomiasis prevalence before and after damming in African watersheds within the range of native river prawns (Dam and prawns) or outside the range of native river prawns (Dam only), compared to undammed watersheds as a control group. The dashed line shows an odds ratio of 1, indicating no change from before to after. Schistosomiasis increased in dammed watersheds, especially those where prawn migration was likely blocked by the dam, and schistosomiasis decreased in 'control', undammed watersheds. The vertical lines are 95% confidence intervals.
Figure created with data from Sokolow et al. (2017).

schistosomiasis slightly decreased in watersheds without dams (Sokolow et al. 2017). Furthermore, schistosomiasis increased *most* in watersheds with dams built along the coast and inside the historical range of migratory prawns. This suggests that prawns independently affect schistosomiasis, as do other environmental changes associated

with dams (e.g. reduced salinity and increased vegetation).

Schistosomiasis affects more than 200 million people, with about 800 million people at risk, and it is a critical challenge for future disease control efforts (see also Box 11.2), but schistosomiasis isn't the only water-associated disease impacted

---

### Box 11.2  Listed by WHO, neglected by whom?

Schistosomiasis is widespread in tropical and subtropical regions and affects millions of people, but many people have never heard of it. In fact, it is one of the most important of 20 **neglected tropical diseases (NTDs)** listed by the World Health Organization. We would like to provide a precise definition of NTD, but that is surprisingly difficult to find. Here are some complications:

Diseases *neglected* by whom? We can agree that NTDs are given too little attention relative to the global burden of disease that they cause. But whom should we blame for this neglect? Researchers? Funders? Policymakers? Drug companies? A combination of these actors is implied by current definitions.

NTDs can often be prevented, controlled, or treated using existing, effective, and relatively cheap methods (e.g. post-exposure treatment, education, or water sanitation), but these known preventatives or cures are not available to afflicted populations. Conversely, the development of treatments for NTDs are often neglected by drug research and development efforts because it isn't profitable to develop drugs for people or countries who won't be able to pay for them.

*Tropical* diseases? The diseases on the NTD list occur mostly in tropical regions, but they also occur in less tropical, lower income nations. Furthermore, some NTDs occur in high income, temperate countries, yet are still 'neglected'. For example, 5–6% of people living in the United States' subtropical Gulf Coast—that's millions of people—are affected by at least one neglected tropical disease (e.g. Chagas disease, cysticercosis), and most of those people are living in poverty (Hotez and Lee 2017).

*Diseases*? The WHO's list of 20 NTDs contains ecologically diverse infectious agents, including viruses, bacteria, protozoa, and helminths. It also contains one non-communicable disease: snake bite envenomation. Snake bite envenomation was removed from the NTD list in 2013, which led to fewer resources being provided globally to prevent and treat snake bites, including reduced produc-

tion and distribution of expensive antivenoms. Following the antivenom shortage and resulting human suffering and mortality, snake bite envenomation was added back to the NTD list in 2017 (Chippaux 2017).

Though they might be difficult to define, the 20 WHO-listed neglected tropical diseases share many similarities. They are all 'poverty associated' and often disproportionately affect women and children. They may even be associated with creating or sustaining poverty traps (Section 11.9). They cause variable burdens of morbidity and mortality globally, but they are all likely to be under-reported when they do occur. And they all lack public and political visibility and discourse because they affect people with limited economic and political power, or they are associated with stigma/shame.

Bacterial NTDs (disease agent) comprise Buruli ulcer (*Mycobacterium ulcerans*), leprosy or Hansen's disease (*Mycobacterium leprae*), yaws (*Treponema pallidum*) and other endemic treponematoses, and trachoma (*Chlamydia trachomatis*). Viral NTDs include the viruses that cause dengue and chikungunya, and rabies. NTDs caused by protists include Chagas disease (*Trypanosoma cruzi*), human African trypanosomiasis or sleeping sickness (*Trypanosoma brucei gambiense* and *Trypanosoma brucei rhodesiense*), and leishmaniasis (*Leishmania* spp.). A whole suite of NTDs are caused by parasites: dracunculiasis or Guinea-worm disease (*Dracunculus medinensis*), echinococcosis (*Echinococcus multilocularis*), food-borne trematodiases or parasitic flukes (e.g. *Fasciola* and *Paragonimus*), lymphatic filariasis or elephantiasis (e.g. *Wucheria bancrofti*), onchocerciasis or river blindness (*Onchocerca volvulus*), schistosomiasis (*Schistosoma* spp.), scabies (*Sarcoptes scabei*) and other ectoparasitoses, soil-transmitted helminthiases, and taeniasis/cysticercosis caused by tapeworms (*Taenia* spp.). Fungal NTDs are represented by mycetoma (caused by bacteria or fungi) as well as chromoblastomycosis and other deep mycoses. And finally, there is one non-infectious NTD: snakebite envenomation (Fig. 11.17).

**Box 11.2** *Continued*

**Figure 11.17** (Top) Black mamba (*Dendroaspis polylepis*), a large (up to 4.5 m) snake from sub-Saharan Africa. The black mamba only bites humans if threatened or cornered, but its venom contains neurotoxins that can be fatal if antivenom is not administered. (Bottom) According to the WHO, snakebite envenomation causes a substantial public health burden: 1.8 to 2.7 million cases of envenomation every year, causing between 81,410 and 137,880 deaths, and approximately three times that number of amputations and other permanent disabilities. Snake bites can be prevented by educating the public about risks and how to seek appropriate medical care, training medical workers how to provide appropriate medical care, and improving access to affordable antivenoms (Lancet 2017).

Photo of black mamba by Tim Vickers, published in the public domain, https://commons.wikimedia.org/w/index.php?curid=11360973. Map of snakebite envenomations by Eightofnine, public domain: https://commons.wikimedia.org/w/index.php?curid=7117176

by dams! On the Narmada River in India, the Koka Reservoir in Ethiopia, and the Gilgel-Gibe dam in Ethiopia, damming increased *Anopheles* mosquito breeding habitat, larval and adult mosquito densities, and/or human malaria incidence because dams and irrigation schemes create pools of stagnant or slow-flowing water that is ideal habitat for some mosquito species (Singh et al. 1999, Kibret et al. 2009, Yewhalaw et al. 2009). In other cases, dams might decrease the incidence of particular infectious diseases, because as with any ecological change, some species may 'win' while others may 'lose'. This reminds us that we must carefully consider how each proposed land use change will impact the specific local ecologies of infectious diseases and the people living nearby.

Dams are just one form of hydrological or water use change that is associated with agriculture. Agriculture can also lead to increasing numbers of cattle near or in waterbodies. Cattle faeces in water is a notorious problem for human diarrhoeal disease, which can be especially deadly for children who lack access to rehydration therapy and other healthcare.

Cattle can also amplify the number of schistosome eggs reaching waterbodies used by humans. As mentioned above, cattle are one possible definitive host for one schistosome species that infects humans (*S. japonicum*). Cattle are also the primary definitive host for *Schistosoma bovis*. Furthermore, *S. bovis* is hybridizing with *S. haematobium* and leading to the emergence of new lineages that can infect people (Borlase et al. 2021). One such hybrid schistosome is responsible for the recent emergence of schistosomiasis in Corsica (southern Europe) in 2013, after endemic schistosomiasis hadn't been seen in Europe for 50 years (Bisoffi et al. 2016). People who had visited West Africa became infected with the hybrid parasite, returned to Corsica, urinated in the Cavu River, and infected local snails (Bisoffi et al. 2016). This is one more example of how livestock, local environmental changes, and globalization interact to lead to infectious disease emergence.

Finally, we will note one last time that not all land use change causes disease emergence. For example, extensive draining and bulldozing of marshes and wetlands to create farmland and

spaces for development greatly reduced vector populations in many places around the world, subsequently decreasing incidences of vector-borne diseases. We do not advocate for destroying wetlands, because this causes many other ecological and human problems, such as loss of species and fish nurseries, and increased risk of flooding. Instead, we want to emphasize thatecological knowledge is often required to understand and predict how land use change will affect infectious disease dynamics.

## 11.8 Urbanization

Livestock populations aren't the only populations increasing globally. Human populations increased from 2.5 billion in 1950 to nearly 8 billion in 2022 (Fig. 11.12). All these people are increasingly concentrated in cities, which are sprawling outward to support burgeoning populations. Cities provide ample chances for pathogen transmission because they contain dense populations of susceptible humans and are often encroaching into wildlife habitat. This gives pause for thought: could spillovers of pathogens that currently tend to sputter out have dramatically different destinies in urban conditions?

Prior to 2013, the largest Ebola virus outbreak totalled 318 cases and occurred in the Democratic Republic of Congo. Until that point, outbreaks involved an initial spillover in a remote community that caused stuttering chains of infection that could be contained by medical interventions. The 2013 to 2016 Ebola virus outbreak in West Africa's Guinea, Liberia, and Sierra Leone demonstrated how new settings—large urban centres with substantial spatial connectivity to other population centres—impacted epidemic growth (Munster et al. 2018). The West Africa Ebola virus outbreak resulted in >28,600 reported cases, >11,300 deaths, an estimated $2.2 billion loss in the gross domestic products of Liberia, Sierra Leone, and Guinea, and international response costs that exceeded $3.9 billion. Minimal public health infrastructure and a slow response from international partners contributed to this tragedy, which was initially exacerbated by urbanization and connectivity of populations.

There are several other reasons why conditions in cities are ripe for zoonotic disease (Wilke et al.

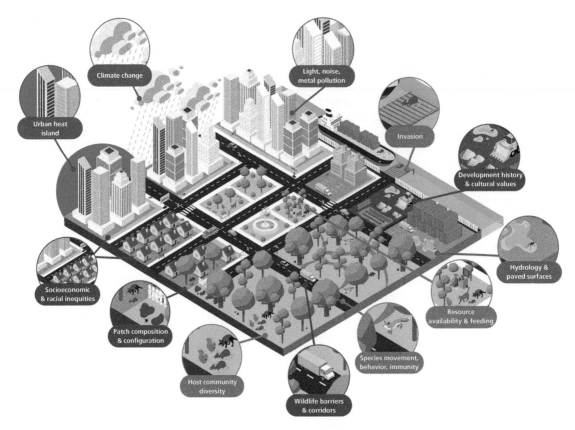

**Figure 11.18** Conceptual diagram illustrating several potential mechanisms by which urbanization can influence zoonotic disease hazards.
Figure reproduced without changes from (Combs et al. 2022) under the Attribution 4.0 International license (CC BY 4.0).

2021, Combs et al. 2022) (Fig. 11.18). For example, urban adapted species like raccoons, foxes, domestic dogs, and rodents are reservoirs for many zoonotic diseases, including leptospirosis, rabies, and hantavirus. This is especially true in cities or neighbourhoods where older buildings and poor sanitation infrastructure (e.g. waste dumping) create ideal conditions for disease-spreading animals. Even green spaces in cities, which tend to be more available near affluent neighbourhoods, can be hot spots for zoonotic disease hazards, because they concentrate interactions between mammals, vector species, and humans (Combs et al. 2022).

Urbanization both increases and decreases zoonotic disease hazards. For example, while highly dense human populations increase the risk of large epidemics, health infrastructure in cities is usually better than in rural regions and

may prevent spillover events from turning into full blown epidemics (Munster et al. 2018). And while urbanization favours some human-centric species, like *Aedes* mosquitoes and rats, it also destroys habitats for many other potential wildlife reservoirs and disease vectors (Baker et al. 2022). These examples illustrate how complex disease systems are, and how difficult it can be to predict how human and wildlife health may be impacted by anthropogenic change.

## 11.9  Poverty traps

This chapter has focused on several **proximate drivers** of disease emergence; but why do people cut down forests, move to cities, consume wild meat, and do other activities that risk exposure to zoonotic diseases for themselves or others?

One such indirect or **ultimate driver** of land use change is poverty. In regions with extreme poverty, people often rely on subsistence livelihoods (e.g. agriculture and wildlife hunting) or are otherwise more often exposed to pathogens in the environment (e.g. because they don't have access to clean water or don't have window screens in their houses). At the same time, poverty makes people less likely to be able to access and afford preventative healthcare, such as vaccines. Thus, a vicious cycle where sick and malnourished people are less able to engage in physically intensive labour, and poverty is exacerbated. This may also affect how people use natural resources. For example, near Lake Victoria, Kenya, a region with high HIV/AIDs prevalence, fishers who are sick are more likely to use illegal fishing practices that are ecologically damaging because those practices are less physically demanding (Fiorella et al. 2017). Environmental degradation can also be caused by **external forces**, such as companies responding to global demand for resources by exploiting cheap labour in poor regions. These processes interact to create feedback loops of poverty, disease, and environmental degradation, where poverty begets more poverty— a disease-driven **poverty trap** (Bonds et al. 2010).

Poverty traps are often studied or discussed in reference to low-income nations. But poverty can also drive disease emergence in relatively wealthy nations because substantial portions of those populations may still live in unsanitary or risky environments. And even in countries where equality is relatively high, environmental change can rapidly shift otherwise functioning systems towards destructive feedback loops.

For example, from 2007 to 2020, a mortgage crisis in the US that was prompted by massive depreciation of home prices triggered a sudden increase in home sales, notices of delinquency of payment, declarations of bankruptcy, and home abandonment. Kern County in southern California was particularly affected, with a 300% increase in notice of delinquency in the spring quarter of 2007 compared with that of 2006. Because of the home abandonments, swimming pools, hot tubs, and ornamental ponds became neglected and permitted algal proliferations; growing numbers of green pools were documented by aerial surveys. These algae-rich water sources were exploited by *Culex pipiens quinquefasciatus*, the common house mosquito, which is a vector of West Nile virus. Coincidentally, there was also a population boom in house sparrows, a competent West Nile virus reservoir. Virus infection incidence in the house mosquitoes increased rapidly in June 2007, a month earlier than normal. This caused a cluster of 140 laboratory-confirmed human cases of West Nile virus in one town in Kern County (Bakersfield), representing a 205–280% increase compared to previous years (Reisen et al. 2008).

If you lived in an affluent neighbourhood, away from abandoned homes and neglected pools, you might have avoided increased West Nile virus exposure. But maybe not! In our increasingly connected world, human infectious diseases can emerge and spread anywhere.

Furthermore, addressing the proximate driver by treating abandoned pools with mosquito larvicides *might* have reduced risks for people in the Bakersfield area, but targeting proximate drivers is often a short-term, 'band aid' solution. To more efficiently intervene, abandoned pools would need to be drained and covered to prevent re-filling. But that would require resources that were already stretched by the financial crisis, and efforts would need to continue as more and more pools were becoming abandoned. A better solution might be to address the ultimate drivers of the pool problems (i.e. financial insecurity and a malfunctioning housing sector), thereby breaking the potential feedback loop between poverty, degradation, and disease.

## 11.10 Summary

Land use change is a catch-all term for ways that humans modify abiotic and biotic environments: paving surfaces, converting forests into agricultural fields, using grasslands to graze livestock, fragmenting habitat, etc. These activities are important for human well-being (e.g. food security, access to social services), but they can also cause emerging infectious disease events. Land can be divided into core wildlife habitats, edge habitats, and matrix, and the ways that people and animals move through and interact in these habitats can affect spillover risks. Changes to water can also affect disease dynamics, as we saw in examples regarding how irrigation and dams affect schistosomiasis and malaria. Disease emergence does not just depend on physical changes to environments, but also on how interactions in those environments change. For example, farming wildlife and domestic species creates large and often highly susceptible host populations that may interact with both wildlife and people, creating frequent opportunities for spillover and spillback of emerging pathogens. Similarly, urbanization can create conditions where some reservoir species, like rodents, flourish and have unusually high interaction rates with humans. These mechanisms are all proximate drivers of disease emergence or disease burdens, which may or may not be the best targets for disease control interventions. For example, efforts to reduce deforestation may be highly beneficial for conservation and may reduce spillover risks for some diseases, but these efforts are unlikely to be successful without addressing the ultimate drivers of deforestation, such as poverty and global demand for forest products.

## 11.11 Notes on sources

Information on snakebite envenoming: https://www.who.int/news-room/fact-sheets/detail/snakebite-envenoming, accessed 27/2/2023.

Information on neglected tropical diseases: https://www.who.int/news-room/questions-and-answers/item/neglected-tropical-diseases, accessed 27/2/2023.

## 11.12 References

Abolnik, C., Olivier, A., Reynolds, C., et al. (2016). Susceptibility and status of avian influenza in ostriches. Avian Diseases, 60, 286–95.

Baker, R.E., Mahmud, A.S., Miller, I.F., et al. (2022). Infectious disease in an era of global change. Nature Reviews Microbiology, 20, 193–205.

Barnosky, A.D. (2008). Megafauna biomass tradeoff as a driver of Quaternary and future extinctions. Proceedings of the National Academy of Sciences, 105(Supplement 1), 11,543–8.

Bar-On, Y.M., Phillips, R., Milo, R. (2018). The biomass distribution on Earth. Proceedings of the National Academy of Sciences, 115(25), 6506–11.

Beaubien, J. (2017). Hippos, anthrax and hunger make a deadly mix. NPR Goats & Soda: https://www.npr.org/sections/goatsandsoda/2017/08/16/543900930/hippos-anthrax-and-hunger-make-a-deadly-mix, accessed 11/11/2022.

Becker, D.J., Eby, P., Madden, W., et al. (2023). Ecological conditions predict the intensity of Hendra virus excretion over space and time from bat reservoir hosts. Ecology Letters, 26, 23–36.

Bell, T., Darwin, C., Owen, R., et al. (1839). The Zoology of the Voyage of H.M.S. Beagle, Under the Command of Captain Fitzroy, R.N., During the Years 1832 to 1836. Published with the approval of the Lords Commissioners of Her Majesty's Treasury. Smith, Elder and Co. Retrieved from the Library of Congress, https://www.loc.gov/item/06016152/.

Benavides, J.A., Paniagua, E.R., Hampson, K., et al. (2017). Quantifying the burden of vampire bat rabies in Peruvian livestock. PLOS Neglected Tropical Diseases, 11, e0006105.

Bisoffi, Z., Buonfrate, D., Beltrame, A. (2016). Schistosomiasis transmission in Europe. Lancet Infectious Diseases, 16, 878–80.

Bonds, M.H., Keenan, D.C., Rohani, P., Sachs, J.D. (2010). Poverty trap formed by the ecology of infectious diseases. Proceedings of the Royal Society B: Biological Sciences, 277, 1185–92.

Bonwitt, J., Dawson, M., Kandeh, M., et al. (2018). Unintended consequences of the 'bushmeat ban' in West Africa during the 2013–2016 Ebola virus disease epidemic. Social Science & Medicine, 200, 166–73.

Borlase, A., Rudge, J.W., Léger, E., et al. (2021). Spillover, hybridization, and persistence in schistosome transmission dynamics at the human–animal interface. Proceedings of the National Academy of Sciences, USA, 118, e2110711118.

Brashares, J.S., Golden, C.D., Weinbaum, K.Z., et al. (2011). Economic and geographic drivers of wildlife consumption in rural Africa. Proceedings of the National Academy of Sciences, USA, 108, 13,931–6.

Chippaux, J.-P. (2017). Snakebite envenomation turns again into a neglected tropical disease! Journal of Venomous Animals and Toxins Including Tropical Diseases, 23, 38.

Combs, M.A., Kache, P.A., VanAcker, M.C., et al. (2022). Socio-ecological drivers of multiple zoonotic hazards in highly urbanized cities. Global Change Biology, 28, 1705–24.

Cronin, D.T., Woloszynek, S., Morra, W.A., et al. (2015). Long-term urban market dynamics reveal increased bushmeat carcass volume despite economic growth and proactive environmental legislation on Bioko Island, Equatorial Guinea. PLOS ONE, 10, e0134464.

Darwin, C. (1890). Journal of Researches into the Natural History and Geology of the Countries Visited During the Voyage Round the World of H.M.S. 'Beagle' Under the Command of Captain FitzRoy. 2nd ed. D. Appleton and Company.

Eby, P., Peel, A., Hoegh, A., et al. (2023). Pathogen spillover driven by rapid changes in bat ecology. Nature, 613, 340–44.

Fenollar, F., Mediannikov, O., Maurin, M., et al. (2021). Mink, SARS-CoV-2, and the human-animal interface. Frontiers in Microbiology, 12, 663815.

Fiorella, K.J., Milner, E.M., Salmen, C.R., et al. (2017). Human health alters the sustainability of fishing practices in East Africa. Proceedings of the National Academy of Sciences, USA, 114, 4171–6.

Friant, S., Paige, S.B., Goldberg, T.L. (2015). Drivers of bushmeat hunting and perceptions of zoonoses in Nigerian hunting communities. PLOS Neglected Tropical Diseases, 9, e0003792.

Giles, J.R., Eby, P., Parry, H., et al. (2018). Environmental drivers of spatiotemporal foraging intensity in fruit bats and implications for Hendra virus ecology. Scientific Reports, 8, 9555.

Golden, C.D., Fernald, L.C.H., Brashares, J.S., et al. (2011). Benefits of wildlife consumption to child nutrition in a biodiversity hotspot. Proceedings of the National Academy of Sciences, USA, 108, 19,653–6.

Guth, S., Mollentze, N., Renault, K., et al. (2022). Bats host the most virulent—but not the most dangerous—zoonotic viruses. Proceedings of the National Academy of Sciences, USA, 119, e2113628119.

Haggerty, C.J.E., Bakhoum, S., Civitello, D.J., et al. (2020). Aquatic macrophytes and macroinvertebrate predators affect densities of snail hosts and local production of schistosome cercariae that cause human schistosomiasis. PLOS Neglected Tropical Diseases, 14, e0008417.

Halpin, K., Hyatt, A.D., Fogarty, R., et al. (2011). Pteropid bats are confirmed as the reservoir hosts of henipaviruses: a comprehensive experimental study of virus transmission. American Journal of Tropical Medicine and Hygiene, 85, 946–51.

Hammer, A.S., Quaade, M.L., Rasmussen, T.B., et al. (2021). SARS-CoV-2 transmission between mink (Neovison vison) and humans, Denmark. Emerging Infectious Diseases, 27, 547–51.

Hayman, D.T.S. (2019). Bat tolerance to viral infections. Nature Microbiology, 4, 728–9.

Hayman, D.T.S., Bowen, R.A., Cryan, P.M., et al. (2013). Ecology of zoonotic infectious diseases in bats: current knowledge and future directions. Zoonoses & Public Health, 60, 2–21.

Henley, J. (2020). Culled mink rise from the dead to Denmark's horror. The Guardian, 25 November: https://www.theguardian.com/world/2020/nov/25/culled-mink-rise-from-the-dead-denmark-coronavirus, accessed 11/11/2022.

Hoover, C.M., Rumschlag, S.L., Strgar, L., et al. (2020). Effects of agrochemical pollution on schistosomiasis transmission: a systematic review and modelling analysis. Lancet Planetary Health, 4, e280–91.

Hotez, P.J., Lee, S.J. (2017). US Gulf Coast states: the rise of neglected tropical diseases in 'flyover nation'. PLOS Neglected Tropical Diseases, 11, e0005744.

Huong, N.Q., Nga, N.T.T., Long, N.V., et al. (2020). Coronavirus testing indicates transmission risk increases along wildlife supply chains for human consumption in Viet Nam, 2013-2014. PLOS ONE, 15, e0237129.

Janoušková, E., Clark, J., Kajero, O., et al. (2022). Public health policy pillars for the sustainable elimination of zoonotic schistosomiasis. Frontiers in Tropical Diseases, 3, 826501.

Johnson, N., Aréchiga-Ceballos, N., Aguilar-Setien, A. (2014). Vampire bat rabies: ecology, epidemiology and control. Viruses, 6, 1911–28.

Jones, I., Lund, A., Riveau, G., et al. (2018). Ecological control of schistosomiasis in Sub-Saharan Africa: restoration of predator-prey dynamics to reduce transmission. 236–51 in Ecology and Evolution of Infectious Diseases, Oxford University Press.

Kibret, S., McCartney, M., Lautze, J., Jayasinghe, G. (2009). Malaria Transmission in the Vicinity of Impounded Water: Evidence from the Koka Reservoir, Ethiopia (Vol. 132). IWMI.

Koopmans, M. (2021). SARS-CoV-2 and the human-animal interface: outbreaks on mink farms. The Lancet Infectious Diseases, 21, 18–19.

Lancet. (2017). Snake-bite envenoming: a priority neglected tropical disease. Lancet, 390, 2.

Lee, D.N., Papeş, M., Bussche, R.A.V.D. (2012). Present and potential future distribution of common vampire bats in the Americas and the associated risk to cattle. PLOS ONE, 7, e42466.

Lehman, M.W., Craig, A.S., Malama, C., et al. (2017). Role of food insecurity in outbreak of anthrax infections among humans and hippopotamuses living in a game reserve area, rural Zambia. Emerging Infectious Diseases, 23, 1471–7.

Loh, E.H., Zambrana-Torrelio, C., Olival, K.J., et al. (2015). Targeting transmission pathways for emerging zoonotic disease surveillance and control. Vector Borne & Zoonotic Diseases, 15, 432–7.

Luis, A.D., Hayman, D.T.S., O'Shea, T.J., et al. (2013). A comparison of bats and rodents as reservoirs of zoonotic viruses: are bats special? Proceedings of the Royal Society B: Biological Sciences, 280, 20122753.

Lund, A.J., Lopez-Carr, D., Sokolow, S.H., et al. (2021). Agricultural innovations to reduce the health impacts of dams. Sustainability, 13, 1869.

Martin, G., Yanez-Arenas, C., Chen, C., et al. (2018). Climate change could increase the geographic extent of Hendra virus spillover risk. Ecohealth, 15, 509–25.

McKee, C.D., Bai, Y., Webb, C.T., Kosoy, M.Y. (2021). Bats are key hosts in the radiation of mammal-associated *Bartonella* bacteria. Infection, Genetics and Evolution, 89, 104719.

Mendes, W. da S., Silva, A.A.M. da, Neiva, R.F., et al. (2009). An outbreak of bat-transmitted human rabies in a village in the Brazilian Amazon. Revista de Saúde Pública, 43, 1075–7.

Mollentze, N., Streicker, D.G. (2020). Viral zoonotic risk is homogenous among taxonomic orders of mammalian and avian reservoir hosts. Proceedings of the National Academy of Sciences, USA, 117, 9423–30.

Munster, V.J., Bausch, D.G., de Wit, E., et al. (2018). Outbreaks in a rapidly changing Central Africa—lessons from Ebola. New England Journal of Medicine, 379, 1198–201.

Murray, K., Rogers, R., Selvey, L., et al. (1995). A novel morbillivirus pneumonia of horses and its transmission to humans. Emerging Infectious Diseases, 1, 31–3.

Olival, K.J., Hosseini, P.R., Zambrana-Torrelio, C., et al. (2017). Host and viral traits predict zoonotic spillover from mammals. Nature, 546, 646–50.

Olson, S.H., Bounga, G., Ondzie, A., et al. (2019). Lek-associated movement of a putative Ebolavirus reservoir, the hammer-headed fruit bat (*Hypsignathus monstrosus*), in northern Republic of Congo. PLoS ONE, 14, e0223139.

Oreshkova, N., Molenaar, R.J., Vreman, S., et al. (2020). SARS-CoV-2 infection in farmed minks, the Netherlands, April and May 2020. Eurosurveillance, 25, 2001005.

O'Shea, T.J., Cryan, P.M., Cunningham, A.A., et al. (2014). Bat flight and zoonotic viruses. Emerging Infectious Diseases, 20, 741–5.

Oude Munnink, B.B., Sikkema, R.S., Nieuwenhuijse, D.F., et al. (2021). Transmission of SARS-CoV-2 on mink farms between humans and mink and back to humans. Science, 371, 172–7.

Plowright, R.K., Field, H.E., Smith, C., et al. (2008). Reproduction and nutritional stress are risk factors for Hendra virus infection in little red flying foxes (*Pteropus scapulatus*). Proceedings of the Royal Society B: Biological Sciences, 275, 861–9.

Poland, J.D., Barnes, A.M., Herman, J.J. (1973). Human bubonic plague from exposure to a naturally infected wild carnivore. American Journal of Epidemiology, 97, 332–7.

Reisen, W.K., Takahashi, R.M., Carroll, B.D., Quiring, R. (2008). Delinquent mortgages, neglected swimming pools, and West Nile virus, California. Emerging Infected Diseases, 14, 1747–9.

Ribeiro, L.P., Carvalho, T., Becker, C.G., et al. (2019). Bullfrog farms release virulent zoospores of the frog-killing fungus into the natural environment. Scientific Reports, 9, 13,422.

Ribeiro Prist, P., Reverberi Tambosi, L., Filipe Mucci, L., et al. (2022). Roads and forest edges facilitate yellow fever virus dispersion. Journal of Applied Ecology, 59, 4–17.

Rohr, J.R., Barrett, C.B., Civitello, D.J., et al. (2019). Emerging human infectious diseases and the links to global food production. Nature Sustainability, 2, 445–56.

Rudge, J.W., Webster, J.P., Lu, D.-B., et al. (2013). Identifying host species driving transmission of schistosomiasis japonica, a multihost parasite system, in China. Proceedings of the National Academy of Sciences, USA, 110, 11,457–62.

Rulli, M.C., Santini, M., Hayman, D.T.S., D'Odorico, P. (2017). The nexus between forest fragmentation in Africa and Ebola virus disease outbreaks. Science Reports, 7, 41,613.

Scott, T.P., Fischer, M., Khaiseb, S., et al. (2013). Complete genome and molecular epidemiological data infer the maintenance of rabies among kudu (*Tragelaphus strepsiceros*) in Namibia. PLOS ONE, 8, e58739.

Singh, N., Mehra, R.K., Sharma, V.P. (1999). Malaria and the Narmada-river development in India: a case study of the Bargi dam. Annals of Tropical Medicine & Parasitology, 93, 477–88.

Smil, V. (2011). Harvesting the Biosphere: What We Have Taken From Nature. MIT Press.

Sokolow, S.H., Huttinger, E., Jouanard, N., et al. (2015). Reduced transmission of human schistosomiasis after restoration of a native river prawn that preys on the snail intermediate host. Proceedings of the National Academy of Sciences, USA, 112, 9650–55.

Sokolow, S.H., Jones, I.J., Jocque, M., et al. (2017). Nearly 400 million people are at higher risk of schistosomiasis because dams block the migration of snail-eating river prawns. Philosophical Transactions of the Royal Society B: Biological Sciences, 372, 20160127.

Sokolow, S.H., Wood, C.L., Jones, I.J., et al. (2018). To reduce the global burden of human schistosomiasis, use 'old fashioned' snail control. Trends in Parasitology, 34, 23–40.

Sow, S., de Vlas, S.J., Engels, D., Gryseels, B. (2002). Water-related disease patterns before and after the construction of the Diama dam in northern Senegal. Annals of Tropical Medicine & Parasitology, 96, 575–86.

Stears, K., Schmitt, M.H., Turner, W.C., et al. (2021). Hippopotamus movements structure the spatiotemporal dynamics of an active anthrax outbreak. Ecosphere, 12, e03540.

Streicker, D.G., Recuenco, S., Valderrama, W., et al. (2012). Ecological and anthropogenic drivers of rabies exposure in vampire bats: implications for transmission and control. Proceedings of the Royal Society of London B: Biological Sciences, 279, 3384–92.

Tucker Lima, J.M., Vittor, A., Rifai, S., Valle, D. (2017). Does deforestation promote or inhibit malaria transmission in the Amazon? A systematic literature review and critical appraisal of current evidence. Philosophical Transactions of the Royal Society B: Biological Sciences, 372, 20160125.

UNEP. (2016). UNEP Frontiers 2016 Report: Emerging Issues of Environmental Concern. United Nations Environmental Programme, Nairobi.

WHO. (2019). Schistosomiasis and soil-transmitted helminthiases: number of people treated in 2018 (No. 50). Weekly Epidemiological Record. World Health Organization, Geneva, Switzerland.

Wilke, A.B.B., Benelli, G., Beier, J.C. (2021). Anthropogenic changes and associated impacts on vector-borne diseases. Trends in Parasitology, 37, 1027–30.

Wilkinson, D.A., Marshall, J.C., French, N.P., Hayman, D.T.S. (2018). Habitat fragmentation, biodiversity loss and the risk of novel infectious disease emergence. Journal of The Royal Society Interface, 15, 20180403.

Yewhalaw, D., Legesse, W., Van Bortel, W., et al. (2009). Malaria and water resource development: the case of Gilgel-Gibe hydroelectric dam in Ethiopia. Malaria Journal, 8, 21.

# SECTION 4

# Conservation, Ecology, and Control

Emerging wildlife pathogens can cause dramatic declines or extinctions for host species. Sometimes, the host species was already rare or threatened and a disease agent exacerbates extinction risk; other times, the infectious disease creates the conservation threat. In this section (Chapters 12–15), we explore why and how some infectious diseases seriously impact host populations, while others do not. We then move on to examples where these impacts have rippled across the host community or even the whole ecosystem. With these catastrophic impacts in mind, we overview methods and controversies for controlling emerging infectious diseases in complex systems. We conclude with a chapter about the COVID-19 pandemic to remind us why 'an ounce of prevention is worth a pound of cure' and how to prevent future pandemics. Also included: a recipe for a hallucinogenic brandy.

Harlequin frogs (*Atelopus* species) have experienced horrendous declines in the past two decades, associated with outbreaks of chytrid fungus acting in concert with climate change (Jaynes et al. 2022).
Photo by Gena Steffens.
*Source:* Jaynes, K.E., Páez-Vacas, M.I., Salazar-Valenzuela, D., et al. (2022). Harlequin frog rediscoveries provide insights into species persistence in the face of drastic amphibian declines. Biological Conservation, 276, 109784.

SECTION

# CHAPTER 12

# Impacts of emerging infectious disease on wildlife populations

## 12.1 Introduction—Arctic foxes and otodectic mange

*Bering Islands, Alaska*—The volcanic Commander Islands can be found in the Bering Sea, at the western end of the Aleutian Islands (USA) and not far east of Kamchatka (Russia). These human-free islands are inhabited by northern fur seals (*Callorhinus ursinus*) and Arctic fox (*Alopex lagopus*) (Fig. 12.1). In 1741, the naturalist Georg Wilhelm Steller (1709–1746) was shipwrecked on Bering Island, the largest of the Commander Islands. If Steller sounds familiar, it might be because he was the first European to describe North America's Steller's jay (*Cyanocitta stelleri*), Steller's sealion (*Eumetopias jubatus*), and the now extinct Steller's sea cow (*Hydrodamalis gigas*), which was a giant member (up to 9 m long) of the dugong family. Impressive work, but it would seem that naturalists' ethos in the 1700s differed from the ethos that predominates now:

> The foxes who were found there in uncountable numbers became more and more daring and aggressive … They stole our baggage, gnawed soles, boots, trousers, coats … We became so embittered with them that we killed them, young and old, played dirty tricks on them and tortured them in the most cruel ways. In the course of my stay on the island I have killed more than 200 of these filthy creatures. On the third day I killed more than 70 foxes in 3 hours with an axe, and we made the roof of our house out of their skins.
> —Georg Steller, 10 November 1741 (Goltsman et al. 1996).

The foxes' fate has worsened further since their discovery. They formerly existed at high densities but populations were steadily reduced and harvested (43,000 foxes were killed on the Commander Islands for their pelts in 1745–1785). In 1976, there was a calamitous decline in numbers following the introduction of otodectic mange, which was brought by dogs (*Canis lupus familiaris*) accompanying seafaring trappers.

In 1976, when the mange mite, *Otodectes cynotis*, was first observed, fewer than 200 foxes survived on Mednyi Island. Only one cub survived of 28 cubs born that year, with most deaths occurring acutely during a few days in August. All cubs were seriously afflicted—'seething'—with mites. Otodectic mange, also known as ear canker, was not previously considered a cause of canid mortality; however, many dying cubs were observed repeatedly shaking their heads, indicating that canker mites had damaged their tympanic membranes and caused inflammation of the brain. Prior to symptoms, the fox cubs appeared to be in good condition, but once one cub was infected, the entire litter was usually dead within 20 to 25 days. Two years later, the population had declined to fewer than 120 animals (Goltsman et al. 1996).

Scientists decided to intervene based on the interpretation that low fox populations 'could be attributed to no other cause than the introduction of canine disease, and such a small and fragmented population was imperiled by stochastic misadventure and possibly by the demographic and genetic hazards of small numbers' (Goltsman et al. 1996). In 1994, mangy fox cubs were captured and treated with Alugan spray and Ivermectin; none of the treated cubs died that year.

*Emerging Zoonotic and Wildlife Pathogens*. Dan Salkeld, Skylar Hopkins, and David Hayman, Oxford University Press.
© Oxford University Press (2023). DOI: 10.1093/oso/9780198825920.003.0012

**Figure 12.1** Arctic fox, Svalbard, Norway.
Photo by Peter Hudson.

Survivorship also appeared to have increased the following year (though the effect was not statistically significant). The population persists in the 2020s but remains at numbers far reduced from the 1770s.

Though parasites and pathogens can regulate host populations, i.e. affecting both births and survival, traditionally they have not usually been associated with dramatic host population declines, population extirpation, or species extinctions (De Castro and Bolker 2005, Smith et al. 2009). For example, only ~8% of species considered critically endangered by the IUCN Red List have an infectious disease listed as one of the factors threatening conservation (Smith et al. 2006), and most of those are amphibians hypothesized to be threatened by a few pathogens (e.g. the chytrid fungus, see below). So how do we explain the sorry story of the Mednyi Arctic fox, or other examples where emerging infectious diseases *did* devastate host populations?

First, remember that host population declines do not necessarily result in extirpations or extinctions; in many cases, abundant or widespread host species can be hit hard by a pathogen during an initial epidemic of an emerging infectious disease before settling at a new **endemic equilibrium** where the host and pathogen co-exist (Table 12.1). The host population might not be as abundant as before, but nevertheless it persists (though it may now be more vulnerable to extinction driven by other factors). But in some cases, the initial decline due to disease is followed by host extirpation (Fig. 12.2). Simple population models (like the SIR models described in Chapter 5) predict that there are three mechanisms by which pathogens can extirpate host populations, assuming that the hosts aren't rescued by evolution or other complexity.

**Mechanism 1: the host population is already small and/or stressed, such that even a small perturbation due to disease leads to host extirpation.**

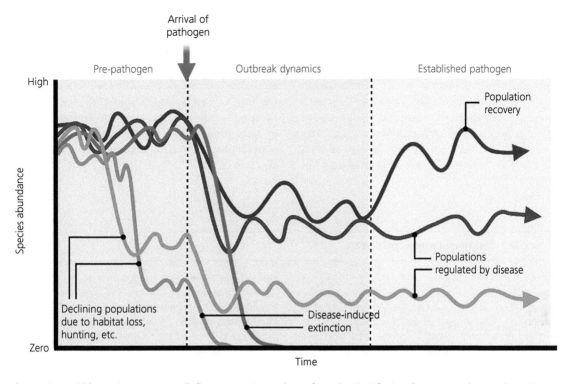

**Figure 12.2** Wildlife population sizes naturally fluctuate over time, and many factors besides infectious diseases can reduce population sizes ('pre-pathogen' period). Emerging infectious diseases can cause declines after their arrival in a naive population ('outbreak dynamics' period). Host population declines may end in extirpation or extinction, which may be especially likely for populations and species that already had reduced population sizes in the pre-pathogen period. Other host populations may initially decline but then persist at smaller sizes in a new endemic equilibrium with their pathogens, or even recover to pre-pathogen sizes as the host and pathogen evolve (changes in behaviour, density, resistance). Inspired by and adapted from Collins (2018), using BioRender.com.

Pathogen spillover is an existential threat to rare, endangered, or fragmented host populations. For example, before mites were introduced, the Arctic fox population was already diminished due to fur-trapping and an axe-happy naturalist explorer. When new pathogens invade stressed or declining populations, even a slight increase in mortality, decrease in birth rate, or change in host behaviour can be the straw that breaks the camel's back. For example, a new disease could reduce host population sizes to such low numbers that the remaining individuals can no longer find mates (an example of an **allee effect**, where host fitness is positively influenced by host abundance). And as the Arctic fox example shows, emerging wildlife pathogens may cause unexpected impacts when they enter naive populations, especially if those populations are already stressed; in the fox example, mites were causing mortality through otodectic mange, which had previously not been considered as a possible cause of death.

**Mechanism 2: the pathogen has density-independent transmission, such that transmission rates do not decline as the host population dwindles.** For many pathogens, we expect transmission rates to be relatively high when host density is high, and relatively low when host density is low (Chapter 5). In this scenario, transmission rates should decline with host density, and the pathogen may be lost from the host population due to limited transmission before the host population is extirpated. But this might not be the case for some pathogens; for example, if transmission rates saturate with host density, there could be very little change in transmission rates across a broad range of host densities, and transmission rates might not decline until the host population is too small to persist anyway.

**Mechanism 3: the pathogen has an abiotic or biotic reservoir outside of the impacted host population, which maintains high transmission rates to the host population even as it declines towards extirpation.** As mentioned above, for many pathogens, we expect transmission rates to be relatively high when host density is high and relatively low when host density is low. But

that won't necessarily occur if the pathogen is maintained in the environment (e.g. white nose syndrome, Chapter 6) or by other reservoir host species (e.g. rabies maintained by domestic dogs, see below). In the next section, we will return to the topic of pathogens that spill over from domestic species into wildlife and that are maintained in large domestic populations even as wildlife populations dwindle towards extinction (see also Box 12.1).

### Box 12.1 Metapopulations and disease

When populations become small, their probability of extinction typically goes up, because demographic and environmental stochasticity (e.g. random unfavourable weather events) are more likely to set the population on an irreversible decline (Heard et al. 2015). However, when one *population* is locally extirpated, a *species* is not necessarily lost; the area might be re-colonized by migrants from a different population later if other populations still exist

nearby. **Metapopulations** are populations spread across a landscape that are linked by dispersal (immigration and emigration). We can represent metapopulations with mathematical models, which show that a balance between population extirpations and re-colonizations can allow a species to persist regionally even as single populations keep blinking out.

**Figure 12.3** Growling grass frog (*Litoria raniformis*), Donnybrook, Victoria, Australia. Photo by Geoffrey Heard.

**Box 12.1** *Continued*

What happens when an infectious disease is introduced into a host metapopulation? You might expect connectivity to be detrimental to regional host persistence when infectious diseases are introduced, because pathogens in one population can invade the other populations via dispersal of infected hosts. This might be the case during an initial outbreak, where control efforts to prevent pathogen spread to other populations are critical. But after a pathogen is already established throughout a region, population connectivity may be required for host persistence (Heard et al. 2015).

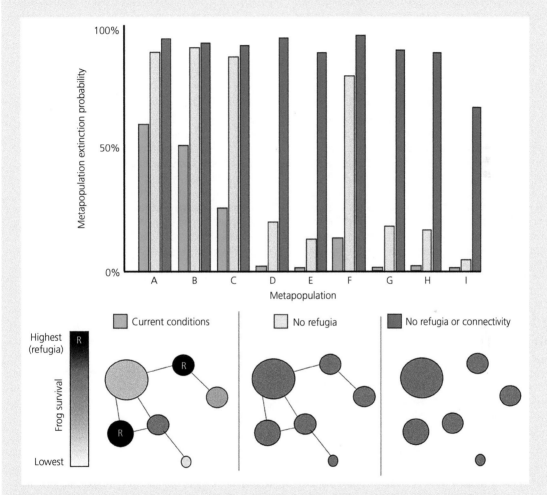

**Figure 12.4** Under current conditions, where the prevalence of chytridiomycosis varies across wetlands due to variable water temperature and salinity, most of the nine growling grass frog *metapopulations* (A to I) are unlikely to be extirpated in any given year, even though some of the populations within each metapopulation may be extirpated (Heard et al. 2015). In contrast, if the warm and relatively salty refugia wetlands were lost from the metapopulations, the metapopulations would be more likely to be extirpated, especially if connectivity among wetlands was also reduced.

---

**Box 12.1** *Continued*

For example, growling grass frogs (*Litoria raniformis*) (Fig. 12.3), named for their distinctive call, were once abundant in south-eastern Australia. Populations live in slow-moving wetlands, including ponds or pools within streams. When populations in these habitats are separated by more than 1 km, they are unlikely to be connected by dispersal, but many populations are close enough to be connected and thus form metapopulations. In a study population near Melbourne, there were nine metapopulations that each contained 6–48 connected wetlands (Heard et al. 2015).

When chytridiomycosis (caused by *Batrachochytrium dendrobatidis*, see below) reached the area, growling grass frogs declined to the point that the species is now considered endangered. Individual populations often winked out due to a combination of chytrid infections and other stressors. But often, these same sites would be re-colonized by growling grass frogs from nearby source populations.

All the wetlands harbour the chytrid fungus—it persists in the environment and on other frog species that appear more **resistant** or **tolerant** (Box 12.4) to chytridiomycosis. But disease outcomes also vary for growling grass frogs across heterogeneous habitats because chytrid fungus cannot reproduce (and may die) at temperatures above 28°C

and at relatively high salinities. Warm, salty ponds/pools are less conducive for the fungus (prevalence <30%), and thus more favourable for the frogs (annual probability of frog population persistence >90%), and may therefore serve as **refugia** for growling grass frogs—habitats where the frogs are relatively safe from severe disease impacts (Fig. 12.4). If populations in refugia do well enough, they may even serve as **source populations** whose emigrating individuals enable population recovery in other sites and habitats. In contrast, relatively cool wetlands with low salinity had higher chytrid prevalence (>60%) and annual population persistence was low (<30%).

In fact, in metapopulation simulations, the growling grass frog was more likely to persist if habitats were heterogeneous (with variable survival) rather than homogeneous (an over-simplification of reality). And if simulations did not include the warm and salty refugia (e.g. if we imagine that they were destroyed by human development) and connections between populations, growling grass frog metapopulations were almost always extirpated. These metapopulation simulations show that the probability of host persistence is highest when there are refuges from disease impacts and populations can be 'rescued' by immigration from other populations.

---

## 12.2 Small carnivore populations threatened by pathogens from domestic dog reservoirs

Domestic dogs have spread across the world with humans to live on every continent, including Antarctica. (However, huskies are no longer used as sled dogs for Antarctic exploration and research, partly because they might introduce pathogens to local seal populations (Holden 1994)). Like most species, dogs host a variety of parasites and pathogens, but unlike most wildlife species, dogs are often hyperabundant because they have access to human-provided resources. Therefore, when dogs live near wildlife—and especially when they are free-ranging or feral, sharing environments and interacting with wildlife—there is high potential for them to act as reservoirs of pathogens that spill over into wildlife populations (Fig. 12.5).

*The Ethiopian Highlands*—The Ethiopian wolf (*Canis simensis*), also known as the Simien jackal or Abyssinian wolf (Fig. 12.6), is the world's rarest canid and Africa's most threatened carnivore. Fewer than 500 of the wolves remain in their natural range in the Ethiopian Highlands. Already vulnerable due to disappearing habitat, Ethiopian wolves are highly susceptible to rabies; four major rabies outbreaks have been recorded in the last three decades, killing dozens of wolves each time (e.g. Randall et al. 2004). The rabies virus doesn't persist within Ethiopian wolf populations between outbreaks, because Ethiopian wolf populations are too small and isolated (Box 12.2). Instead, the Ethiopian wolves are exposed to rabies by interactions with feral domestic dogs.

To complicate matters, dog populations are too large and mobile to easily achieve broad enough vaccination coverage to completely protect wolves

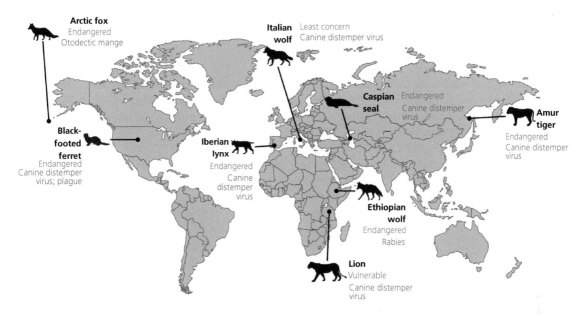

**Figure 12.5** Carnivore species, their conservation status (from IUCN red list or publication), and one important pathogen affecting populations. See the main chapter text for further details.
Made using BioRender.com.

**Figure 12.6** Ethiopian wolf (*Canis simensis*) and cub.
Photo by Rebecca Jackrel and provided by the Ethiopian Wolf Conservation Programme.

given the resources available. Vaccinations are now also targeted at the wolves themselves. The rabies vaccine baits are deployed inside pack territories using goat meat baits, delivered from horseback during the night to maximize successful bait delivery (Carrington 2018). As we will see in Chapter 14, vaccination of wildlife is rarely a viable control strategy for wildlife disease control, but a huge effort has been made to protect endangered Ethiopian wolves.

Another endangered African canid is the African wild dog (*Lycaon pictus*). Always uncommon due to competition (bullying?) by lions and hyaenas and needing large ranges, a combination of new threats has caused further decline in numbers: habitat loss; depletion of wild prey; direct persecution and eradication campaigns involving shooting, snaring, and poisoning; road accidents; population fragmentation; and, of course, disease (Woodroffe and Ginsberg 1999). Outbreaks of rabies virus and canine distemper virus have caused extirpations of entire wild dog packs and possibly even local populations. Just like the case of the Ethiopian wolves, domestic dog populations are likely operating as pathogen reservoirs (Kat et al. 1995, Woodroffe and Ginsberg 1999).

*The Mediterranean*—Of course, examples of small carnivore populations threatened by invasive pathogens shared by domestic dogs are not limited to Africa. The Iberian lynx (*Lynx pardinus*) is one of the most endangered felid species in the world, with fewer than 200 individuals surviving in southwestern Spain. In 2005, an adult female Iberian lynx was found dead in Doñana National Park; that lynx and 15% of (88) living lynxes tested positive for canine distemper virus (Meli et al. 2010). And in Spain, just like the Serengeti, evidence of infection has been observed in a broad range of species: domestic dogs and cats (*Felis catus*), ferrets (*Mustela putorius furo*), mink (*Mustela lutreola*), red foxes (*Vulpes vulpes*), genets (*Genetta genetta*), and Spanish wolves (*Canis lupus*). Similarly, in Italy, canine distemper virus has been held responsible for deaths of Apennine wolves in the Abruzzi region of Italy at a time when local domestic dogs also exhibited increased incidence of infection. Other wildlife species (red foxes and badgers (*Meles meles*)) also appeared to be dying, demonstrating once again that canine distemper virus can cause multi-host epizootics/outbreaks (Di Sabatino et al. 2014).

---

**Box 12.2 Infectious disease and inbreeding**

Small, fragmented, and isolated populations may suffer reduced genetic variability, particularly if isolation and inbreeding have existed for multiple generations. The impacts of inbreeding may persist even after the population increases. Reduced genetic variation may impact populations' abilities to evolve in the face of environmental change, such that instead of evolving, populations are extirpated. This has received much attention with regards to infectious diseases, because inbreeding is thought to reduce heterozygosity of the **major histocompatibility complex** (MHC), which is linked to effective immune responses (Penn 2002). For example, inbreeding among Californian sea lions (*Zalophus californianus*) has led to greater susceptibility to infectious disease and cancer risk (Acevedo-Whitehouse et al. 2003). This suggests that when populations become small or fragmented due to anthropogenic stressors like habitat destruction, their risk of disease-induced extinction may increase.

The unfortunate poster child for inbreeding and disease may be Tasmanian devils (*Sarcophilus harrisii*), who have been horribly afflicted by Tasmanian devil facial tumour disease (see also Chapters 2 and 13) (Fig. 12.7). Now limited to the island of Tasmania (the arrival of dingoes extirpated them from mainland Australia), they were persecuted in the 1800s and poisoned in the 1900s, causing the species to experience multiple population bottlenecks before being protected in 1941. Numbers rebounded—up to 150,000ish in the 1990s—but the reduction in genetic diversity imposed by the historic declines in number remains. Reduced genetic diversity has led to Tasmanian devils being susceptible to a range of infectious diseases (Jones et al. 2004, Pyecroft et al. 2007, Woods et al. 2007). Most bizarrely, the Tasmanian devil's immune system is unable to recognize tumour cells as 'non-self' (Pearse and Swift 2006). 'It's not that evolution has made the devil tumour invisible. It's that genetic impoverishment has made the immune system blind'

**Box 12.2** *Continued*

(Quammen 2008). Devil facial tumour disease caused such severe population declines that Tasmanian devils were predicted to become extinct within 20 years (McCallum and Jones 2006, McCallum et al. 2007, 2009). However, more recent analyses suggest that there is some population recovery in these defiantly resilient animals, and that coexistence between host and pathogen is a more likely outcome than Tasmanian devil extinction (Patton et al. 2020).

**Figure 12.7** A Tasmanian devil leaving a trap after being surveyed by researchers. Photograph by David Hamilton.

## 12.3 Small carnivore populations threatened by pathogens from wildlife reservoirs

Domestic dogs can't be blamed for all infectious diseases affecting wildlife populations.

*Far East Asia*—Much, much bigger than Iberian lynx, the Amur tiger (*Panthera tigris altaica*, formerly known as the Siberian tiger) comprises fewer than 3500 individuals in two discrete populations in Russia and China along the Amur River. This is another cat species susceptible to canine distemper virus (suggesting a pathogen misnomer (Terio and Craft 2013)). Prior to 2000, seroprevalence for canine distemper virus antibodies in Amur tigers was zero (0/18) but increased subsequently (20/54; 37%), suggesting that the pathogen invaded the population at that time. Death from canine distemper virus was first witnessed in Amur tigers in 2003 and then again in 2010, raising alarm about the potential impact on the tiger species (Gilbert et al. 2020). Even sporadic mortality from infectious disease can impact population viability, because female Amur tigers reproduce slowly; they have their first litters at roughly 42–54 months, and then reproduce at intervals of nearly two years. Model predictions generate high likelihoods of extinction of Amur tigers in the next 50 years in scenarios without canine distemper virus control efforts (Gilbert et al. 2014).

To intervene in tiger conservation, it is vital to understand the local ecology of disease dynamics, including which species serve as reservoir hosts. In the Russian far east, the tigers' habitat encompasses 17 other wild carnivore species and domestic dogs. But unlike our previous examples regarding canine distemper virus, domestic dogs do not appear to be the main culprit. Most seropositive (i.e. previously exposed) dogs came from more remote human settlements, and those dog populations were likely too small and isolated to maintain the pathogen. Therefore, interventions focused on domestic dog populations would have failed to address the issue of viral transmission in the tiger populations (Gilbert et al. 2014, 2020). Instead, interventions must grapple with infection dynamics among and between wildlife carnivore species. This includes attempting to vaccinate as many tigers as is feasible, so that populations are more resilient to future outbreaks. Unfortunately, there is no good oral delivery system for canine distemper virus vaccines, so the vaccine needs to be given individually; a big ask for a large, rare, cryptic carnivore that can potentially eat you.

## 12.4 Environmental reservoirs and resistant reservoir hosts drive amphibian species to extinction

At 7am on January 13, 1984, ranger Keith McDonald was driving along a dirt road deep in the rainforest of Eungella National Park, 80km west of Mackay [Australia], when one of the most implausible events in the natural world occurred on the passenger seat beside him.

Inside a container of water, a female frog he had collected from a stream the night before opened her mouth and spat out a fully formed juvenile frog. Over the next half-hour, 14 more froglets were born through their mother's mouth. As any child will tell you, frogs don't give birth through their mouth. They don't give birth at

all. They lay eggs, which hatch into tadpoles and metamorphose into frogs underwater. It was the first and last time anyone would see the unique birthing approach of the northern gastric-brooding frog. By March of 1985 the frogs, endemic to this one area on Earth, were gone, never to be seen again.

—Ricky French, 'Case of the missing frogs' (2019)

*Canterbury, England, 1989*—At the first World Congress of Herpetology, amphibian researchers gathered together and realized something concerning: rapid and severe declines in amphibian populations were being observed in countries around the world (Fisher and Garner 2020). It was unclear why these declines were happening; in many cases, dead frogs were never actually seen. Populations just declined or disappeared in a way that researchers called 'enigmatic'. Though amphibians had known conservation threats, like wetland destruction and other land use change, many of the observed declines were happening in relatively remote and 'pristine' habitats where anthropogenic influences were minimal (Blaustein and Wake 1990, Lips 2016). In the next few years, researchers would compile the data around the world to confirm that amphibian population declines were widespread and that no single known threat could explain them all. The hunt for the cause continued.

Today, we recognize several pathogens that cause widespread amphibian mortality, including Ranaviruses and *Batrachochytrium dendrobatidis*, the 'chytrid fungus'. *Batrachochytrium dendrobatidis* wasn't named until 1999 (Berger et al. 1998, Longcore et al. 1999), and though it has clearly devastated many amphibian populations, it is difficult to say exactly how many local extirpations and species extinctions can be attributed to the fungus (see also Box 12.3). One estimate suggests that dozens of amphibian species extinctions and more than a hundred population declines greater than 90% were caused by this single fungal pathogen (Scheele et al. 2019).

Further complicating our understanding, *Batrachochytrium dendrobatidis* does not affect all amphibian species in the same way. When individual hosts are infected, the worst symptoms result from damage to the skin, which for amphibians means a disruption of respiration, osmoregulation, and other functions. But not all infected hosts develop chytridiomycosis, and not all diseased hosts die from their infections. For example, *Batrachochytrium dendrobatidis* is less likely to cause disease in salamander species than in frog species, and some frog species (e.g. American bullfrogs) show few signs of disease when they are infected. The most heavily infected amphibians are usually the most likely to experience the most extreme pathology, morbidity, and mortality (Lips 2016), which we have previously defined as **intensity-dependent disease outcomes**. In turn, infection intensity is driven by several processes related to the pathogen, host(s), and environmental characteristics, as suggested by the **disease triangle** (Clare et al. 2016, Bosch et al. 2018, Lips 2016, Figure 12.8).

The *environmental conditions* (Fig. 12.8) that can exacerbate chytridiomycosis include temperature, humidity, and host stressors like UV radiation and pesticide exposure (Kriger et al. 2007, Walker et al. 2010, Puschendor et al. 2011, Garner et al. 2011, Rohr et al. 2013). For example, chytrid has had especially devastating impacts on montane frogs in the Neotropics, where frog diversity is high and the cool, moist conditions are ideal for the fungus. The *host factors* (Fig. 12.8) that influence disease

outcomes include host community composition; for example, some frog communities contain 'carrier' host species that experience limited pathology but can spread the pathogen to other species. Another host factor is the host's microbiome; some skin bacteria produce antifungal metabolites that make hosts more resistant to chytrid infection (Harris et al. 2009). And finally, the *pathogen characteristics* (Fig. 12.8) that influence disease outcomes include the strain of the pathogen and how long it has existed in a particular region.

The 'panzootic' strain of *Batrachochytrium dendrobatidis* that has caused epizootics around the world seems to have originated in Asia (O'Hanlon et al. 2018, Fisher and Garner 2020). It likely spread repeatedly to other countries through the pet trade and other human-assisted movement (Chapter 9). In Asia, this panzootic strain does not seem to cause mass mortality events and population declines (Fisher and Garner 2020). Similarly, there are other 'endemic' strains of *Batrachochytrium dendrobatidis* that do not cause obvious amphibian declines in particular regions, perhaps because those strains have co-existed with their regional hosts for a long time. But note that *Batrachochytrium dendrobatidis* is not done evolving; co-infection of hosts with

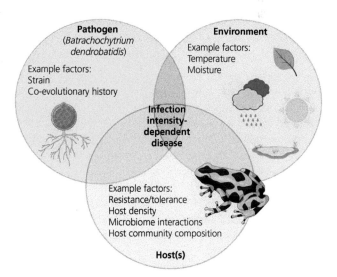

**Figure 12.8** The disease triangle for chytridiomycosis, showing environment, host, and pathogen characteristics that lead to intensity-dependent disease outcomes for hosts.
Figure adapted from Lips (2016) and made with Biorender.com.

**Figure 12.9** Fire salamanders (*Salamandra salamandra*) can grow up to 35 centimetres long and live more than 40 years. In European mythology, fire salamanders were associated with, um, fire. This false association may have derived from hibernating salamanders scampering out of logs that had been put on a fire. The vivid coloration warns predators of an alkaloid poison, samandarin, around their head and back, which triggers muscle convulsions, hypertension, and hyperventilation in any vertebrates who persist in trying it. The samandarine family of compounds comprises the all-important ingredient in a Slovenian salamander brandy: 'Throw a couple of salamanders (one every ten litres) into a barrel with fruit (soaked before distilling), seal everything with a wooden cover and leave it to the will of the gods for a couple of months' (Kozorog 2003). Salamander brandy is credited with being both hallucinogenic *and* intensely aphrodisiac, which could make for a strange and memorable evening, potentially culminating in carnal relations with a beech tree (*Fagus* spp.) (Kozorog 2003).
Photo by Petar Milošević, CC BY-SA 4.0, https://commons.wikimedia.org/w/index.php?curid=96146618

multiple strains can lead to recombination and new variants, which then may spread widely and negatively impact naive host species.

And unfortunately, *Batrachochytrium dendrobatidis* isn't the only amphibian fungus. Another species, *Batrachochytrium salamandrivorans*, was discovered in 2010 when it extirpated fire salamanders (*Salamandra salamandra*, Fig. 12.9) in the Netherlands (Martel et al. 2013, Fitzpatrick et al. 2018). *Batrachochytrium salamandrivorans* differs from *Batrachochytrium dendrobatidis* in several ways: it has a somewhat different infection cycle and mechanisms for causing pathology; it tends to be more detrimental for salamanders than for frogs; and it was

discovered before it was globally distributed. But like *Batrachochytrium dendrobatidis*, *Batrachochytrium salamandrivorans* is known to cause high mortality rates in some species; it originated in Asia (Martel et al. 2014, Laking et al. 2017); and it could easily be spread by humans to areas of high salamander diversity via the pet trade. Therefore, many *Bsal*-free countries (i.e. United States, Canada, and several European countries) have proactively banned importation of salamanders to prevent accidental importation of the fungus (Garner et al. 2016, EFSA Panel AHAW 2018, US Fish and Wildlife Service 2016, Canada Border Services Agency 2018).

## Box 12.3  Fungal pathogens

In the past two centuries, many new fungal diseases have emerged in plant and animal populations. These include chytridiomycosis in amphibians; white nose syndrome in bats; Dutch elm disease; chestnut blight; aspergillosis in endangered kakapos (*Strigops habroptilus*, a flightless, nocturnal, ground-dwelling parrot in New Zealand); ophidiomycosis or snake fungal disease; and late or potato blight (Fig. 12.10), the disease associated with the fungus-like oomycete which caused the Irish Potato Famine. These fungal diseases have decimated and extinguished wildlife populations; altered plant communities and entire landscapes; released massive amounts of carbon dioxide through tree death; threatened food security; and driven historically important events of human immigration and mortality (Fisher et al. 2012). Therefore, understanding and controlling fungal pathogens is important, but it is also difficult!

Fungal pathogens may be especially deadly and difficult to control for several reasons (Fisher et al. 2012). Fungi can reproduce quickly, and their rapid growth on hosts can lead to high host fatality ratios and rapid spread through host populations. Highly resilient environmental stages of fungal

pathogens (spores) can persist in environmental reservoirs and continue to expose dwindling host populations, such that the force of infection remains high even as the host population declines (Chapter 5). And free-living, saprophytic stages or durable spores may be especially adept at getting moved around by humans, such as through soil and water transported for agriculture and the pet trade. These problems might explain why fungal pathogens seem to cause more host extinctions than other types of disease agents, on average (Fisher et al. 2012).

It is unlikely that climate change is universally making all fungal pathogens worse—some probably benefit while others do not. But many fungal pathogens are highly sensitive to environmental conditions like temperature and soil moisture, and these conditions often predict whether host disease will be especially severe. Therefore, our changing climate might exacerbate many fungal diseases, and it certainly complicates our ability to predict and control outbreaks.

Though individual hosts can often be treated and cured of their fungal infections, fungal diseases are extremely difficult to control due to their environmental reservoirs. Fungal pathogens therefore remind us of the adage that 'an ounce

**Figure 12.10**  The oomycete that causes potato late blight (*Phytophthora infestans*) is still a major crop pathogen today, and it is not confined to Ireland or potatoes! The pathogen causes diseases in several crops on all continents except Australia and Antarctica. Australia has strong biosecurity protocols to prevent the accidental importation of the pathogen. In these photos, late blight is shown on a ripe tomato.
Photograph by Downtowngal, CC BY-SA 3.0, https://commons.wikimedia.org/w/index.php?curid=15991500

**Box 12.3** *Continued*

of prevention is worth a pound of cure'. Biosecurity efforts to prevent fungal pathogens of plants and animals from spreading to new areas are costly, but prevention costs are a mere drop in the ocean compared to the huge costs of trying to control emerging fungal pathogens unleashed on naive plant and animal populations.

## 12.5 Infectious diseases can make common species rare

Most of the preceding examples described what happens when an emerging infectious disease spills over into a population that is already small or stressed. In those scenarios, infectious diseases are like the 'straw that broke the camel's back' for threatened species. Though such scenarios may be increasingly common as anthropogenic change threatens more wildlife species, it is also important to remember that emerging infectious diseases can also decimate species with large and widespread populations.

In Chapter 6, we described the spread of *Pseudogymnoascus destructans*, a fungal pathogen, from hibernating bats in New York in 2006 to across the United States and Canada by 2022. As it spread, *Pseudogymnoascus destructans* caused mass mortality events in bat hibernacula, but how much did this impact bat populations? A lot.

Several cave-hibernating bat species have experienced dramatic population declines of 95–99% across their ranges, with many hibernacula being completely extirpated (Cheng et al. 2021). The impacted species include those that were most common before the epizootic began in 2006, such as the little brown bat (*Myotis lucifugus*), tri-coloured bat (*Perimyotis subflavus*), and northern long-eared bat (*Myotis septentrionalis*), which are now considered vulnerable to future extinction. The millions of bats that died during the initial spread of *Pseudogymnoascus destructans* also included bat species that were already threatened, like the Indiana bat (*Myotis sodalis*). Because these species all reproduce relatively slowly, it will take decades for bat communities to recover (if species manage to persist that long). Therefore, it is critical to protect bats as much as possible during this recovery phase, such as by conserving their summer foraging habitats and maternal roosting locations.

However, like amphibians infected with chytridiomycosis, not all North American bats were impacted to the same extent by *Pseudogymnoascus destructans*. This is another example of the disease triangle concept: some host, environment, and pathogen conditions create a deadly combination that leads to high infection intensities and disease, while others do not. For example, *Pseudogymnoascus destructans* does not seem to have impacted bat species that do not hibernate in caves and mines, likely because trees and other roosts do not create ideal environmental reservoirs for the fungus. But even bat species that commonly roost together in the same caves may have quite different responses to infection. For example, big brown bats (*Eptesicus fuscus*) tend to roost in relatively cold locations near cave and mine entrances and have experienced only mild to moderate population declines. In contrast, little brown bats (*Myotis lucifugus*) and tri-coloured bats (*Perimyotis subflavus*) prefer relatively warmer roosting locations that are more amenable to fungal growth and have experienced moderate to severe population declines (Fig. 12.11) (Verant et al. 2012, Langwi et al. 2016, Hopkins et al. 2021). Figuring out these species-level differences at the start of a wildlife disease outbreak is challenging, but it is critical for understanding how to allocate limited resources for monitoring and control.

On a brighter note, even populations that experienced extreme declines during the initial white nose syndrome epizootic may be stabilizing over time. These populations may have become more resistant or tolerant to *Pseudogymnoascus destructans* infections (Frick et al. 2017, Langwig et al. 2017, Box 12.4). For example, in hibernacula where *Pseudogymnoascus destructans* has been present the longest, bat populations may have experienced

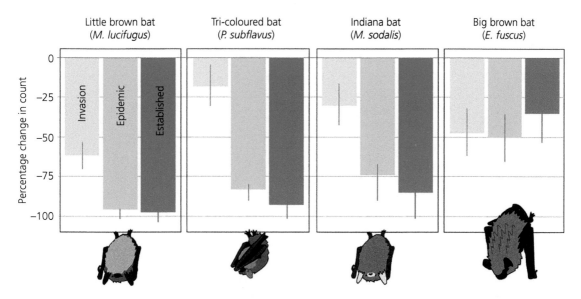

**Figure 12.11** Average percentage declines in bat populations from four bat species that hibernate in caves and mines. The bars show progress through time, from the years when *Pseudogymnoascus destructans* first appeared at a site (invasion) to when it spread and caused widespread mortality (epidemic) to several years after invasion (established). Note that some species, like the big brown bat, experienced moderate declines, whereas others experienced severe declines. On this graph, a −100% change would indicate complete extirpation.
Graph was created with data from Cheng et al. (2021).

some microevolution, where genes related to thermoregulation, torpor, and fat production are more common now than they were before *Pseudogymnoascus destructans* invaded (Auteri et al. 2020, Lilley et al. 2020). This is to be expected, because this disease and the mass mortality it causes is a very strong selection pressure! Furthermore, bat survival may be increasing over time. For example, Grimaudo et al. (2022) did two transplant experiments where bats were translocated to a hibernaculum

where all the bats had been extirpated by white nose syndrome. In the experiment, conducted in 2009, during the initial epizootic, the transplanted bats all died. In the experiment in 2018, 80% of the transplanted bats survived. If North American bats can evolve fast enough, we may eventually end up in a situation similar to that in Eurasia, where bats are persisting with lower average fungal loads and pathology than bats in present-day North America (Zukal et al. 2016, Hoyt et al. 2020).

---

### Box 12.4  Resistance vs. tolerance

Though often used loosely or interchangeably, **resistance** and **tolerance** mean different things in disease ecology (and in immunology!). To see the difference, imagine that four little brown bats and three big brown bats were exposed to the same amount of *Pseudogymnoascus destructans* and their (imaginary) pathogen loads and disease severity are plotted in Fig. 12.12.

**Resistance** refers to a host's ability to limit parasite burden, and resistance can vary among individuals of the same species and between species. In the plot, little brown bats

A and B have different fungal loads, even though they were exposed to the same pathogen dose. Since little brown bat A was better able to reduce pathogen replication and infection loads, it is more resistant than little brown bat B. (Little brown bats A and B are also both more resistant than bats C and D, which have high loads and low resistance.) We also see that big brown bats have lower maximum and average pathogen loads than little brown bats, and thus big brown bats are the more resistant species in this thought experiment.

**Box 12.4** *Continued*

**Tolerance** refers to a host's ability to limit the disease severity induced by a given parasite burden, and it can also vary among individuals of the same species and between species. In the plot, little brown bats C and D have the same pathogen loads and are thus equally resistant, but little brown bat C is better able to tolerate that load and remain healthy (and alive) than little brown bat D, who died from disease symptoms. We also see that on average, big brown bats had better average health for equivalent pathogen loads, such that big brown bats were more tolerant than little brown bats in this thought experiment.

Note that unless we know pathogen exposure or load, we cannot differentiate between resistant versus tolerant individuals or populations; a very sick host could have high pathogen loads, or low tolerance to a low pathogen load.

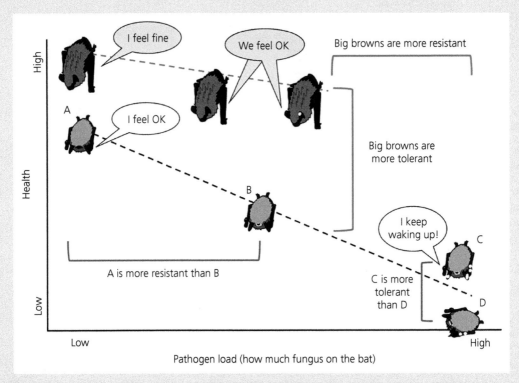

**Figure 12.12** A conceptual diagram comparing the resistance and tolerance of individuals of two bat species based on a thought experiment where all bats were exposed to the same amount of *Pseudogymnoascus destructans* and then their fungal loads and health were later quantified.

## 12.6 Summary

Parasites and pathogens can regulate, and even decimate, host populations by affecting birth rates, death rates, and other demographic processes. Without long-term or careful data, these impacts can go undocumented, yet infectious disease dynamics—whether among endangered populations, from environmental reservoirs, or due to spillover from more abundant, animal reservoir populations—is a critical aspect of conservation biology. Indeed, these impacts may ripple through ecosystems and communities—the topic of the next chapter!

**Table 12.1** Some additional examples of wildlife populations impacted by infectious diseases.

| Host species | Pathogen | Outbreak information | Comments | Reference |
|---|---|---|---|---|
| Wild bearded pig (*Sus barbatus*) | African swine fever virus (Asfarviridae) | First confirmed in Borneo in early February 2021. Outbreak caused a rapid decline in wild bearded pigs. | Vaccines (for domestic pigs) are not currently available. | Ewers et al. 2021 |
| Gorilla (*Gorilla gorilla*); chimpanzee (*Pan troglodytes*) | Ebola virus (Filoviridae) | Republic of Congo. Estimated 50% decline in gorilla observations; 8 groups (143 individuals) disappeared; 88% decline in chimpanzee observations. | Declines coincident with human outbreaks. | Leroy et al. 2004 |
| American crows (*Corvus brachyrhynchos*) | West Nile virus (WNV) (Flaviviridae) | USA. More than two-thirds of crows died of WNV infection in local outbreaks. | | Yaremych et al. 2004 |
| Caspian seals (*Phoca capsica*) | Canine distemper virus (CDV) (Paramyxoviridae) | Spring 2000, Black Sea: estimated 10,000 seals died of CDV infection. | | Kennedy et al. 2000 |
| Prairie dogs (*Cynomys* spp.) | *Yersinia pestis* (causes plague) | Invasive pathogen periodically causes local extirpations of prairie dog colonies. | | Salkeld et al. 2016, Tripp et al. 2017 |
| Bird species (gannets, penguins, waterfowl, etc.) and sea mammals | Avian H5N1 influenza | 2022 outbreaks globally causing die-offs of birds and seals. | | Sidik 2022, Weston 2022, Puryear et al. 2023 |
| Harbour seals (*Phoca vitulina*) and grey seals (*Halichoerus grypus*) | Phocine distemper virus | Massive epidemic in 1988 in north-western Europe, killed >18,000 seals. | | Duignan et al. 2014 |
| Black abalone (*Haliotis cracherodii*) | *Candidatus Xenohaliotis californiensis* (causes abalone withering syndrome) | California coast. Black abalone started declining in 1985, especially during warm years, and are now federally endangered. | Commercially important abalone fishery was closed. Disease is mediated by a hyperparasite of *C. X. californiensis*. | Lafferty and Kuris 1993, Friedman et al. 2002, Crosson et al. 2014 |

## 12.7 References

Acevedo-Whitehouse, K., Gulland, F., Greig, D., Amos, W. (2003). Disease susceptibility in California sea lions. Nature, 422, 35.

Auteri, G.G., Knowles, L.L. (2020). Decimated little brown bats show potential for adaptive change. Scientific Reports, 10, 3023.

Berger, L., Speare, R., Daszak, P., et al. (1998). Chytridiomycosis causes amphibian mortality associated with population declines in the rain forests of Australia and Central America. Proceedings of the National Academy of Sciences, USA, 95, 9031–6.

Blaustein, A.R., Wake, D.B. (1990). Declining amphibian populations: a global phenomenon? Trends in Ecology and Evolution, 5, 203–04.

Bosch, J., Fernandez-Beaskoetxea, S., Garner, T.W.J., Carrascal, L.M. (2018). Long-term monitoring of an amphibian community after a climate change- and infectious disease-driven species extirpation. Global Change Biology, 24, 2622–32.

Canada Border Services Agency. (2018). Environment and Climate Change Canada (ECCC)'s Import Restrictions on Salamanders: Customs Notice 17-17. https://www.cbsa-asfc.gc.ca/publications/cn-ad/cn17-17-eng.html

Carrington, D. (2018). Ethiopia deploys hidden rabies vaccine in bid to protect endangered wolf. The Guardian, 22 August: https://www.theguardian.com/environment/2018/aug/22/ethiopia-deploys-hidden-rabies-vaccine-in-bid-to-protect-endangered-wolf, accessed 11/18/2021.

Cheng, T.L., Reichard, J.D., Coleman, J.T., et al. (2021). The scope and severity of white-nose syndrome on hibernating bats in North America. Conservation Biology, 35, 1586–97.

Clare, F.C., Halder, J.B., Daniel, O., et al. (2016). Climate forcing of an emerging pathogenic fungus across a montane multi-host community. Philosophical Transactions of the Royal Society B: Biological Sciences, 371, 20150454.

Collins, J.P. (2018). Change is key to frog survival. Science, 359, 1458–9.

Crosson, L.M., Wight, N., VanBlaricom, G.R., et al. (2014). Abalone withering syndrome: distribution, impacts, current diagnostic methods and new findings. Diseases of Aquatic Organisms, 108, 261–70.

De Castro, F., Bolker, B. (2005). Mechanisms of disease-induced extinction. Ecology Letters, 8, 117–26.

Di Sabatino, D., Lorusso, A., Di Francesco, C.E., et al. (2014). Arctic lineage-canine distemper virus as a cause of death in Apennine Wolves (*Canis lupus*) in Italy. PLoS ONE, 9, e82356.

Duignan, P.J., Van Bressem, M.F., Baker, J.D., et al. (2014). Phocine distemper virus: current knowledge and future directions. Viruses, 6, 5093–134.

EFSA Panel on Animal Health and Welfare (AHAW), More, S., Angel Miranda, M., et al. (2018). Risk of survival, establishment and spread of *Batrachochytrium salamandrivorans* (Bsal) in the EU. EFSA Journal, 16, e05259.

Ewers, R.M., Nathan, S.K.S.S., Lee, P.A.K. (2021). African swine fever ravaging Borneo's wild pigs. Nature, 593, 37.

Fisher, M., Henk, D., Briggs, C., et al. (2012). Emerging fungal threats to animal, plant and ecosystem health. Nature, 484, 186–94.

Fisher, M.C., Garner, T.W. (2020). Chytrid fungi and global amphibian declines. Nature Reviews Microbiology, 18, 332–43.

Fitzpatrick, L.D., Pasmans, F., Martel, A., Cunningham, A.A. (2018). Epidemiological tracing of *Batrachochytrium salamandrivorans* identifies widespread infection and associated mortalities in private amphibian collections. Scientific Reports, 8, 1–10.

French, R. (2019). Case of the missing frogs. The Weekend Australian Magazine, 8 June 2019.

Frick, W.F., Cheng, T.L., Langwig, K.E., et al. (2017). Pathogen dynamics during invasion and establishment of white-nose syndrome explain mechanisms of host persistence. Ecology, 98, 624–31.

Friedman, C., Biggs, W., Shields, J., Hedrick, R.P. (2002). Transmission of withering syndrome in black abalone, *Haliotis cracherodii* Leach. Journal of Shellfish Research, 21, 817–24.

Garner, T.W., Schmidt, B.R., Martel, A., et al. (2016). Mitigating amphibian chytridiomycoses in nature. Philosophical Transactions of the Royal Society B: Biological Sciences, 371, 20160207.

Garner, T.W.J., Rowcliffe, J.M., Fisher, M.C. (2011). Climate change, chytridiomycosis or condition: an experimental test of amphibian survival. Global Change Biology, 17, 667–75.

Gilbert, M., Miquelle, D.G., Goodrich, J.M., et al. (2014). Estimating the potential impact of canine distemper virus on the Amur Tiger population (*Panthera tigris altaica*) in Russia. PLoS ONE, 9, e110811.

Gilbert, M., Sulikhand, N., Uphyrkin, O., et al. (2020). Distemper, extinction, and vaccination of the Amur tiger. Proceedings of the National Academy of Sciences, USA, 117, 31,954–62.

Goltsman, M., Kruchenkova, E.P., Macdonald, D.W. (1996). The Mednyi Arctic foxes: treating a population imperilled by disease. Oryx, 30, 251–8.

Grimaudo, A.T., Hoyt, J.R., Yamada, S.A., et al. (2022). Host traits and environment interact to determine persistence of bat populations impacted by white-nose syndrome. Ecology Letters, 25, 483–97.

Harris, R.N., Brucker, R.M., Walke, J.B., et al. (2009). Skin microbes on frogs prevent morbidity and mortality caused by a lethal skin fungus. The ISME Journal, 3, 818–24.

Heard, G.W., Thomas, C.D., Hodgson, J.A., et al. (2015). Refugia and connectivity sustain amphibian metapopulations afflicted by disease. Ecology Letters, 18, 853–63.

Holden, C. (Ed.) (1994). Last bark for Antarctic huskies. Science, 263, 606.

Hopkins, S.R., Hoyt, J.R., White, J.P., et al. (2021). Continued preference for suboptimal habitat reduces bat survival with white-nose syndrome. Nature Communications, 12, 166.

Hoyt, J.R., Langwig, K.E., Sun, K., et al. (2020). Environmental reservoir dynamics predict global infection patterns and population impacts for the fungal disease white-nose syndrome. Proceedings of the National Academy of Sciences, 117, 7255–62.

Jones, M.E., Paetkou, D., Geffen, E., Moritz, C. (2004). Genetic diversity and population structure of Tasmanian devils, the largest marsupial carnivore. Molecular Ecology, 13, 2197–209.

Kat, P.W., Alexander, K.A., Smith, J.S., Munson, L. (1995). Rabies and African wild dogs in Kenya. Proceedings of the Royal Society of London B, 262, 229–33.

Kennedy, S., Kuiken, T., Jepson, P.D., et al. (2000). Mass die-off of Caspian seals caused by canine distemper virus. Emerging Infectious Diseases, 6, 637–9.

Kozorog, M. (2003). Salamander brandy: 'A Psychedelic Drink' between media myth and practice of home alcohol distillation in Slovenia. Anthropology of East Europe Review, 21, 63–71.

Kriger, K.M., Hero, J.M. (2007). Large-scale seasonal variation in the prevalence and severity of chytridiomycosis. Journal of Zoology, 271, 352–9.

Lafferty, K., Kuris, A. (1993). Mass mortality of abalone *Haliotis cracherodii* on the California Channel Islands: tests of epidemiological hypotheses. Marine Ecology Progress Series, 96, 239–48.

Laking, A.E., Ngo, H.N., Pasmans, F., et al. (2017). *Batrachochytrium salamandrivorans* is the predominant chytrid fungus in Vietnamese salamanders. Scientific Reports, 7, 1–5.

Langwig, K.E., Frick, W.F., Hoyt, J.R., et al. (2016). Drivers of variation in species impacts for a multi-host fungal disease of bats. Philosophical Transactions of the Royal Society B: Biological Sciences, 371, 20150456.

Langwig, K.E., Hoyt, J.R., Parise, K.L., et al. (2017). Resistance in persisting bat populations after white-nose syndrome invasion. Philosophical Transactions of the Royal Society B: Biological Sciences, 372, 20160044.

Leroy, E.M., Rouquet, P., Formenty, P., et al. (2004). Multiple Ebola virus transmission events and rapid decline of central African wildlife. Science, 303, 387–90.

Lilley, T.M., Wilson, I.W., Field, K.A., et al. (2020). Genome-wide changes in genetic diversity in a population of Myotis lucifugus affected by white-nose syndrome. G3 Genes | Genomes | Genetics, 10, 2007–20.

Lips, K.R. (2016). Overview of chytrid emergence and impacts on amphibians. Philosophical Transactions of the Royal Society B: Biological Sciences, 371, 20150465.

Longcore, J.E., Pessier, A.P., Nichols, D.K. (1999). *Batrachochytrium dendrobatidis* gen. et sp. nov., a chytrid pathogenic to amphibians. Mycologia, 91, 219–27.

Martel, A., Blooi, M., Adriaensen, C., et al. (2014). Recent introduction of a chytrid fungus endangers Western Palearctic salamanders. Science, 346, 630–31.

Martel, A., Spitzen-van der Sluijs, A., Blooi, M., et al. (2013). *Batrachochytrium salamandrivorans* sp. nov. causes lethal chytridiomycosis in amphibians. Proceedings of the National Academy of Sciences, USA, 110, 15,325–9.

McCallum, H., Jones, M. (2006). To lose both would look like carelessness: Tasmanian devil facial tumour disease. PLoS Biology, 4, e342.

McCallum, H., Jones, M., Hawkins, C., et al. (2009). Transmission dynamics of Tasmanian devil facial tumor disease may lead to disease-induced extinction. Ecology, 90, 3379–92.

McCallum, H., Tompkins, D.M., Jones, M., et al. (2007). Distribution and impacts of Tasmanian devil facial tumor disease. EcoHealth, 4, 318–25.

Meli, M.L., Simmler, P., Cattori, V., et al. (2010). Importance of canine distemper virus (CDV) infection in free-ranging Iberian lynxes (*Lynx pardinus*). Veterinary Microbiology, 146, 132–7.

O'Hanlon, S.J., Rieux, A., Farrer, R.A., et al. (2018). Recent Asian origin of chytrid fungi causing global amphibian declines. Science, 360, 621–7.

Patton, A.H., Lawrance, M.F., Margres, M.J., et al. (2020). A transmissible cancer shifts from emergence to endemism in Tasmanian devils. Science, 370, eabb9772.

Pearse, A.M., Swift, K. (2006). Transmission of devil facial-tumour disease. Nature, 439, 549.

Penn, D.J. (2002). The scent of genetic compatibility: sexual selection and the major histocompatibility complex. Ethology, 108, 1–22.

Puryear, W., Sawatzki, K., Hill, N., et al. (2023). Highly pathogenic Avian Influenza A(H5N1) virus outbreak in New England seals, United States. Emerging Infectious Diseases, 29, 786–91.

Puschendor, R., Hoskin, C.J., Cashins, S.D., et al. (2011). Environmental refuge from disease-driven amphibian extinction. Conservation Biology, 25, 956–64.

Pyecroft, S.B., Pearse, A.M., Loh, R., et al. (2007). Towards a case definition for Devil Facial Tumour Disease: what is it? EcoHealth, 4, 346–51.

Quammen, D. (2008). What's killing the Tasmanian Devil? Yale Environment 360: https://e360.yale.edu/features/whats_killing_the_tasmanian_devil, accessed 14/1/2022.

Randall, D.A., Williams, S.D., Kuzmin, I.V., et al. (2004). Rabies in endangered Ethiopian wolves. Emerging Infectious Diseases, 10, 2214–17.

Rohr, J.R., Raffel, T.R., Halstead, N.T., et al. (2013). Early-life exposure to a herbicide has enduring effects on pathogen-induced mortality. Proceedings of the Royal Society B: Biological Sciences, 280, 20131502.

Salkeld, D.J., Stapp, P., Tripp, D.W., et al. (2016). Ecological traits driving the outbreaks and emergence of zoonotic pathogens. BioScience, 66, 118–29.

Scheele, B.C., Pasmans, F., Skerratt, L.F., et al. (2019). Amphibian fungal panzootic causes catastrophic and ongoing loss of biodiversity. Science, 363, 1459–63.

Sidik, S.M. (2022). Why is bird flu so bad right now? Nature: https://www.nature.com/articles/d41586-022-03322-22, accessed 15/12/2022.

Smith, K.F., Acevedo-Whitehouse, K., Pedersen, A.B. (2009). The role of infectious diseases in biological conservation. Animal Conservation, 12, 1–12.

Smith, K.F., Sax, D.F., Lafferty, K.D. (2006). Evidence for the role of infectious disease in species extinction and endangerment. Conservation Biology, 20, 1349–57.

Terio, K.A., Craft, M.E. (2013). Canine distemper virus (CDV) in another big cat: should CDV be renamed carnivore distemper virus? mBio, 4, e00702–13.

Tripp, D.W., Rocke, T.E., Runge, J.P., et al. (2017). Burrow dusting or oral vaccination prevents plague-associated prairie dog colony collapse. EcoHealth, 14, 451–62.

US Fish and Wildlife Service. (2016). Listing salamanders as injurious due to risk of salamander chytrid fungus. US Fish and Wildlife Service, Washington, DC: https://www.fws.gov/injuriouswildlife/salamanders.html.

Verant, M.L., Boyles, J.G., Waldrep, W. Jr., et al. (2012). Temperature-dependent growth of *Geomyces destructans*, the fungus that causes bat white-nose syndrome. PLoS ONE, 7, e46280.

Walker, S.F., Bosch, J., Gomez, V., et al. (2010). Factors driving pathogenicity vs. prevalence of amphibian panzootic chytridiomycosis in Iberia. Ecology Letters, 13, 372–82.

Weston, P. (2022). The scale is hard to grasp: avian flu wreaks devastation on seabirds. The Guardian: https://www.theguardian.com/environment/2022/jul/20/avian-flu-h5n1-wreaks-devastation-seabirds-aoe, accessed 15/12/2022.

Woodroffe, R., Ginsberg, J.R. (1999). Conserving the African wild dog *Lycaon pictus*. I. Diagnosing and treating causes of decline. Oryx, 33, 132–42.

Woods, G.M., Kreiss, A., Belov, K., et al. (2007). The immune response of the Tasmanian devil (*Sarcophilus harrisii*) and Devil Facial Tumour Disease. EcoHealth, 4, 338–45.

Yaremych, S.A., Warner, R.E., Mankin, P.C., et al. (2004). West Nile virus and high death rate in American crows. Emerging Infectious Diseases, 10, 709–11.

Zukal, J., Bandouchova, H., Brichta, J., et al. (2016). White-nose syndrome without borders: *Pseudogymnoascus destructans* infection tolerated in Europe and Palearctic Asia but not in North America. Scientific Reports, 6, 19,829.

# CHAPTER 13

# Infectious diseases in ecosystems

## 13.1 Communities and ecosystems

Ecologists have a system for organizing processes by scale. At the lowest ecological scale, there are individuals—single living organisms of any type. A group of individuals of the same species living in the same place at the same time is a **population**. All the populations of all the species living in the same place at the same time are a **community**. And the biotic community combined with the abiotic conditions in that place are the **ecosystem**. In this chapter, we will focus on processes occurring at the community and ecosystem scales.

Most disease ecology research focuses on a single pathogen in a single host species or a single pathogen in a multi-host community. However, every host is a little world for microbes, so most hosts are simultaneously **co-infected** with multiple parasites and beneficial symbionts. Furthermore, host species interact with many other species besides parasites, including predators, prey or resources, competitors, and vectors. This reminds us that while it may be easier to focus ecological studies on single hosts and pathogens, a myopic view might cause us to miss important details in complex systems.

Before diving into those complex details, let's define a few more terms commonly used to describe ecological interactions between species. **Direct interactions** occur when one species impacts a second species, and the impact is not mediated or transmitted by a third species (solid lines in Fig. 13.1). For example, prey abundance often directly impacts predator foraging success, reproduction, and population sizes. In contrast, **indirect interactions** occur when the impact of one species on a second species is mediated by a third species (dashed lines in Fig. 13.1). For example, if a pathogen that only infects the prey population reduces the size of the prey population, predators might experience an indirect negative effect of the pathogen: the pathogen causes a reduction in predator's foraging success, reproduction, or survival, even though the pathogen and predator never directly interact. The following two sections will contain many specific examples of pathogens' indirect effects within ecosystems.

## 13.2 Bottom-up effects

As we've discussed in previous chapters, infection and disease do not always occur when pathogens and hosts overlap; disease dynamics often depend on environmental conditions like temperature and precipitation (i.e. the disease triangle concept; Chapters 2 and 12). Furthermore, environmental conditions don't just affect hosts and pathogens, but *all* species and their interactions—plants, herbivores, carnivores, etc. For example, mild weather conditions in the arid south-western US increase primary plant production. Plants are not part of the hantavirus disease system, but plants do provide food for deer mice, the reservoir host for hantavirus (Chapter 8). Therefore, environmental conditions that increase primary productivity also increase rodent population density, hantavirus prevalence in mice, and risk and incidence of Hantavirus Pulmonary Syndrome in humans living in rural areas (Parmenter et al. 1999, Yates et al. 2002). This is an example of a **bottom-up effect**—when a lower trophic level (plants) impacts a higher trophic level (rodents). Bottom-up effects like these can ripple all the way up food chains, such that primary produces

*Emerging Zoonotic and Wildlife Pathogens.* Dan Salkeld, Skylar Hopkins, and David Hayman, Oxford University Press.
© Oxford University Press (2023). DOI: 10.1093/oso/9780198825920.003.0013

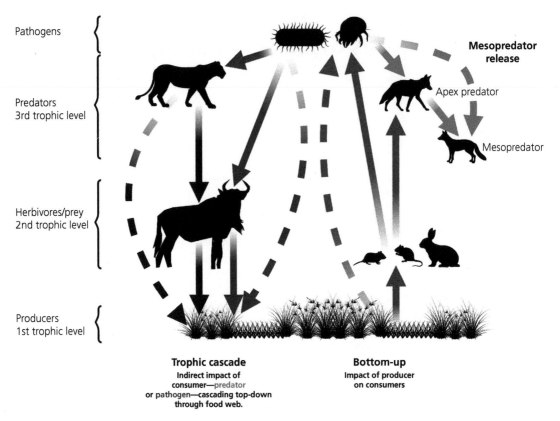

**Figure 13.1** Stylized examples of top-down and bottom-up effects in ecosystems. The solid lines indicate direct interactions between adjacent trophic levels, such as predators consuming prey. The dashed lines indicate indirect interactions that 'skip' a trophic level, such as when pathogens of a consumer indirectly impact the consumer's resource population. Note that pathogens are not always the highest trophic level in ecosystems; they are depicted this way for simplicity here.
This figure was created using BioRender.com.

affect primary consumers, which in turn affect secondary and tertiary consumers.

*Parque Nacional G.D. Omar Torrijos Herrera, Panama*—The amphibian chytrid fungus, *Batrachochytrium dendrobatidis* (see Chapter 12), was first observed in a national park in central Panama in 2004 in mature, undisturbed secondary forest. Fortunately for science, nigh-on 600 surveys for amphibians and reptiles had occurred on seven permanent transects during the seven years before the chytrid's arrival. After the invasion began, more than 500 surveys were conducted on the same transects during the next six years. Before the epizootic, the amphibian community at the study site comprised >70 species. During the epizootic,

more than 30 species were extirpated, and overall amphibian abundance declined by >75% (Crawford et al. 2010).

The impacts didn't stop there. Many snake species prey on frogs and frog eggs, and without their frog prey, snakes suffered. Snakes can often survive without meals for long periods, but snakes that preferred amphibian diets became increasingly emaciated during the five years post-invasion, suggesting that those snakes were struggling to find prey. Correspondingly, snake abundance declined, likely as starving snakes began to die or move to new areas (if possible). As just one example, observation rates for a snake called the Argus snail sucker (Fig. 13.2) were only a third of what they had

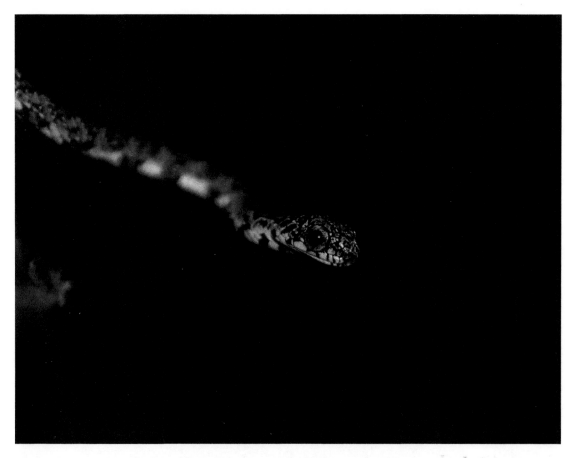

**Figure 13.2**  An Argus snail sucker (*Sibon argus*) in Costa Rica, one snake species that consumes frog eggs.
Photograph by Lucas Vogel, used under a CC BY-SA 4.0 license.

been pre-chytrid. Overall, the total number of snake species observed on the transects declined from 30 to 21 (Zipkin et al. 2020). Several rare species may now be locally extirpated, but it is difficult to be sure, because snakes are sneaky and cryptic and thus difficult to find even when they are abundant (Chapter 9).

Although most Panamanian snake species were negatively affected by the decimation of amphibians, some species fared better and even increased in abundance or body condition. In other words, the fungus outbreak indirectly produced many 'loser' snake species but also a few 'winners', an ecological phenomenon frequently observed after disturbance. The impact on snake ecology likely also has further knock-on effects throughout the local food

web, but they remain undescribed; for example, snakes are prey for many other species, like birds of prey, and it is unclear how those apex predators have been impacted. This sad episode demonstrates the indirect and cascading bottom-up effects of infectious agents upon ecological communities.

## 13.3 Top-down effects: mesopredator release

**Apex predators**—predators at the 'top' of a food web—can disappear because of hunting, extermination campaigns, land use change, and of course, deadly infectious diseases. These changes to food webs can have effects that trickle from 'the top' (i.e. predators) down, which are helpfully called **top-down effects**.

*Tasmania, 1970s to 2000s*—Ever since the thylacine or Tasmanian tiger (*Thylacinus cynocephalus*) went extinct, the apex mammalian predator in Tasmania has been the pugnacious Tasmanian devil (*Sarcophilus harrisii*, which translates to Harris's flesh-lover). Tasmanian devils traditionally prey upon Tasmanian pademelons (*Thylogale billardierii*, Fig. 13.3) and the common brushtail possum (*Trichosurus vulpecula*). However, these ecosystems also contain a variety of other mammalian species (Fig. 13.4),

including the domestic cat (*Felis catus*), which is sometimes consumed by Tasmanian devils. Considering the many predator and prey species interactions, we would expect a sudden decline in the apex Tasmanian devils to have direct or indirect effects on the rest of the community.

Starting in 1996, devil facial tumour disease (DFTD) caused a progressive disease-induced decline in Tasmanian devils that spread from the north-east to the south and west. To study whether this had cascading impacts on the mammal community, researchers needed a data set that

**Figure 13.3** A Tasmanian pademelon (*Thylogale billardierii*) at Mount Field National Park, Tasmania, Australia.
Photo by JJ Harrison, on Wikimedia, under a CC BY-SA 3.0 license. https://commons.wikimedia.org/w/index.php?curid=6976776

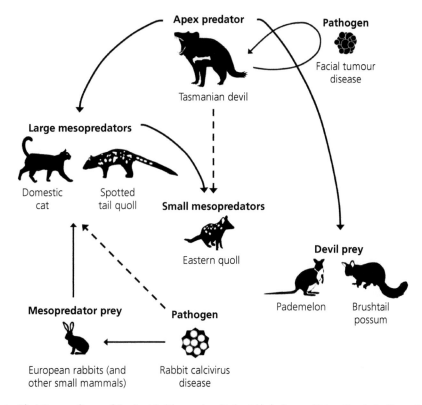

**Figure 13.4** A simplified diagram of some of the direct (solid arrows) and indirect (dashed arrows) interactions in the Tasmanian ecosystem (simplifying a similar diagram from Hollings et al. (2014)). Tasmanian devil facial tumour disease caused a top-down effect, where reduced devil populations caused mesopredator release, indirectly reducing eastern quoll populations. Rabbit calicivirus disease caused a bottom-up effect, where the virus indirectly negatively impacted mesopredators through impacts on the rabbit population. The impacts of environmental conditions are not shown here.

could disentangle the effects of DFTD from other variables that affect mammal community ecology across Tasmania, such as vegetation, rainfall, and invasive species. Therefore, researchers compared mammal communities across north-eastern Tasmania, where the disease has been present the longest and where rainfall and forest cover are high, with mid-Tasmania, which has less rainfall, more agricultural land, denser human settlement, and higher feral cat populations.

Of course, such a comparison would not have been possible without long-term data, which fortuitously existed. Annual 'spotlighting surveys' were initiated in Tasmania in 1975 to assess the effects of harvesting Tasmanian pademelons, brushtail possums, and Bennett's wallabies for meat, fur, and crop protection. Spotlighters drove along 10 km transects at a constant speed (25 km/h), scanning

with spotlights to detect glowing eyeshine of the marsupial herbivores. Thankfully, spotlighters also recorded sightings of other mammal species spotted incidentally, and thus there are (admittedly imperfect) data on mammal abundance from more than a decade before the emergence of devil facial tumour disease. The 1985–2008 data covered regions where DFTD arrived earliest (north-east), where DFTD arrived later (mid-Tasmania), and where devils were still disease-free.

We might have expected pademelons and brushtail possum populations to directly increase as their predators' populations declined. But instead, slight decreases of both followed devil facial tumour disease's arrival, and that probably had nothing to do with DFTD; rainfall patterns were more influential and there wasn't enough harvesting/culling data to rule out harvesting/culling as a cause of the population dips. Impacts of DFTD also weren't obvious for several of the smaller mammals (bandicoots, potoroos, rabbits), perhaps in part because spotting them from a moving vehicle was less likely in the first place.

In contrast, the spread of devil facial tumour disease clearly impacted feral cats and, in turn, eastern quolls (*Dasyurus viverrinus*). Previously, on individual transects, eastern quolls were strongly positively associated with devils and were detected less frequently when feral cats were present. As devil populations declined early in north-eastern Tasmania, feral cats were 'released' from predation pressures and their populations increased. In turn, eastern quoll numbers declined, and these impacts on the quoll population lagged somewhat behind Tasmanian devil declines. In mid-Tasmania, feral cats had higher abundance before DFTD spread to the region, and feral cats were less impacted by the disappearance of Tasmanian devils; feral cat abundance was better explained by rainfall, vegetation type, prey abundance, and human presence. But quoll numbers still declined, and this time they dropped simultaneously with DFTD's arrival. Therefore, it seems that eastern quolls are indirectly protected from feral cats by Tasmanian devils. This is an example of **mesopredator release**, a type of top-down effect where the decline or removal of a top predator increases populations of small or medium-sized predators (also called mesopredators), creating indirect, knock-on impacts for prey species of the mesopredators (Hollings et al. 2014).

Meanwhile, invasive European rabbits (*Oryctolagus cuniculus*) were involved in their own disease dynamics, where populations are declining due to rabbit calicivirus disease (RCD). In the 'mid region' where this was most obvious, feral cat populations also declined, because rabbits serve as cat prey. So, in addition to top-down effects, this system also has bottom-up effects: disease affecting prey affecting predators! (This effect was not evident in the early region, which had lower densities of rabbits to start with.)

## 13.4 Top-down effects: trophic cascades

Traditionally, and strictly, **trophic cascades** describe indirect top-down regulation of productivity,

abundance, or biomass at one trophic level by higher-level consumers at least one trophic level removed. For example, the impact of predators upon the density or behaviour of herbivores, which trickles down to indirectly impact plant abundance (Stapp 2007, Buck and Ripple 2017, Monk et al. 2022). Most ecological research regarding trophic cascades has focused on similar scenarios and free-living consumers, like apex predators. However, like predators, pathogens can also occupy high trophic levels and may therefore cause trophic cascades.

*Serengeti, Tanzania*—Rinderpest (German for 'cattle plague') is a disease caused by a morbillivirus (family Paramyxoviridae), like the human measles virus and canine distemper virus. Rinderpest virus (RPV) can infect a wide range of hosts—predominantly cattle, but also more than 40 other artiodactyl species (artiodactyls are the 'even-toed ungulates' and include pigs, camels, deer, giraffes, antelopes, sheep, goats, and cattle). Most cattle (80–90%) die within 10 days of RPV infection, which causes fever, discharges from the eyes and nose, diarrhoea, and dehydration. Outbreaks devastated cattle herds and thus caused awful famines in livestock-dependent human populations before 2011.

To control rinderpest, local and global vaccination campaigns focused on cattle, because cattle were important to humans and were the main reservoir host facilitating RPV spillovers into wildlife species. Rinderpest virus was officially declared globally **eradicated** on 25 May 2011—only the second disease that humans have deliberately eradicated (smallpox was declared vanquished in 1980). In the Serengeti, control efforts had locally extirpated rinderpest in the 1960s, and the ecosystem changed dramatically after that.

In the pre-1960s Serengeti, rinderpest was particularly deadly to blue wildebeest calves (*Connochaetes taurinus*, also known as common wildebeest, white-bearded gnu, or brindled gnu; Fig. 13.5). In fact, wildebeest populations were regulated by the virus, where host abundance was impacted by virus dynamics. After rinderpest was extirpated, wildebeest populations boomed from ~200,000 beests to more than a million (Fig. 13.5). That's a lot of large animals (140–290 kg each) grazing grass.

**Figure 13.5** (top) Wildebeest in Serengeti National Park, Tanzania. (bottom) As the seroprevalence of rinderpest in Serengeti wildebeest declined from 1958–1963 (open boxes), the wildebeest population increased (filled circles). The open triangles are rinderpest seroprevalence estimates (all 0%) from wildebeest as measured in a separate study during a small outbreak in buffaloes in the 1980s. (Left) Photo by Kai Pütter under an Unsplash license. (Right) Graph by Holdo et al. (2009) under a CC BY 3.0 license.

As wildebeest ate more grass, they reduced the likelihood of fires by reducing available fuel loads (Holdo et al. 2009). Grass abundance and fire regimes were further influenced by intra-annual variation in rainfall. As fire frequency declined, more saplings could be recruited and more trees could persist, leading to increased tree density. The changing landscape was brilliantly documented by serial photographs of landscape plots spanning half a century (1960–2003) (Holdo et al. 2009), which could be statistically correlated to fire frequency (Fig. 13.6). No alternative hypotheses—e.g. did declining elephant populations allow trees to flourish?—received the same statistical or logical support as the wildebeest–grass–fire–tree cascade that occurred after rinderpest was eliminated.

This example is about wildebeest, but we can imagine how it might play out in many systems: a pathogen controls herbivore (grazer) population size, which controls plant biomass, which affects fire patterns, which affects tree population dynamics. As Holdo et al. (2009) note: 'seemingly small ecological perturbations such as disease outbreaks have

the potential to profoundly affect ecosystem function'. Indeed, after RPV was eradicated, increased tree density may have shifted the Serengeti from being a net source for carbon to a net carbon sink! In other words, for tree density, no gnus is not good gnus.

*The Andes, Argentina*—The high alpine Andes wildlands are inhabited by elegant vicuña (*Vicugna vicugna*) (Fig. 13.7) that graze grasses, rushes, and sedges, and are preyed upon by pumas (*Puma concolor*). Vicuña carcasses also constitute an important food resource for Andean condors (*Vultur gryphus*)—gigantic scavengers with wingspans of 3.3 metres that can reach 15 kg in weight, making them one of the largest and heaviest flying birds in the world. Until recently, there were robust populations of vicuña, pumas, and condors at San Guillermo National Park, a remote, undisturbed, protected area in north-western Argentina.

Vicuña used to heavily graze the park's vast open plains, in part because the open environment made them safer from puma predation. Exclosure experiments (fencing out the vicuña) showed that they

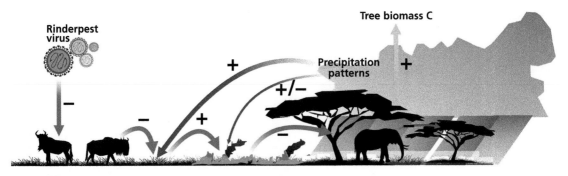

**Figure 13.6** Red arrows denote a four-step pathway of causality linking rinderpest virus with tree population dynamics. Rinderpest virus negatively impacted wildebeest populations. Wildebeest abundance and rainfall impact grass biomass. When wildebeest populations increased after rinderpest was eradicated, the reduced (grazed) grass provided less fuel for wildfires. Fewer wildfires allowed more saplings to grow and increased tree density, which can affect tree biomass carbon.
Created in BioRender based on a similar diagram in Holdo et al. (2009).

**Figure 13.7** Vicuña posing in front of Chimborazo volcano, Ecuador (not the location where the mange outbreak occurred, which was in Argentina).
Photograph by David Torres Costales, from Wikimedia under a CC BY-SA 3.0 license.

reduce vegetation biomass by 85% and plant cover by 50%. The vicuña also use canyons and meadows to access water, but this is dangerous big cat territory, with rocky terrain and tall vegetation that makes predators hard to see. Imminent death lurking behind every rock makes the vicuña nervously vigilant in these habitats, and they seem to lose their appetites, grazing less in the canyons and meadows and having less impact upon vegetation. Therefore, the pumas drive vicuña behaviour and foraging, concentrating their grazing in open plain environments.

These details had all been worked out by researchers who radio-collared pumas and condors and used regular vehicular surveys to monitor patterns of vicuña density from 2004 to 2020. (Where are these fun jobs advertised?!) During their long-term study, in August 2015, sarcoptic mange (*Sarcoptes scabiei*) began to ravage the vicuña herds (Fig. 13.8). Sarcoptic mange is caused by a dastardly mite that parasitizes the host's skin, and these mites probably (not definitely) spilled over into the vicuña herds from domestic llamas (*Lama glama*) (Ferreyra

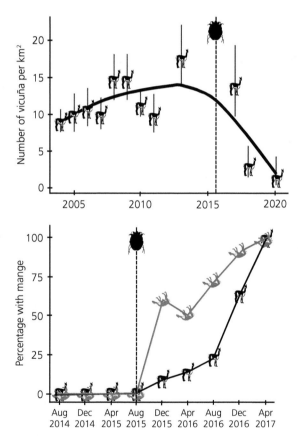

**Figure 13.8** After the mange epizootic began in August 2015, the mean vicuña population density declined (top) as the percentage of living (black) and dead (red) vicuña with mange increased (bottom). The vertical purple bars are 95% confidence intervals for the mean. This figure was modified from Monk et al. (2002).

et al. 2022). The mites are extremely itchy and cause lesions, alopecia (hair loss), and hyperkeratosis (thickening of the outer layer of the skin, which is made of keratin), which can lead to secondary infections, weakened immune systems, and thermal stress. In severe cases, infestations can result in death. Vicuña numbers in the park declined from the thousands, to fewer than 100.

Local condor activity also plummeted after the mange outbreak. This effect lagged behind vicuña declines, because condors probably took advantage of mange-killed vicuña carcasses in the early stages of the disease outbreak. But fresh carcasses were rare after 2017, so condors sought pastures new.

Astonishingly, after the mange outbreak, field measurements and remote sensing data revealed that plant biomass (as well as cover and height) increased dramatically in the plains where herbivory had previously been high (Fig. 13.9). Grass biomass increased 900%! In contrast, plant biomass was relatively constant in the puma-risky canyons and meadows—where grazing pressure had never been high. This suggests that the temporal change in plant biomass in the plains was not simply a response to unusual patterns of rainfall and could be attributed to altered vicuña abundance and grazing habits.

This chapter contains several dramatic and clear examples of infectious diseases having cascading bottom-up or top-down effects in food webs. These examples only exist because scientists happened to be conducting long-term studies *before* epizootics began, such that there were well-established baseline data to compare to the post-epizootic ecological community (Zipkin et al. 2020). For example, the world may never have known that chytrid in amphibians had a dramatic impact on frog-eating snakes if snakes hadn't been monitored before and during the chytrid epizootic in Panama (Section 13.2). Furthermore, not just any long-term data set will be sufficient; the data set must allow researchers to observe the knock-on trophic effects *and* to rule out alternative hypotheses (e.g. changing abiotic conditions, anthropogenic disturbance), confirming that observed trophic effects were caused by infectious disease dynamics (Monk et al. 2022). Data sets that can do all of this are both rare and extremely valuable in our understanding of the

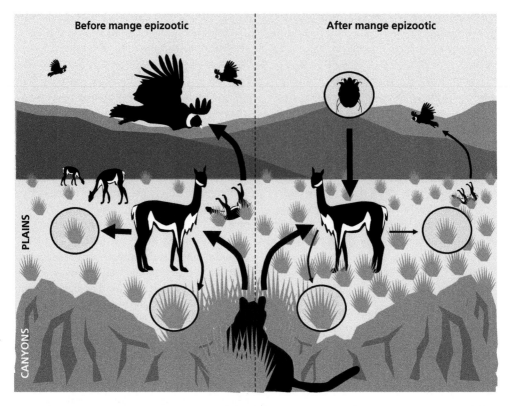

**Figure 13.9** Before the mange epizootic, vicuñas heavily grazed plains habitats, where puma predation pressures were low, and lightly grazed canyon habitats, where rocks provided better cover for pumas. Given the abundant vicuña population, condors had a consistent supply of carcasses. During and after the mange epizootic, there were few surviving vicuñas, causing the grass biomass to greatly increase in the plains habitats (but not in the canyons and meadows). Vulture sightings also declined after the mange epizootic. The thickness of the arrow indicates the strength of the impact of species from one trophic level to another.
Modified from a figure published by Monk et al. (2022).

impacts of pathogens on ecosystems! This reminds us that while most scientific studies are relatively short (e.g. one field season or the time it takes a student to get their PhD), more long-term studies are needed to understand how communities and ecosystems are responding to the changing world.

It is also worth noting that all the examples above involved infectious diseases that caused high mortality upon their host organisms. Dramatic wildlife mortality events and their effects on ecosystems may be relatively easy to observe (though still not easy!), but trophic cascades may also be triggered by infectious diseases that do not typically cause the deaths of their hosts (Buck and Ripple 2017, Koltz et al. 2022). For example, sublethal

infections of parasitic worms can reduce appetite in their hosts—parasite-mediated anorexia. If sublethal infections are common in herbivorous hosts, they may add up to a landscape-sized effect. For example, mathematical models parameterized with data from reindeer (*Rangifer tarandus*—referred to as caribou in North America) infected with gastrointestinal helminth worms (*Ostertagia* spp.) suggested that sublethal infections could reduce reindeer feeding rates enough to trigger a trophic cascade (Koltz et al. 2022). Is a world with more sublethal parasitic infections a greener world? Have sublethal parasitic infections been overlooked as candidate drivers of ecosystem-level processes? If so, this might be an argument for conserving, not just expunging, some parasite species.

## Box 13.1  Nematomorphs and trophic cascades

Nematomorphs, also called Gordian worms or horsehair worms, are worm-like parasites in the phylum Nematomorpha. Like many other parasites, nematomorphs have complex life cycles, where individuals must successfully move through the world as eggs, larvae, cysts, and then reproductive adults (Fig. 13.10). But unlike many of the worm-like parasites that we have previously illustrated in life cycle diagrams, adult nematomorphs spend time as free-living organisms outside of a host. In particular, the adult worms leave their hosts and swim through aquatic environments, where they find mates and reproduce.

Many adult nematomorphs find themselves in a position *worse* than up the creek without a paddle; they aren't in a creek at all. Before emerging as adults, nematomorphs can often be found living inside a *terrestrial*, ground-dwelling insect, like a cricket or beetle. This alone is amazing, because the adult worms are relatively huge and can take up most of the space inside the insect's body! But perhaps more amazingly, nematomorphs appear to have the ability to manipulate their host's behaviour. For example, crickets and grasshoppers infected with nematomorphs are 20 times more likely to jump in Japanese headwater streams than uninfected crickets and grasshoppers (Sato et al. 2011). (Note that uninfected crickets do sometimes jump or fall into streams, but this seemingly maladaptive behaviour is *more* likely in infected crickets.) After an infected cricket enters the water, the unbelievably long nematomorph exits the cricket, like a scene from a horror movie.

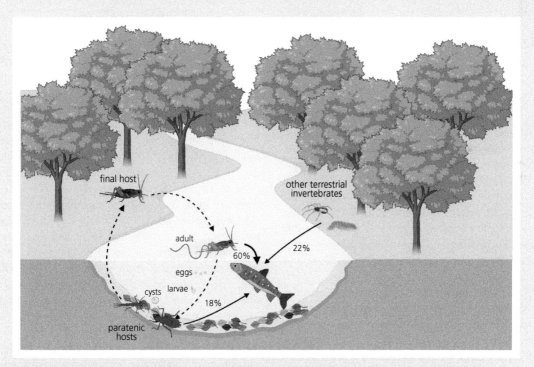

**Figure 13.10** Nematomorphs use crickets and other terrestrial insects as final hosts. After manipulating the host to jump into water, nematomorphs exit, find a mate, reproduce, and lay eggs. Later, the larvae encyst in aquatic insect larvae, which eventually emerge as adults, die, and are eaten by crickets and other insects, completing the life cycle (dashed arrows). Infected crickets that jump into Japanese headwater streams are a substantial diet component for Kirikuchi char—more so than other terrestrial insects that fall in or benthic invertebrates (solid arrows and diet percentages).
Figure remade from Sato et al. (2011) using BioRender.com and some animals drawn by Skylar Hopkins.

---

**Box 13.1** *Continued*

Remarkably, infected crickets *can* survive having a huge worm stuffed inside their body. But they *cannot* survive being eaten by hungry char (a fish). This is the fate of many of the crickets that so unwisely jump into Japanese headwater streams. In fact, infected crickets comprised 60% of the annual energy intake of the endangered Kirikuchi char (*Salvelinus leucomaenis japonicus*) in Japanese headwaters! It is even possible that this endangered fish species could not persist without the extra crickets that jump into streams because of parasite manipulation (Sato et al. 2011). (Note that the nematomorphs often get eaten along with the crickets, as described by Sato et al. in 2012: 'The worms, together with their cricket hosts, are vulnerable to aquatic predators, such as fishes and frogs, although the worms usually survive attacks on the crickets by squirming out the mouth, gills or anus'.)

During 'nematomorph season', when crickets and grasshoppers are abundant inputs to headwater streams, char eat fewer stream insects, like stonefly or mayfly larvae (Sato et al. 2011). This trophic cascade continues to trickle down further in the food web. When there are more algae and leaf-eating benthic stream insects, algae and leaves break down slightly faster. These details were all carefully documented in a field experiment that either excluded all terrestrial insects from entering streams, excluded only infected crickets from entering streams, or added crickets to the stream at levels consistent with an abundant nematomorph site (Sato et al. 2012). This is a clear example of how parasites that change host behaviour can have dramatic impacts on ecosystems.

## 13.5 Parasites in food webs

In the preceding sections, we mostly considered the impacts of a single parasite or pathogen (RPV, DFTD, mange mites, chytrid fungus) on hosts at a single trophic level (e.g. herbivores, apex predators), and how those impacts subsequently rippled through the rest of the food web. But of course, infectious agents are not confined to hosts in single trophic levels in food webs; they are ubiquitous! In fact, parasitism is the most common trophic strategy for animals (Dobson et al. 2008) and has independently evolved in animals more than 200 times, so diverse and abundant parasites occur across 15 phyla (Weinstein and Kuris 2016) and across food webs.

Food webs are usually visually represented as a diagram or 'ecological map' where each species is represented by a node (represented by a circle, point, square, or label) and trophic relationships between species are represented by links (sometimes called edges and represented by lines). For example, a simplified version of the vicuña food web would have nodes for the vicuñas' favourite grass to eat, vicuñas, pumas, and Andean condors, and there would be a line from grass to vicuñas, a line from vicuñas to pumas, and a line from vicuñas to condors. This simple food web already contains

several trophic levels—primary producers (grass), herbivores (vicuñas), predators (pumas), and detritivores (condors). We could also include the parasites that infect all these species, like the mange mites infesting vicuñas.

Parasites were historically excluded from early food web studies, and they continue to be left out of some food web studies in the present day (Fig. 13.11). For example, only 7% of marine food webs constructed by researchers—of which there are more than 100!—contain parasite species (McLaughlin et al. 2020). Parasites are neglected for several reasons: they are small and thus easy to overlook; they often cannot be observed without dissecting hosts, which is not always a feasible study design for an entire food web; they can be difficult to identify without specialized taxonomic expertise; and they are often assumed to be less important than other trophic groups, like predators. But as we've seen in the preceding examples, parasites *can* have strong influences on ecosystems!

When we compare food webs that exclude versus include parasites, we see different views of ecological systems. Two things are immediately obvious: (1) parasites are involved in many food web links, such that food webs with parasites included are far more connected than food webs

(a)                                          (b)

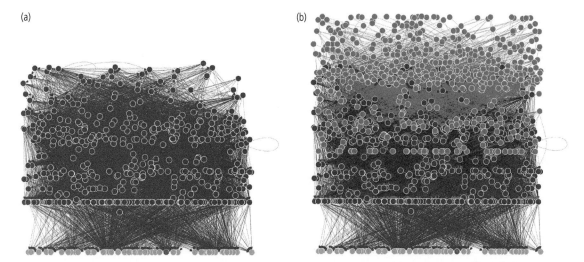

**Figure 13.11** Two detailed kelp-forest food webs, where circles represent species and lines represent trophic relationships between species. The left food web excludes parasites, and the right food web includes parasites. Point colours indicate each species' trophic strategy: red = parasites, blue = free-living consumers, and green = autotrophs (primary producers). The brown points indicate detritus.
This figure and all the food web data were published by Morton et al. (2021) under a creative Commons Attribution 4.0 International License.

that exclude parasites (Lafferty et al. 2006); and (2) when the abundance of each species is quantified, it becomes clear that parasites have a large total biomass. For example, in three estuaries in California, there was more parasite biomass than top predator biomass, and trematode parasites had especially high biomass—more than birds, fish, and other abundant groups (Kuris et al. 2008)! Therefore, when we exclude parasites from food webs, we underestimate the diversity, biomass, and connectedness of ecosystems (Box 13.1).

Food webs that include versus exclude parasites also vary in **stability**. Stability is most simply quantified by removing one species from the food web and counting how many other species will be lost when their resources disappear. This method makes a few big assumptions: consumers do not go extinct until their resource is completely gone; consumers can persist if any resource species remains, even if it was previously a small part of the diet; and consumers cannot 'switch' to other resources that they have never been observed consuming before. As long as we remember that these assumptions are unlikely to hold up in real ecosystems, extinctions in food webs can be simulated to understand how stable they might be given ecological change. Simulations like these reveal that food webs that include

parasites are *less* stable, which might seem counterintuitive! Parasites decrease stability because parasites themselves are often reliant on several specific host species to complete their life cycles. When one host species is lost, several host-specific parasite species may be lost with it. If parasites are really so prone to co-extinction in a changing world, efforts to conserve parasites along with their threatened host species may be important.

## 13.6 Ecosystem functions performed by parasites

In this book about emerging infectious diseases, we spend many chapters describing the big problems that a few pathogens cause and how to prevent or control their spread. This might lead you to believe that all infectious agents are 'bad' and need to be controlled. But beyond the few thousand control-worthy pathogen species that infect humans and our domestic species or cause major wildlife population declines (Carlson et al. 2020a), there are perhaps millions of parasite species that are natural parts of ecosystems.

In fact, a 'healthy' ecosystem is probably one that contains many diverse parasite species. For example, in a salt marsh in California, there were some degraded regions that were targeted for a restoration project—restoring tidal flow by constructing new channels and planting native vegetation. Before restoration began in 1997, snails in the degraded regions had lower trematode parasite prevalence and species richness than nearby 'control' sites that were relatively undisturbed (Huspeni and Lafferty 2004). During the six years of marsh restoration, trematode prevalence and richness in snails increased in the restored sites and were even higher than the control site in the sixth year. This increase likely occurred because as the salt marsh was restored, birds increasingly used the habitat, and birds are definitive hosts for many trematodes that use snails as intermediate host species. Perhaps then, parasites can be bioindicators of ecosystem health, as their presence demonstrates that diverse and trophically connected host species are also in the ecosystem (Box 13.2).

As we have shown, parasites and pathogens play important roles in ecosystems (Box 13.2). They can also provide **ecosystem services**—ecological functions provided by a species that benefit humans.

---

### Box 13.2 Parasites as bioindicators

Parasites may be useful **bioindicators** of ecosystem health—the species richness, prevalence, or abundance of parasites can be measured to distinguish between degraded ecosystems and ecosystems with intact food webs. Parasites may also be useful bioindicators for the presence or ecology of individual host species.

The diamondback terrapin (*Malaclemys terrapin*) (Fig. 13.12) is a small turtle species that is vulnerable to

**Figure 13.12** A diamondback terrapin on land—the only easy way to photograph them!
Photograph by Ryan Hagerty, published on Wikipedia in the public domain.

---

**Box 13.2** *Continued*

extinction, with declining populations in the eastern United States (Maerz et al. 2018). To monitor and conserve this species, researchers and managers need to be able to find terrapins and estimate how many are present in various habitats. This is quite tricky for a turtle that spends much of its time underwater! Fortunately, diamondback terrapins have a species-specific trematode parasite, *Pleurogonius malaclemys*, that only uses the terrapins as final hosts and can thus be used as a reliable bioindicator of terrapin presence. The trematode uses mud snails, *Tritia obsolete*, as first and second intermediate hosts; the larval parasites first multiply in the snail's gonads and then move outside the snail, where they encyst on the snail's shell and operculum. Later, the snail and these 'metacercarial pearls' may be consumed by a terrapin, completing the life cycle. Researchers can eavesdrop on this life cycle to learn more about terrapins and have demonstrated a strong correlation between the abundance of terrapins derived from mark-recapture surveys and the abundance of *Pleurogonius malaclemys*

'pearls' (Byers et al. 2011). In fact, it is so easy to sample snails and look for the trematode 'pearls' that there is a citizen science programme that uses this method to inform estimates of terrapin abundance in New York (Eugene et al. 2021).

> Operculum (/ōˈpərkyəl(ə)m/)—a structure that closes or covers an aperture, including a secreted plate that closes the aperture of a gastropod mollusc's shell when the animal is retracted.

Not all parasites will be useful bioindicators and not all host species need parasites as bioindicators for their monitoring or conservation. Parasites are most likely to be useful bioindicators for hosts that are difficult to monitor (e.g. terrapins) and ecological interactions that are difficult to observe. Whereas ecological interactions are usually brief, parasites may persist in the host for weeks to years, preserving a record for those brave enough to peek inside the guts of a dead animal.

---

For example, parasitoid wasps lay their eggs inside insect hosts, including major agricultural pests, and this function is worth *billions* of dollars in pest control annually just in the United States. Pest control by parasites suggests that losing parasites from ecosystems could have large and unexpected consequences on ecosystem dynamics (Wood and Johnson 2015).

Infectious agents can also be important for individual hosts in the development of immunity. Exposure to relatively benign parasites may be important for hosts developing their immune systems, and may be critical for preparing hosts raised in captivity for exposure to parasites upon eventual release into the wild, where parasites are common (Spencer and Zuk 2016). Or, host species may actually benefit from **co-infection**, when hosts are simultaneously infected by multiple parasite species (Section 13.7). For example, one pathogen species may be able to prevent another, more virulent infectious agent from establishing, or reduce its ability to replicate, providing some protection to the host from the second species. Though these scenarios are still poorly understood, they suggest that trying to eliminate all the parasites from endangered wildlife or captive

populations may have unexpected negative consequences for the hosts.

Since pathogens are highly diverse, important in ecosystems, and vulnerable to extinction (or co-extinction with hosts), parasite/pathogen conservation is a growing subfield within invertebrate conservation. To conserve parasites, we first need to learn more about them, because at present, most parasite species have yet to be described by science, and even among the described species, we often lack even the most basic details about their life cycles, ecology, and functions (Carlson et al. 2020b). Addressing those data gaps will take a lot of work, but will be important for recognition of the ecological role of pathogen diversity, and the importance of parasite conservation, in the natural world.

## 13.7 Co-infection

When you imagine an ecosystem, you probably imagine something vast, like an ocean full of marine creatures or a jungle full of diverse plants, animals, and fungi. But you may be biased by your relatively large mammalian form, which determines how far you can travel and how much you can observe in an hour or day. From a different perspective—say, the perspective of a single bacterium—an individual host could be an entire ecosystem, complete with variable environmental conditions and interacting species.

For hosts, **co-infection** is the normal state. Humans, animals, plants, fungi, protists, and even bacteria and archaea are simultaneously colonized by a range of **symbionts**: parasites, beneficial mutualists, and commensal symbionts that seem to neither benefit nor harm the host. For example, most people worldwide have herpes simplex virus-1, herpes simplex virus-2, or both (Wald and Corey 2007); these are lifelong infections that often cause few symptoms, but some people have extremely uncomfortable symptoms, including genital lesions. Similarly, in the US, most people (80%) will become infected by a human papillomavirus by the age of 45 years (Chesson et al. 2014). Given these high prevalences of infection, it is nearly guaranteed that whenever we are infected by another virus (e.g. a cold virus), we will be *co-infected* by at least two potentially pathogenic viruses (e.g. the cold virus *and* herpes simplex virus-1). And that's just viruses! Every host is an ecosystem full of symbionts.

Given that so many species are living on and in the host simultaneously, it is no surprise that many of them interact. Interactions between co-infecting symbionts may benefit the host, harm the host, or have no noticeable impact at all (Griffiths et al. 2011, Buffie and Pamer 2013). However, since we cannot see most parasites and pathogens, it is difficult to observe and categorize their direct or indirect interactions (Pflughoeft and Versalovic 2012). It can be especially difficult to try to deduce interaction strengths or directions from co-infection patterns in nature. For example, if we never saw Parasite B co-infecting hosts with Parasite A, we might hypothesize that when Parasite A colonizes a host first, it prevents Parasite B from establishing through antagonistic interactions (i.e. priority effects). But there could also be many alternative hypotheses. For example, perhaps hosts that are exposed to Parasite A never visit the habitats where they could be exposed to Parasite B, and thus the lack of co-infections has nothing to do with parasite interactions. In many cases, interactions between

co-infecting parasites can only be disentangled using careful experiments.

Some of the best-known interactions between co-infecting species come from human infectious diseases, which tend to receive the most research effort. For example, around 14 million people worldwide are co-infected with *Mycobacterium tuberculosis* (the bacterium that causes tuberculosis) and HIV (Pawlowski et al. 2012). HIV compromises the immune system and thereby increases the risk of TB infection (and infections by other pathogens, like *Histoplasma*). Either pathogen can make people very sick, but co-infection is especially harmful.

Whether co-infecting parasites and pathogens interact often depends on where they reside within or on the host. For example, when viruses infect different cell types, interactions between viral species are unlikely. In contrast, co-infections of the same tissues can lead to competition among infections, just like different free-living species occupying the same habitat and consuming the same resources. This occurs during co-infection among children with rhinovirus and other common respiratory viruses (e.g. respiratory syncytial virus, metapneumovirus, or parainfluenza virus): co-infections lead to lower viral loads compared to infections with rhinovirus alone (Waghmare et al. 2019). Sometimes interactions in the same tissues are direct and caused by limited physical space. One can only fit so many worms in the large intestine! In other cases, parasite interactions are indirectly mediated by the host's immune response (Box 13.3). For example, if a relatively immune-resistant parasite species causes an elevated immune reaction, it may persist while the host's immune system eliminates less immune-resistant parasite species. Alternatively, pathogens that suppress the host's immune system may indirectly benefit other species (e.g. the HIV and TB example).

The co-infection of a single cell by two or more virus particles can also lead to viral recombination or reassortment. For example, during COVID-19 surveillance efforts in Malaysia, an alphacoronavirus was isolated from a human pneumonia patient and identified as a novel canine–feline recombinant (Vlasova et al. 2022). This virus discovery suggests that a host (human, dog, cat, pig, something else) was co-infected with both a cat virus and a dog virus, and those two viruses recombined to make a new virus inside a cell. The new recombinant virus does not appear to be spreading and causing widespread morbidity or mortality, which is fortunate. In many other cases, recombination creates a virus novel enough that it is not as recognizable by hosts' immune systems, creating new and potentially devastating infectious diseases (e.g. highly pathogenic avian influenza).

## 13.8 Summary

It has been said that 'No man is an island entire of itself; every man is a piece of the continent, a part of the main' (Donne 1624). The same could be said of a vicuña. Or a hookworm. Yet we often view host–parasite interactions through the lens of single, simple relationships, even though it ought to be well understood that the interactions between pathogens, hosts, and vectors are played out within ecological communities and ecosystems. Long-term

---

### Box 13.3  Co-infection in endangered sea lions

Most mammals, including humans, can be 'infected' by *Klebsiella pneumoniae*, a commensal bacterium that lives in the mucosal lining of the intestines. This bacterium is normally harmless, but it sometimes goes rogue and causes host morbidity and mortality. In humans, this possibility is most likely to occur in medical settings, but hypervirulent *Klebsiella pneumoniae* can also occur in wildlife in areas as remote as the subantarctic.

The subantarctic Auckland Islands are home to the endangered New Zealand sea lion (*Phocarctos hookeri*), the rarest sea lion species in the world. As with any endangered species, successful reproduction and survival of young is critical for the persistence of New Zealand sea lions, and thus much research and monitoring focuses on the pup population (Fig. 13.13). In one monitoring study that ran from the 1998–1999 breeding season to the 2004–2005 breeding season, more than 400 dead pups were examined by researchers (Castinel et al. 2007). The single most common cause of death (35%) was assumed to be trauma caused by adult and subadult males that kill pups. But this was rivalled by the combined impacts of a few infectious diseases that also caused substantial pup mortality (at least 37% of pup deaths, and perhaps more if some of the trauma cases were misclassified and actually caused by infectious disease).

New Zealand sea lion pups are commonly infected with hookworm (*Uncinaria* spp.) (Castinel et al. 2007). These infections are often mild, but severe infections may cause pup mortality. Hookworm infections may also make sea lion pups more vulnerable to other sources of mortality. For example, in 2001–2002, pup mortality increased, and many pup deaths were associated with infection by hypervirulent *Klebsiella pneumoniae*; this was the first season where this pathogen was identified in the population. It seems that hypervirulent *Klebsiella pneumoniae* infections might

be facilitated by hookworm infections, because hookworms might damage the mucosal lining in the host's intestines. And both parasites might make sick pups more vulnerable to attacks from male sea lions. But disentangling parasite interactions and their impacts on hosts is difficult from co-infection and mortality data alone.

In 2016–2017, hypervirulent *Klebsiella pneumoniae* was firmly established in the sea lion population on Enderby Island, causing 70% of pup mortality events (Michael et al. 2019). So, researchers started a two-year randomized controlled clinical treatment trial to treat pups for hookworm using ivermectin and determine how that impacted pup survival (Michael et al. 2021). Roughly half of the 698 captured pups were given an ivermectin injection when they were approximately one week old, and all pups were individually marked using a PIT tag. As the season went on, dead pups could be examined to determine whether they had a PIT tag, if they'd received ivermectin, and their cause of death. Additionally, every time a dead pup was found, one or two living pups were sampled to compare the risk factors for dead and living pups (a 'case control study'). As you might imagine, this careful and extensive study was incredibly time consuming, but studies like these are critical for our understanding of interactions between co-infecting parasites.

Compared to control pups, the pups that received ivermectin injections had significantly lower mortality (24.1% control, 11.1% ivermectin treatment; Michael et al. 2021, Fig. 13.3). When ivermectin-treated pups were found dead, they also tended to be less likely to have hypervirulent *Klebsiella pneumoniae* infections than dead pups from the control group. This suggests that hookworm infections *do* facilitate *Klebsiella pneumoniae* infections, though the exact mechanisms underlying this interaction are still unclear (e.g. mucosal disruption).

**Box 13.3** *Continued*

**Figure 13.13** (Top) Adorable New Zealand sea lion pup nursing on Enderby Island. (Bottom) Proportion of surviving New Zealand sea lion pups in ivermectin treatment and control groups due to mortality only from hypervirulent *Klebsiella pneumoniae* (left) and all causes of mortality (right). The shaded regions depict 95% confidence intervals.

(Top) Photo by Kimberley Collins, published on Wikimedia Commons under a CC BY-SA 4.0 license. (Bottom) This figure was published by Michael et al. (2021) under a CC BY license.

data and careful experiments have sometimes been able to illustrate how and when pathogens' impacts occur (bottom-up, top-down, etc.). These insights can then be used to decide how best to control pathogen transmission (for the benefit of hosts) or to conserve parasite biodiversity (for the benefit of hosts and parasites).

## 13.9 References

Buck, J.C., Ripple, W.J. (2017). Infectious agents trigger trophic cascades. Trends in Ecology and Evolution, 32, 681–94.

Buffie, C.G., Pamer, E.G. (2013). Microbiota-mediated colonization resistance against intestinal pathogens. Nature Reviews Immunology, 13, 790–801.

Byers, J.E., Altman, I., Grosse, A.M., et al. (2011). Using parasitic trematode larvae to quantify an elusive vertebrate host. Conservation Biology, 25, 85–93.

Carlson, C.J., Hopkins, S., Bell, K.C., et al. (2020a). A global parasite conservation plan. Biological Conservation, 250, 108596.

Carlson, C.J., Dallas, T.A., Alexander, L.W., et al. (2020b). What would it take to describe the global diversity of parasites? Proceedings of the Royal Society B, 287, 20201841.

Castinel, A., Duignan, P.J., Pomroy, W.E., et al. (2007). Neonatal mortality in New Zealand sea lions (*Phocarctos hookeri*) at Sandy Bay, Enderby Island, Auckland Islands from 1998 to 2005. Journal of Wildlife Diseases, 43, 461–74.

Chesson, H.W., Dunne, E.F., Hariri, S., Markowitz, L.E. (2014). The estimated lifetime probability of acquiring human papillomavirus in the United States. Sexually Transmitted Diseases, 41, 660–64.

Crawford, A.J., Karen, R., Lips, K.R., Bermingham, E. (2010). Epidemic disease decimates amphibian abundance, species diversity, and evolutionary history in the highlands of central Panama. Proceedings of the National Academy of Sciences, USA, 107, 13,777–82.

Dobson, A., Lafferty, K.D., Kuris, A., et al. (2008). Homage to Linnaeus: how many parasites? How many hosts? Proceedings of the National Academy of Sciences, USA, 105(S1), 11,482–9.

Donne, J. (1624). No man is an island. Meditations XVII, Devotions Upon Emergent Occasions.

Eugene, A., Burke, R.L., Williams, J.D. (2021). Of mudsnails, terrapins, and flukes: use of trematodes as a field-based project in parasitology research. Invertebrate Biology, 140, e12326.

Ferreyra, H.dV., Rudd, J., Foley, J., et al. (2022). Sarcoptic mange outbreak decimates South American wild camelid populations in San Guillermo National Park, Argentina. PLoS ONE, 17, e0256616.

Griffiths, E.C., Pedersen, A.B., Fenton, A., Petchey, O.L. (2011). The nature and consequences of coinfection in humans. Journal of Infection, 63, 200–06.

Holdo, R.M., Sinclair, A.R.E., Dobson, A.P., et al. (2009). A disease-mediated trophic cascade in the Serengeti and its implications for ecosystem C. PLoS Biology, 7, e1000210.

Hollings, T., Jones, M., Mooney, N., McCallum, H. (2014). Trophic cascades following the disease-induced decline of an apex predator, the Tasmanian Devil. Conservation Biology, 28, 63–75.

Huspeni, T.C., Lafferty, K.D. (2004). Using larval trematodes that parasitize snails to evaluate a saltmarsh restoration project. Ecological Applications, 14, 795–804.

Koltz, A.M., Civitello, D.J., Becker, D.J., et al. (2022). Sublethal effects of parasitism on ruminants can have cascading consequences for ecosystems. Proceedings of the National Academy of Sciences, USA, 119, e2117381119.

Kuris, A., Hechinger, R., Shaw, J., et al. (2008). Ecosystem energetic implications of parasite and free-living biomass in three estuaries. Nature, 454, 515–18.

Lafferty, K.D., Dobson, A.P., Kuris, A.M. (2006). Parasites dominate food web links. Proceedings of the National Academy of Sciences, USA, 103, 11211–16.

Maerz, J.C., Seigel, R.A., Crawford, B.A. (2018). Conservation in terrestrial habitats: mitigating habitat loss, road mortality, and subsidized predators. 201–20 in W.M. Roosenburg, V.S. Kennedy (Eds.), Ecology and Conservation of the Diamond-backed Terrapin, Johns Hopkins University Press.

McLaughlin, J.P., Morton, D.N., Lafferty, K.D. (2020). Parasites in marine food webs. 45–60 in D.C. Behringer, B.R. Silliman, K.D. Lafferty (Eds.), Marine Disease Ecology, Oxford University Press.

Michael, S.A., Hayman, D.T., Gray, R., Roe, W.D. (2021). Risk factors for New Zealand sea lion (*Phocarctos hookeri*) pup mortality: ivermectin improves survival for conservation management. Frontiers in Marine Science, 8, 881.

Michael, S.A., Hayman, D.T.S., Gray, R., et al. (2019). Pup mortality in New Zealand sea lions (*Phocarctos hookeri*) at Enderby Island, Auckland Islands, 2013-18. PLoS ONE, 14, e0225461.

Monk, J.D., Smith, J.A., Donadío, E., et al. (2022). Cascading effects of a disease outbreak in a remote protected area. Ecology Letters, 25, 1152–63.

Morton, D.N., Antonino, C.Y., Broughton, F.J., et al. (2021). A food web including parasites for kelp forests of the Santa Barbara Channel, California. Scientific Data, 8, 99.

Parmenter, R.R., Yadav, E.P., Parmenter, C.A., et al. (1999). Incidence of plague associated with increased winter-spring precipitation in New Mexico. American Journal of Tropical Medicine and Hygiene, 61, 814–21.

Pawlowski, A., Jansson, M., Sköld, M., et al. (2012). Tuberculosis and HIV co-infection. PLoS Pathogens, 8, e1002464.

Pflughoeft, K.J., Versalovic, J. (2012). Human microbiome in health and disease. Annual Review of Pathology: Mechanisms of Disease, 7, 99–122.

Sato, T., Egusa, T., Fukushima, K., et al. (2012). Nematomorph parasites indirectly alter the food web and ecosystem function of streams through behavioural manipulation of their cricket hosts. Ecology Letters, 15, 786–93.

Sato, T., Watanabe, K., Kanaiwa, M., et al. (2011). Nematomorph parasites drive energy flow through a riparian ecosystem. Ecology, 92, 201–07.

Spencer, H.G., Zuk, M. (2016). For host's sake: the pluses of parasite preservation. Trends in Ecology and Evolution, 31, 341–3.

Stapp, P. (2007). Trophic cascades and disease ecology. EcoHealth, 4, 121–4.

Vlasova, A.N., Diaz, A., Damtie, D., et al. (2022). Novel canine coronavirus isolated from a hospitalized patient with pneumonia in East Malaysia. Clinical Infectious Diseases, 74, 446–54.

Waghmare, A., Strelitz, B., Lacombe, K., et al. (2019). 2626. Rhinovirus in children presenting to the emergency department: role of viral load in disease severity and co-infections. Open Forum Infectious Diseases, 6, S915–16.

Wald, A., Corey, L. (2007). Persistence in the population: epidemiology, transmission. Chapter 36 in A. Arvin, G. Campadelli-Fiume, E. Mocarski, et al. (Eds.), Human Herpesviruses: Biology, Therapy, and Immunoprophylaxis, Cambridge University Press.

Weinstein, S.B., Kuris, A.M. (2016). Independent origins of parasitism in Animalia. Biology Letters, 12, 20160324.

Wood, C.L., Johnson, P.T. (2015). A world without parasites: exploring the hidden ecology of infection. Frontiers in Ecology and the Environment, 13, 425–34.

Yates, T.L., Mills, J.N., Parmenter, C.A., et al. (2002). The ecology and evolutionary history of an emergent disease: hantavirus pulmonary syndrome. Bioscience, 52, 989–98.

Zipkin, E.F., DiRenzo, G.V., Ray, J.M., et al. (2020). Tropical snake diversity collapses after widespread amphibian loss. Science, 367, 814–16.

# CHAPTER 14

# Infectious disease control

## 14.1 Treating infected wildlife: a tale of scabid wombats

Infectious diseases can cause substantial human morbidity or mortality or devastate wildlife populations. In this chapter, we will cover some general methods for controlling infectious diseases and the challenges inherent to disease control, including relevant case studies from wildlife and human populations.

*Sarcoptes scabiei* is an archetypal host generalist (Escobar et al. 2022). The mite can infect hundreds of mammal species, causing a disease called sarcoptic mange (in wildlife) or scabies (in humans). Mites are transmitted via direct contact transmission between animals or environmental transmission from soil to animals, where mites can typically survive off-host in the environment for up to 19 days (Arlian and Morgan 2017, Browne et al. 2021). The combination of environmental reservoirs (Chapter 6) and reservoir hosts (Chapter 7) makes mange very difficult to control (Escobar et al. 2022).

In Australia, *Sarcoptes scabiei* has been introduced several times by European settlers and their domestic species (Fraser et al. 2017, Fraser et al. 2019), where it now infects the previously naive bare-nosed wombat (*Vombatus ursinus*): a delightfully fluffy and round creature (Fig. 14.1) that produces surprisingly cube-shaped poops (Yang et al. 2021). Wombats suffer especially high morbidity and mortality from mange, perhaps because they eat relatively nutrient-poor foods and have slow metabolisms. Wombats always seem to develop the most severe type of mange, called 'crusted mange' (Fig. 14.1). Symptoms include hair loss (alopecia) and associated heat loss, changed activity

patterns (more time spent outside their burrows), and lethargy (Simpson et al. 2016, Martin et al. 2018).

Given these disease outcomes, mange has devastated some wombat populations. For example, in 2010, a mange outbreak began in a relatively isolated wombat population in Narawntapu National Park on the central-north coast of Tasmania. Mange spread west across the park, causing nearly 100% mortality in the wombats behind the travelling wave of infection (Martin et al. 2018). Across the entire park, wombat abundance declined by 94% over seven years (Figure 14.2). Clearly, mange control methods were urgently needed for wombats.

What kinds of interventions might slow mange transmission and/or reduce the impacts of mange on individual wombats? Wombats are relatively solitary animals with limited contact with other individuals, so efforts to further reduce contacts would likely be ineffective. Instead, mite transmission most likely occurs via wombat burrows (Martin et al. 2018); wombats happen to be the largest burrowing herbivorous mammal (Johnson 1998). Multiple wombats may share burrow networks over time, providing opportunities for mites to 'hop' off an infected wombat one day and hop on to an uninfected wombat using the same burrow a few days later (Skerratt et al. 2004, Martin et al. 2019). In fact, relatively cool and moist wombat burrows may provide ideal survival conditions for mites (Browne et al. 2021). All of this suggests that treating burrows might be highly effective for controlling mange (Beeton et al. 2019), but at present, there is no safe and feasible way to do so! Burrows are too small for most people to enter, and it is unclear if fumigating chemicals pumped into the burrows could reach

*Emerging Zoonotic and Wildlife Pathogens.* Dan Salkeld, Skylar Hopkins, and David Hayman, Oxford University Press.
© Oxford University Press (2023). DOI: 10.1093/oso/9780198825920.003.0014

**Figure 14.1** (Top) Mum and joey (parent and offspring) bare-nosed wombats (*Vombatus urcinus*) in Tasmania. (Bottom) A wombat that is included in a research study (notice the ear tag) that is suffering from mange, as evidenced by missing hair and crusted skin.
Photographs by Scott Carver and Alynn Martin, who research wombat mange.

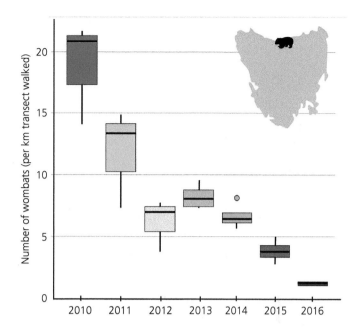

**Figure 14.2** A seven-year time series of wombat counts in Narawntapu National Park, Tasmania, as mange spread through the population.
This figure was modified from Martin et al. (2018).

deep enough to kill mites without also harming the wombats inside. Given these complications, control efforts have instead focused on treating infected wombats (Old et al. 2018, Rowe et al. 2019); the goals are to increase infected wombat survival and to reduce the duration that they are infectious and spreading mites among burrows.

Fortunately for wombats, several drugs (chemotherapeutic interventions) already existed to kill mites and thus cure animals with mange (Rowe et al. 2019, Bains et al. 2022). But there were also several complications with available options. The existing drugs were originally developed for humans and domestic species, like cats and dogs, and it was unclear which dose would be safest and most effective for wombats; this remains a controversial topic even after years of interventions (Mounsey et al. 2022). Furthermore, the early drugs needed to be administered once per week for 12 weeks—somewhat tedious for treating a pet dog, but a logistical nightmare for treating a wild animal that is difficult to find and capture and may be stressed by every interaction with humans. Fortunately, in recent years, new mite-killing drugs that last for months at a time have been developed (Wilkinson et al. 2021).

And then for the *really* difficult part: how do you apply a topical medicine to a wombat? We can rule out capturing every mangy wombat for captive rehabilitative treatment. They generally do not do well in captivity and capturing them is difficult; when researchers do need to handle a wombat, a common technique is for several researchers to form a circle around it and slowly close in, hoping that one person can swing a giant fortified butterfly net down over the wombat before it can sprint away. Though researchers are often successful, their frequent failures cause them to wonder if humans really evolved to hunt in groups using sticks and rocks.

Thus, instead of capturing mangy wombats, drug deployment focuses on treating wombats with longer distance methods (Old et al. 2018). For example, there is the 'pole and scoop' technique, which involves pouring the liquid medicine into a tiny cup at the end of a long pole. A person tries to casually approach a wombat within pole distance when the wombat is outside a burrow, aiming to dump the cup's contents on the wombat's back. This becomes more difficult to do successfully as wombats get healthier and more skittish, but still need additional treatments. Another method, which

**Figure 14.3** A 'burrow flap' in front of a wombat burrow entrance. Here the flap has already been pushed up by an exiting wombat, dumping the medicine to treat mange out of the black reservoir. Burrows are so small that most people cannot enter; much of our knowledge of the burrows comes from a 15-year-old boy named Peter Nicholson who crawled inside and mapped some burrows in the 1960s. These days, researchers are developing a robot called a WomBot for burrow exploration and research (Ross et al. 2021).
Photo by Alynn Martin.

allows for more remote deployment, is called the 'burrow flap' method (Fig. 14.3): a plastic flap with a tiny cup is suspended in front of the burrow by a wire frame, and when the wombat pushes past it, the medicine dumps from the cup, much like the pail-of-water-over-the-door prank in cartoons (Martin et al. 2019).

As you might imagine, these application methods can be somewhat unpredictable and prone to errors. Mathematical models of mange dynamics that compare treatment regimens predict that as the reliability of the treatment delivery method increases, the likelihood of successfully eliminating mange from wombat populations also increases (Martin et al. 2019). Control efforts to reduce mange transmission in wombats are therefore limited by the ability to deploy successful treatments to enough wombats. As we will see below, challenges associated with vaccinating or treating enough individuals to make a difference is a common barrier to controlling infectious diseases in wildlife and human populations.

## 14.2 Vector control and vaccination: conserving the plagued black-footed ferret

*Western US*—As the examples of Arctic foxes and Tasmanian devils in Chapter 12 demonstrated, small and fragmented populations can be vulnerable to disease-induced extirpation, especially if the infectious agent does not require the vulnerable population for its persistence. The posterchild for disease-induced extirpation of small populations might be the black-footed ferret (*Mustela nigripes*), an endangered carnivore in North America that nearly went extinct after plague ravaged ferret and prey populations (Fig. 14.4). Black-footed ferret conservation has required extensive disease control interventions.

The black-footed ferret needs prairie dogs (*Cynomys* species) to survive; prairie dogs serve as ferrets' primary prey and prairie dog burrow complexes also provide shelter for ferrets (Biggins

**Figure 14.4**  A black-footed ferret (*Mustela nigripes*) preying upon a black-tailed prairie dog (*Cynomys ludovicianus*).
Photo: Mike Lockhart.

et al. 2006). Unfortunately, the native prairie dog populations have been persecuted for decades because they are considered a verminous pest that compete for cattle forage, act as an impediment to land settlement, or serve as a convenient target for recreational shooting. Prairie dogs are also impacted by the invasive pathogen *Yersinia pestis*, the flea-vectored bacteria that causes plague. These factors have all contributed to declines in prairie dog populations in the western US, which in turn impacted black-footed ferrets. Things got so bad that black-footed ferrets were listed as endangered under the United States' Endangered Species Act in 1973. But before the US Fish and Wildlife Service approved a recovery plan in 1978 (Biggins et al. 2006), they had seemingly already gone extinct.

Mercifully, in 1981, in Meeteetse, north-western Wyoming, a remnant wild population of about 100 ferrets was discovered when a ranch dog named Shep presented a rancher named John Hogg with a dead ferret. Mr. Hogg remembered: "'We took it down to a taxidermist in town here, and he said, 'Oh my God, you got a ferret'". And I said, "What the hell's that?" Ha, ha, I didn't know what it was' (Preston 2016). Fortunately, other people did know

what ferrets were, and they set out to protect the only known remnant population.

Ferret conservation was not easy. Shortly thereafter, the local prairie dog populations were devastated by plague, and the ferrets were probably afflicted by both plague and canine distemper virus. Fearing that this last-hope population was about to wiped out, the last 18 individuals were brought into captivity in 1987 (Williams et al. 1988, Biggins et al. 2006). Since then, captive breeding programmes have resulted in over 8500 descendants. These ferrets were used to begin reintroduction efforts at 28 locations, such that approximately 300 individuals currently survive in the wild. But because plague still presents a threat by reducing and directly killing wild ferrets (Williams et al. 1994, Matchett et al. 2010), disease control programmes are ongoing. So, take a moment to consider this conundrum: what interventions could be used to maintain a species of rare carnivore in the wild given its predilection for a prey base that succumbs to a mutually deadly disease?

There are many moving parts to the notoriously complex plague system (ferrets, prairie dogs, fleas, *Yersinia pestis*, other flea host species), and some

**Figure 14.5** A Gunnison's prairie dog (*Cynomys gunnisoni*) eating an oral bait containing sylvatic plague vaccine, which is a recombinant raccoon poxvirus engineered to express two protective antigens against *Yersinia pestis* and made attractive by the addition of peanut butter (Rocke et al. 2014). A biomarker (rhodamine B) is added that allows researchers to determine the extent to which animals are actually eating the bait by examining hairs and whiskers under a fluorescence microscope, where the marker fluoresces red.
Photo: Dan Tripp, Colorado Parks & Wildlife.

of them are more viable disease control targets than others. For example, it *wouldn't* be logistically feasible to trap most of the prairie dogs on most prairie dog colonies, so injectable plague vaccines and topical anti-flea medication for prairie dogs cannot be distributed at an ecologically relevant scale. In contrast, since many ferrets are born in captivity, it *is* feasible to give a large proportion of ferrets an injectable vaccine; this vaccine has been available since 2008, and vaccinated captive-reared ferrets have higher survival rates than unvaccinated captive-reared ferrets in the wild (Matchett et al. 2010). To reach the much larger wild prairie dog populations, control programmes deploy oral vaccine baits (Fig. 14.5), which increase survival and abundance in treated prairie dog populations; this preserves the ferret's prey base and reduces *Yersinia pestis* transmission from prairie dogs to ferrets (Biggins et al. 2011, Abbott et al. 2012). Control programmes also target the fleas that transmit *Yersinia pestis* by 'dusting' prairie dog burrows with insecticides (Seery et al. 2003, Tripp et al. 2009, 2016). Neither dusting nor prairie dog bait vaccination is completely successful in preventing prairie dog mortality, but the potential for intervention in plague transmission dynamics to secure sustainable and resilient prairie dog populations is encouraging (Rocke et al. 2017, Tripp et al. 2017). As we will see throughout this chapter, disease control efforts are often most successful when multiple imperfect

interventions target multiple parts of the complex system (e.g. hosts, vectors, pathogen).

Of course, there can also be drawbacks, challenges, or trade-offs associated with disease control interventions (Abbott et al. 2012, Tripp et al. 2016). For example, dusting prairie dog burrows with insecticides indiscriminately kills invertebrates, creates a selection pressure for pesticide-resistant insects, requires substantial human power to successfully pull off, and fails to halt outbreaks if applied too late. As another example, black-footed ferret vaccination does not protect the prairie dog prey base from the disease, and to be most successful, it requires the difficult capture and vaccination of naive, born-in-the-wild ferrets. In both cases, programmes might spend much time and money on an intervention that isn't successful or even creates new problems. This reminds us that when managing wildlife disease, one must always carefully consider whether an intervention can be done successfully at the necessary scale and whether any non-target impacts are justified (Tripp et al. 2017). If not, it is better to focus resources on a different intervention that is more likely to succeed.

The work on prairie dog plague management involves partnerships between federal, state, and local stakeholders, demonstrating that large, collaborative teams can—and need to—work together to achieve innovative solutions to difficult questions (Chapter 15). Without cross-sectoral

---

**Box 14.1 Field work challenges**

You're in a cave in Uganda, surrounded by Marburg and rabies and black forest cobras, wading through a slurry of dead bats, getting hit in the face by live ones … and the walls are alive with thirsty ticks, and you can hardly breathe, and you can hardly see, and … you've got time to be *claustrophobic?*

'It was really unnerving,' Amman told me. 'I'd probably never do it again.' —David Quammen, *Spillover*

Field investigations need to be large enough to capture the dynamic ecological and epidemiological processes at play, covering relevant spatial and temporal scales. This aspect of disease ecology is sometimes under-appreciated and often under-achieved!

In the plague studies described in this chapter, collaborative teams spent three years monitoring the impacts of vaccines and dusting in prairie dog colonies, which involved capturing 10,000 prairie dogs and testing more than 10,000 pools (combined samples) of fleas for *Yersinia pestis*. Those numbers seem large but also rather divorced from the actual field efforts to garner the data. Imagine the effort, trials, and tribulations of carrying vast numbers of heavy metal traps across wind-blown, sunburnt, cactus-infested landscapes filled with prairie dog holes (Fig. 14.6). These landscapes are also home to lurking, petulant, well-camouflaged rattlesnakes (*Crotalus viridis*) and skulking black-widow spiders (*Latrodectus hesperus*). (One of the book's authors wants to

**Box 14.1** *Continued*

**Figure 14.6** Field work often involves long hours of physical labour in remote environments. This collage provides a glimpse into the field lives of five ecologists. Across the top, there is Carissa Turner, Tom Lautenbach (photo by Evan Hockridge), and Mark Spychala, and across the bottom, there is Jasmine Childress and Julia Buck.

sensationalize these venomous species and another wants to sensibly point out that most snake bites occur when people are walking in bare feet and literally step on the snakes or when people are foolishly engaged in trying to kill snakes.) Regardless, a walk in the prairie is no walk in the park!

Before embarking on gruelling field work, disease ecologists must consider whether the efforts will be enough and whether they can be justified. Will surveillance efforts for disease outbreaks really encompass the time, space, and numbers required for the full picture to emerge? Will sample sizes be large enough to make conclusions, or to justify the efforts required to attain them? Will the right host species be targeted for capture? How might the transient or sporadic nature of spatiotemporal disease patterns be reflected by, or hampered by, the surveillance designs? When people outside the team later read the study and interpret the results, they must also consider how the logistic demands of field work may have influenced the study.

For the studies in northern Colorado, plague outbreaks occurred on some of the sites before investigators could even properly measure prairie dog populations. As Tripp et al. (2017) commented: 'Plague's emergence . . . preceded

or coincided with the beginning of our study, confounding survival analyses.' This echoes the sentiments of the mammologist Robert Burns, who wrote in 1785: 'the best-laid plans of mice and men often go awry' (actually, he wrote 'The best laid schemes o' mice an' men/Gang aft a-gley'.). Disease ecologists strive to understand the emergence and spread of wildlife and zoonotic pathogens, but it is worth remembering that field work can be messy and unpredictable, even while Pasteur's old saw 'chance favours the prepared mind' holds true (Salkeld 2017).

partners, management interventions may stutter and fail to be adopted at the necessary scales. And importantly, plague management in the wild shows that even imperfect interventions can be useful when carefully planned and executed with fully engaged teams.

## 14.3 Control interventions

The wombat and prairie dog case studies illustrated some of the interventions that are possible for controlling infectious disease transmission. Attempts to control mange in wombats involved **treatment** of the host, whereas attempts to control plague in prairie dogs included **vector control** and **vaccination**. In this section, we will overview several other types of transmission control and the conditions where they are most effective.

Successful disease control interventions must do one of three things: (1) prevent pathogen invasion in the first place; (2) reduce transmission after invasion to the point where the pathogen is eventually **eliminated** or **eradicated** (Box 14.3); or (3) accept long-term pathogen persistence but try to limit how many hosts become infected or improve disease outcomes for infected hosts and afflicted host populations. The best ways to prevent or reduce pathogen transmission (i.e. $R_0$ or $R_e$; Box 14.2) will depend on the details of the disease system, and we might organize possible interventions in several ways. One way is to organize them into six categories: Barriers, Education, Eradication/Elimination, Treatment, Isolation, and Vaccination. If it's helpful as a mnemonic, you can imagine having intravenous treatment using the extract of the most intense of vegetables (Robbins 1984): BEET-IV (Fig. 14.9).

The basic reproductive number ($R_0$) and the effective reproductive number ($R_e$) of a pathogen—the average number of secondary cases caused by a primary case in a completely susceptible population ($R_0$) or a population with some pre-existing exposure and immunity ($R_e$)—are concepts associated with thresholds (see Chapter 5). When $R_0$ is less than 1, pathogens can't successfully invade a new population; they fizzle out over the long term, even if they initially cause a small outbreak. And when $R_e$ is less than 1

in a population with an established pathogen, transmission tends to wane until the pathogen disappears from the host population(s) over the long term. Therefore, to prevent or reduce pathogen transmission, the goal is often to reduce $R_0$ or $R_e$, and in the best-case scenario, to reduce them below 1. So, what can we do to reduce these metrics?

For any given host–pathogen system, there are several factors that impact $R_0$ and $R_e$, so there may also be many possible targets for interventions. In perhaps the simplest

**Box 14.2** *Continued*

possible example, let's take a system where $R_0 = \beta/\gamma$. Intuitively, interventions that decrease the transmission rate ($\beta$) will decrease $R_0$; the slower a pathogen spreads from one host to the next, the less likely it is to cause an epidemic in a susceptible population. In contrast, *increases* in the recovery rate ($\gamma$), which is the denominator of the $R_0$ equation, will decrease $R_0$; the faster infected hosts recover or die from their infections (the smaller the duration of infectiousness), the less likely a pathogen is to successfully spread from one host to another, and thus the less likely that pathogen is to cause an epidemic. Following this simple example, we can develop a quick rule of thumb: we usually want to reduce the rate parameters in the numerator and increase the rate parameters in the denominator of the $R_0$ equation.

Of course, the equations can become much more complicated! Instead of just showing all possible equations, let's revisit the concept of stock and flow diagrams. In this example, the squares are 'compartments' containing Susceptible (S), Infectious (I), Quarantined (Q), and Resistant (R) host individuals (Fig. 14.7). There is also a compartment for infectious stages in the environment, which we are defining broadly for this example; it could be larval stages of the parasite, viruses floating in the air, or a vector population with its own SIR disease dynamics. When we draw host–parasite systems in this way, we can see each 'pipe' that controls disease dynamics and consider whether we can increase or decrease the flow in that pipe to our advantage. For example, indiscriminate culling of livestock or wildlife hosts could affect the death rates of all free-ranging hosts (S, I, and R), which might not be ideal; if some hosts are already resistant, we might not want to cull them! Instead, some control programmes are '**test and treat**' or '**test and cull**', where only infected/infectious individuals are targeted for treatment or culling. Similarly, we see that some interventions could affect multiple pipes: treating wombats for mange both improves wombat survival and reduces the duration of infectiousness. This diagram will vary among disease systems, and it reminds us that we must carefully consider how hosts, vectors, pathogens, and their environments are interacting to design effective control programmes.

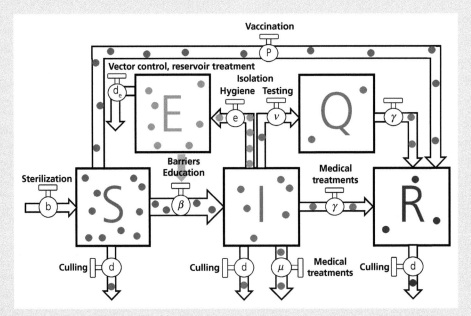

**Figure 14.7** A 'stock and flow' diagram for a pathogen that is transmitted both from direct contacts between susceptible and infectious hosts and from contact with infectious agents in an environmental reservoir. The labels indicate various interventions that could control the flow of individual hosts or pathogen stages through the compartments, thereby affecting the parameters that control $R_0$ or $R_e$. In addition to culling to control host population sizes, we also included sterilization, which is sometimes attempted to reduce host population sizes by reducing birth rates (e.g. trap-neuter-release programmes for free-ranging dogs and cats).

**Barriers** reduce disease spread by preventing pathogen transmission from one host to another, or from one population to another. For example, face masks reduce airborne transmission, mosquito nets and insect repellent reduce vector-borne transmission, and biocontrol efforts at ports reduce the chance of pathogens, vectors, or hosts accidentally moving to new areas on planes or boats.

**Education** can be used to increase awareness of pathogen risks (Fig. 14.10), thereby empowering people (or animals that can be educated) to avoid or reduce pathogen transmission. For example, cook your chicken liver pâté to appropriate temperatures to avert *Campylobacter* infection, avoid mosquitoes at dusk and dawn to reduce likelihood of West Nile virus infection, and use protection to avoid sexually transmitted infections. Education is also an important way to reduce the likelihood that infected humans spread their infections. For example, society can encourage infected people to self-quarantine when they are sick or to only use restrooms/latrines to avoid spreading pathogens in shared environments.

**Elimination/eradication** are terms that have particular meanings when used in the public health arena (see Box 14.3), but both refer to reducing the number of infectious agents, hosts, or vectors over time until there are no more new infections in a defined area. Barriers, education, treatment, and isolation can all contribute to elimination and eradication efforts.

**Treatment** of infected cases can reduce disease severity and/or diminish the window of opportunity for transmission. For example, Arctic foxes (*Alopex lagopus*) were successfully treated for sarcoptic mange with Alugan spray and ivermectin, and this improved pup survival (Goltsman et al. 1996). Similarly, treating wombats for mange improves the survival of individuals *and* reduces spread to uninfected individuals—two separate and important goals.

Antibiotics and antivirals are now commonplace for treating human and livestock disease (though at the risk of the evolution of antibiotic resistance). There are also antifungal treatments, which are becoming increasingly important for treating wildlife diseases like chytridiomycosis, white nose syndrome, and ophidiomycosis. For example, antifungal treatments have been critical for maintaining disease-free captive assurance colonies of amphibians threatened by chytridiomycosis (Bosch et al. 2015). Antifungal treatments have also been used in heroic efforts to save wild amphibian populations, like the Mallorcan midwife toads (*Alytes muletensis*), an amphibian endemic to Mallorca, Spain. Conservationists removed tadpoles from ponds, bathed them in antifungal drugs, and temporarily maintained them in aquaria while the ponds were drained, dried out in the Spanish sun, and treated with disinfectants. Treated and recovered tadpoles were returned (by helicopter!) to the ponds after they had refilled. This was an extraordinary and successful effort, but its success is somewhat idiosyncratic as this was a relatively simple system (no other reservoir hosts for the chytrid) and there were only a few ponds on this isolated island in the Mediterranean Sea.

**Isolation** prevents exposure by keeping sources of infection from contact with susceptible hosts. Strictly, **medical isolation** occurs when confirmed disease cases are isolated. But the philosophy behind this control intervention is the same for **quarantine**, which separates and restricts the movement of people who were (potentially) exposed to a contagious disease. The term quarantine originates from the Italian word for 40, quarantinario—the 40-day enforced waiting period for ships arriving in Venice during the 14th century Black Death, during which no passengers or goods could come ashore. Staying home from work or school when you are sick is a form of quarantine (Fig. 14.8).

**Vaccination** is the process of exposing susceptible individuals to a pathogen in a way that allows them to become fully or partially immune to the pathogen without becoming infectious or without experiencing the same disease severity as an infected host. Vaccination can be a tremendous tool to protect populations and control disease and was central to the eradication of both smallpox and rinderpest viruses, the only infectious diseases that humans have purposefully and successfully eradicated.

The smallpox vaccine was originally developed in 1796 as part of an ethically dubious collaboration between Edward Jenner, an English physician and scientist; James Phipps, the eight-year-old son of Dr Jenner's gardener; Sarah Nelmes, a milkmaid who had recently contracted cowpox; and Blossom, a dairy cow, source of Sarah Nelmes's cowpox, and

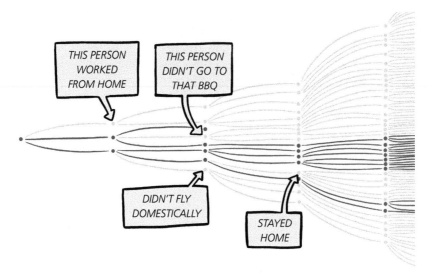

**Figure 14.8** A hypothetical transmission chain for a pathogen with $R_0 = 3$, where, on average, infected hosts (points) pass on infections to three new susceptible hosts depending on their individual circumstances. The pink points and their connections show what would happen with no interventions to reduce transmission and the grey points and lines show how transmission chains can be interrupted by behavioural interventions, like staying home when sick, i.e. isolation.
Created by Siouxsie Wiles and Toby Morris and originally published by thespinoff.co.nz.

bestower of the term *vaccination* from *vacca*, the Latin for cow (Senthilingam 2020).

Aware of rumours that milkmaids exposed to cowpox became naturally protected against smallpox, Jenner removed some pus from Nelmes's pock-marked hand and inoculated it into the arm of Phipps. The lad developed a mild fever and loss of appetite (demonstrating zoonotic transmission of cowpox virus), but he recovered. Two months later, in July, Jenner exposed Phipps directly to smallpox using the practice of variolation, a procedure where the skin is scratched with the skin scab material of someone with a mild form of smallpox. This technique had been described in England in 1721 by Lady Mary Wortley Montagu, who had witnessed the custom being used by women medical practitioners in Turkey (then the Ottoman Empire). Lady Mary had been personally scarred by smallpox, and she had her son inoculated in Constantinople (now Istanbul), and her daughter when back in London (Brumfiel 2021). Phipps did not become sick from smallpox and the adventure constitutes the first 'scientific' validation of vaccination.

The World Health Organization (WHO) orchestrated an 'Intensified Eradication Program' for smallpox in 1967, a time when there were 10 million cases of smallpox (comprising two closely related strains of variola virus: variola major and variola minor) across 43 countries. The programme used a combination of active surveillance for cases and **ring vaccination**, i.e. vaccinating all the contacts and possible exposures in a defined radius of the case patient. For example, the last case of variola major was in Bangladesh in a toddler, Rahima Banu, in 1975, and once located, vaccination teams embarked on a mission to vaccinate everyone living within a 1.5-mile radius of the house, which included more than 18,000 people (Escarce and Jahan 2022).

The last naturally occurring case of variola minor was in a 23-year-old cook, Ali Maow Maalin, in Somalia, in October 1977. Maalin had actually worked as a vaccinator in the smallpox programme but had avoided the vaccine himself due to a fear of needles; fortunately, he recovered from the illness. (The last non-natural cases occurred in Birmingham, England, in August 1978, when Janet Parker, a medical photographer, is presumed to have contracted the virus in a laboratory accident at Birmingham University medical school. She died from the disease, but her mother, who

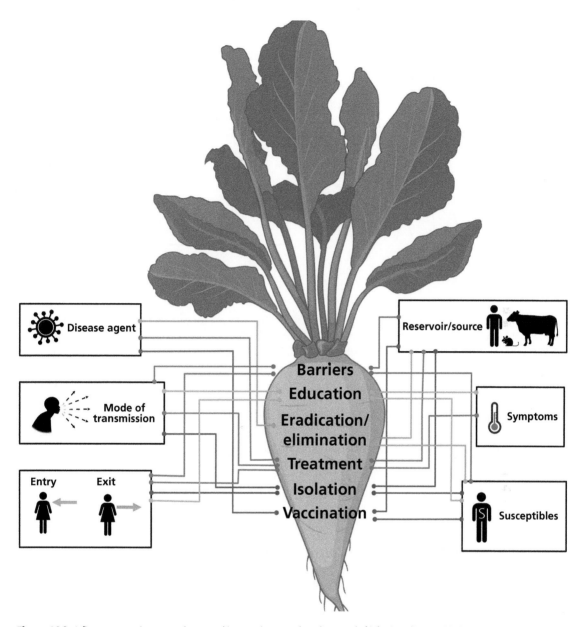

**Figure 14.9** A figure mnemonic to remember control interventions to reduce the spread of infectious diseases: BEET-IV. Made with BioRender.com.

had been caring for her and who also became ill, survived.) Smallpox was officially declared eradicated in 1980.

Vaccines have also been critical in combating wildlife diseases. For example, development of a thermostable vaccine contributed to the eradication of rinderpest (Mariner et al. 2012, Chapter 13). The entire population of endangered California condors, *Gymnogyps californianus*, captive and field-released, received at least one dose of a West Nile vaccine in anticipation of the arrival of West

**Figure 14.10** A lack of control intervention.
'Scientific Briefing' by xkcd, https://xkcd.com/2278/

Nile virus in western regions of the US (Chang et al. 2007). And, as discussed in Chapter 12, vaccinations against rabies and other pathogens have been critical for conserving endangered wild carnivores.

Disease interventions with vaccines also have challenges (also see Chapter 5; Barnett and Civitello 2020). Delivery may require capturing wild animals, potentially multiple times if the vaccine requires boosters, which is always logistically difficult. If oral vaccines can be used, then one must be aware of safety concerns for other animals that might eat the bait. And vaccine campaigns may be expensive; vaccines for humans and domestic species cannot necessarily be used safely with wildlife species, and it requires massive efforts to develop and deploy safe vaccines at the scales necessary to control disease dynamics. Though there are challenges, vaccination also has fewer ethical and welfare issues compared to tactics like culling (Miguel et al. 2020). We'll look at some of these approaches in real-world contexts below.

---

**Box 14.3 Pathogen elimination, eradication, and containment**

Pathogen elimination and pathogen eradication are different goals.

Pathogen **elimination** refers to a scenario where a pathogen no longer persists endemically in a region; some infections may still occur in the region, but the pathogen cannot persist. For example, malaria has been eliminated in the United States since 1949 due to extensive (and often environmentally destructive) mosquito control efforts (Williams 1963). Though malaria infections still pop up in the United States when infected people return from travelling in other countries, these infections have not caused malaria epidemics in the United States for decades. After elimination is achieved, programmes must maintain control efforts indefinitely to prevent re-introduction and establishment of the pathogen in the region.

Pathogen **eradication** refers to zero cases globally, where the pathogen is driven extinct. Though humans have tried to eradicate many pathogens, we have only been successful twice. The virus that caused smallpox in humans was officially declared eradicated in 1980, after hundreds of years of inoculating people to stimulate smallpox immunity. The virus that caused rinderpest in cattle, buffalo, and other wildlife was eradicated in 2011, reducing livestock losses, famine, and resulting human deaths (Morens et al. 2011). For other pathogens and parasites, like the polio virus or the nematode (*Dracunculus medinensis*) that causes Guinea worm

**Box 14.3**  *Continued*

disease, global eradication has been a goal for decades, and though huge strides have been made, only regional eliminations have been achieved so far (Fig. 14.11). Given the scale of these goals, eradication efforts are usually coordinated by international agencies, like the World Health Organization, and require collaborations at multiple levels of government.

When elimination or eradication seem particularly difficult to achieve or the pathogen causes relatively low rates of morbidity or mortality, decision-makers may instead choose **containment** as their disease control goal. Under containment scenarios, the pathogen will remain endemic in the host population, but transmission rates may be reduced and the pathogen will hopefully be less likely to spread to new regions. For example, during the initial SARS-CoV-2 pandemic, some island countries were able to eliminate the virus, but most could not institute this much control, and thus opted to instead reduce transmission as much as possible until a vaccine was developed. This approach aims to reduce morbidity or mortality to an 'acceptable' level (e.g. hospital capacity not overwhelmed), and what is considered 'acceptable' can change over time as control strategies improve or, for wildlife diseases, as wildlife populations decline due to the disease.

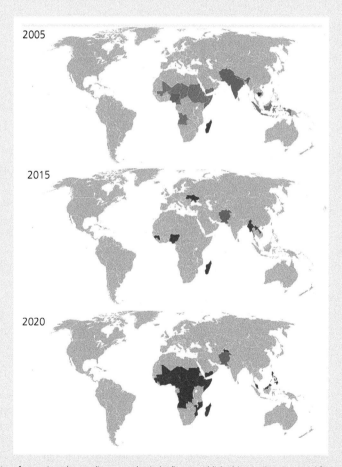

**Figure 14.11**  Time series of countries where polio was endemic (red), re-established (orange), or imported from another region (green), or where cases were only related to vaccine-derived virus (blue).

All maps were published in the public domain on Wikimedia Commons or under a CC BY-SA 4.0 license by users Fvasconcellos, Tobus, and Agricolae.

## 14.4 Culling wildlife to prevent wildlife–livestock disease transmission: the case of badgers and bovine tuberculosis

*Mycobacterium bovis* is the bacterium that causes bovine tuberculosis (referred to here as bovine TB). In cattle, *Mycobacterium bovis* infects the lungs and then either remains localized or worsens to a chronic condition, causing generalized lesions and health declines. Infected cattle can transmit *Mycobacterium bovis* to other herd members or species, including people, though people these days are less likely to become infected through the historical route of drinking contaminated, unpasteurized milk. To prevent disease spread, bovine TB-positive cattle are compulsorily slaughtered, i.e. test and cull.

Bovine TB has been resurging in the United Kingdom and Ireland in the past three decades, especially in the south-west and west of England and the south-west of Wales. In 2013, 32,000 cattle were slaughtered to control bovine TB transmission (out of more than 8 million cattle tested). Each animal's slaughter is followed by replacement of the slaughtered animal, further testing, and the need to quarantine the rest of the herd, which all causes financial and emotional hardship for the farmer (Pérez-Morote et al. 2020). The pecuniary burden of bovine TB surveillance is also borne by the beef and dairy industry. So of course, people want to know why bovine TB incidence is increasing, but the issue is not straightforward. In fact, the rise in infections represents 'one of the most complex, persistent and controversial problems facing the British cattle industry, costing the country an estimated £100 million per year' (Brooks-Pollock et al. 2014).

The bovine TB issue in the UK is complicated by three main factors. First, the testing regimen suffers from inaccuracies and involves time lags between testing and detected infections. Animals are traditionally tested by looking for an immune reaction by injecting them with an antigen mixture of tuberculins derived from *Mycobacterium bovis* and *Mycobacterium avium* (called the 'single intradermal comparative cervical tuberculin (SICCT) test', or more commonly the 'tuberculin skin test'). The skin test *doesn't* directly detect the bovine TB bacteria, but rather whether the animal has previously been exposed and can mount an immune reaction. So, there is a window of time when animals are infected but the immunological skin test is unable to detect the antibodies, and consequently there will be false negatives (Chapter 3). Furthermore, overall the tuberculin skin test's sensitivity (proportion of previously infected animals successfully detected, see Chapter 3 again) is in the range of 70–90% (Brooks-Pollock et al. 2014); this low sensitivity compromises efforts to accurately monitor and interpret bovine TB transmission among and between herds. To recap, one main complication in the UK bovine TB problem is testing inaccuracy and uncertainty, which muddies the waters of scientific, public, and political understanding of the issue.

The second complication is perhaps difficult to believe: vaccinating cattle against TB is prohibited by European Union regulations (Pérez-Morote et al. 2020). The reasons are various, but a partial explanation includes the fact that there is no way to differentiate vaccinated animals from infected animals by the tuberculin skin test.

The third complication is really the crux of the issue: bovine TB transmission dynamics in UK cattle are complicated by the role of wildlife populations as a source of infection. European badgers (*Meles meles*) (Fig. 14.12) appear to be an important wildlife reservoir for bovine TB in the British Isles (Crispell et al. 2019). Badgers can also be found earnestly foraging for luscious earthworms, *Lumbricus terrestris*, in the same idyllic green pastures where cows are happily munching grass. In these areas of overlap, direct badger-to-cow interactions appear rare; instead, spillover of bovine TB appears to occur through indirect environmental transmission (O'Mahony 2014, Woodroffe et al. 2016)—badgers defaecate where cows eat. (Rude? Maybe, but cows also defaecate where badgers eat.) Transmission of bovine TB from badgers to cattle is more commonplace than transmission from cattle to badgers (which also occurs), and intra-species transmission (i.e. within badger populations, or within cattle populations) occurs at a much higher rate than inter-species transmission between badgers and cattle (Crispell et al. 2019). It's worth noting that this system is almost the opposite to the rigmarole of canine distemper or rabies, where the problem is

**Figure 14.12** (Top) European badger (*Meles meles*). (Bottom) Badger being fitted with a proximity collar.
(Top) Photo by Vincent van Zalinge on Unsplash. (Bottom) Photo by Seth Jackson.

spillover from domestic animal populations to vulnerable wildlife populations.

To protect cows from bovine TB, one potential solution would be to reduce exposure to infected badgers, and one proposed way to do that would be to reduce the badger population by **culling** badgers. (Culling is a synonym for killing.) Reducing the population size should theoretically reduce the force of infection from the badger population to the cattle population, as long as transmission is density-dependent (Chapter 5).

The Randomised Badger Culling Trial (RBCT) was an experimental investigation to discern the validity of culling badgers as a means of reducing bovine TB in nearby cows. Badgers were treated in one of three ways: (1) culled proactively (i.e. widespread killing of badgers that were near cow herds, independently of badgers' bovine TB status, i.e. indiscriminate culling, Box 14.4, Fig. 14.14); (2) culled in response to positive TB tests in local cows; or (3) left alone (Donnelly et al. 2005).

As you might have predicted, the number of bovine TB cases in cattle decreased (by 19%) in areas where proactive culling of badgers was carried out (treatment #1). But the number of bovine TB cases actually *increased* (by 29%) in herds located adjacent to the areas where badgers were killed, i.e. doing an intervention *here* seemed bad for cows over *there*. Furthermore, culling of badgers in response to positive bovine TB cases in cows (treatment #2) also seemed to amplify bovine TB transmission (by 25%) and cause more 'herd breakdowns'—positive tests in herds that were previously TB-free and not subjected to movement restrictions. So overall, attempts to reduce reservoir host populations resulted in more disease in the cattle. What on earth was going on?

Badger social groups can have as many as 20 members, and they typically have stable territories containing multiple subterranean dens, called setts. Badgers move from sett to sett, but not necessarily as a single unified group. And individual badgers vary in their sociality and movement on the landscape, which can be represented visually and mathematically using network models (Box 14.4). Network models are usually built using data on who contacts whom, such as which wild badger wearing a proximity logger was near which other wild badger similarly adorned with proximity loggers, and for how long and how often. Some badgers had relatively few within-group contacts, were more socially isolated from their resident group, and spent more time visiting remote setts at the borderlands of their group territories (Weber et al. 2013). Other badgers were more socially connected; they had higher **flow-betweenness** in their network (Box 14.4).

Variation in badger sociality doesn't just affect who got invited to the most tea parties. The more curmudgeonly and distant badgers were also more likely to be infected with bovine TB, perhaps because they spent more time interacting with members of other social groups. Whether these badgers were more likely to be infected because of their behaviours or whether they were more socially distant because they were infected is not known. However, these differences do have implications when trying to understand the impacts of culling.

---

### Box 14.4 An introduction to network theory

Imagine each badger individual, and in the language of **network theory** (also known as **graph theory**), call them a **node** (or **vertex**). Each badger, or node, may interact with another badger and, if they do, there is a connection between them, referred to as an **edge** or **link** (Fig. 14.13). (Yes, for some reason, there are two, or duplicate, possible terms for every concept or idea in network/graph theory.) Links can be **weighted** to reveal the frequency or intensity or quality of each connection. The number of links connected to a node is called the **degree** and reveals how many other individuals one node has interacted with. For example, a node's degree will be small for socially reluctant introverts and high for over-bearing extroverts. If you can draw all the nodes and edges in a population—the full network—you will have a snapshot understanding of who contacts whom. Most networks are **modular**, with several closely-knit groups that frequently interact with each other and only rarely interact with individuals from other groups.

**Box 14.4** *Continued*

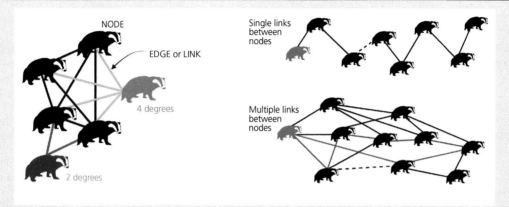

**Figure 14.13** Conceptual diagram to illustrate network (graph) theory terms showing nodes, edges (links), and connectivity in degrees. Top right shows a network of badgers (nodes) where each badger has a maximum of two degrees. Pathogen transmission must therefore be sequential, and an outbreak would be relatively slow. Interventions to break the chain of transmission (dotted line) would prevent subsequent pathogen transmission. Bottom right illustrates a network of badgers that have multiple links, which could potentially result in a larger, faster outbreak. Interventions may be unsuccessful because they fail to disrupt all the potential exposures.
Rather special badger icon by Anthony Caravaggi, PhyloPic.org.Figure created using BioRender.com.

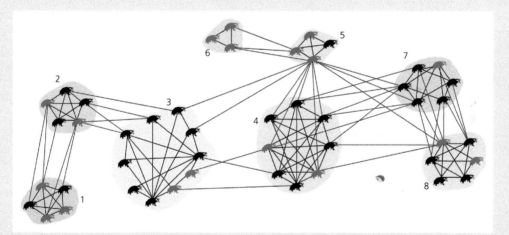

**Figure 14.14** Social network and tuberculosis test outcome in a population of wild badgers based (but adapted somewhat) on real data from Weber et al. (2013). Group territories (setts) are denoted by coloured blobs and approximate the spatial configuration of the social groups. The proximity of badger nodes to one another is simply illustrative and does not represent spatial closeness. The links show close interactions recorded between badgers. In this network, some badger groups are less connected (more distant) in that they interact with only one other sett (setts 1 and 6) whereas other setts are more connected (e.g. sett 4 has interactions with setts 3, 5, 7, and 8). Particular setts (e.g. sett 5) have important roles in connecting other setts. For example, sett 5 acts as a conduit for contacts between sett 6 and setts 3, 4, 7, and 8, and indeed it seems that one particular infected badger (red) acts as a potential superspreader with a high degree. Setts 3 and 4 have similar numbers of badgers (nodes) but the badgers have quite different degrees (sett 3 has badgers with fewer links than the badgers in sett 4). You can imagine the various likelihoods of disease transmission in the two different setts.
Figure made using BioRender.com.

**Box 14.4** *Continued*

If an individual in a network becomes infected, we might want to know how many susceptible individuals are in the infected host's typical social circle (the infected host's degree) or how central the infected host is to the entire network. To answer the second question, we can draw paths through the network connecting all pairs of individuals and ask how many of those paths involve the infected host (a metric called **flow-betweenness**). Or, instead of considering every possible path, we can count how many of the shortest paths between pairs involve the infected host, which is called **betweenness (or closeness) centrality**. These metrics help us quantify whether an individual node is a stepping stone in many pathways or only a few, where a node with a high betweenness centrality is more likely to have a large influence on the transmission of a pathogen through the network.

For the badger population sporting proximity-logging collars, within-group flow-betweenness was significantly lower for badgers infected with bovine TB: the loner badgers weren't central to their group networks. In contrast, those same badgers had significantly higher among-group flow-betweenness in both summer and winter: the loner badgers were more likely to be involved in connections between groups. Again, it's worth remembering that we don't know whether this behaviour was a factor influencing the likelihood of infection, or whether infection was causing the behaviour; the data showed a correlation rather than a causal dynamic. But the inference is that the bovine TB-positive animals are not making a large contribution to bovine TB transmission within their own groups but may be making a disproportionately high contribution to TB spread among groups (Weber et al. 2013).

So far, we have learned that badgers don't move and associate randomly like gas molecules following Brownian mass action (Chapter 5). (The authors hope that you are picturing badgers drifting aimlessly and randomly around British woodlands like grizzled helium balloons!) Instead, badgers have structured social networks, and when those networks are perturbed by culling individuals (**social perturbation**), it can alter how badger interactions and disease dynamics play out across the landscape (Fig. 14.15). In particular, when individual badgers are culled, the surviving neighbouring badgers increase their movement (more ranging), including immigrating into these now-empty territories. Increased movement exacerbates exposures and infections between badger groups and the probability that badgers with bovine TB will roam into new territories and shed TB in cow pastures, resulting in more badger-to-cattle TB spillover (Carter et al. 2007, Vicente et al. 2007). As this example illustrates, understanding the social behaviour of hosts is critical for understanding disease transmission and designing successful control interventions.

Armed with the knowledge that culling might exacerbate bovine TB transmission in cattle and that stable social structure probably reduced TB transmission, the United Kingdom's government decided to... continue with the badger-culling approach. (Yes, really. This is a textbook.) The plan to continue culling was derided as misguided because there wasn't even good baseline data on badger populations: how do you know that you're reducing a badger population sufficiently if you don't even know the original population size?! And there were not sufficient resources to improve density estimates or cull sufficient badgers to reduce herd breakdowns.

Nonetheless, more than 150,000 badgers have been killed since culling was initiated, and the annual government expenditure on compensation and control measures is roughly £100 million/$130 million. Despite all this effort, there are no clear indications that badger culling has reduced the bovine TB problem. The UK's Environment Secretary, Owen Paterson, defending the government's culling decisions after being accused of incompetence, bemoaned that 'the badgers moved the goalposts. We're dealing with a wild animal, subject to the vagaries of the weather and disease and breeding patterns' (BBC 2013). (One of the authors, the smooth talking, adventuresome one, is considering titling their next ecological paper 'The vagaries of weather and disease and breeding patterns'.)

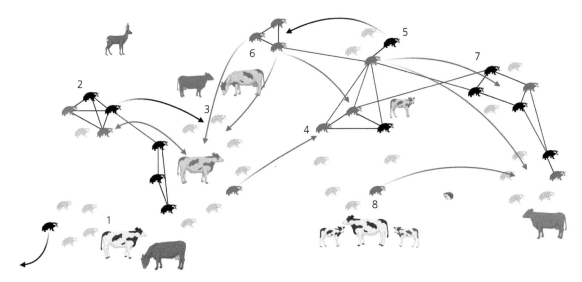

**Figure 14.15** Conceptual diagram showing impact of badger culling on bovine TB movement. In an unaltered badger population (see Fig. 14.14), movement of badgers has evolved to a status quo. Culling removes some but not all badgers (grey badger shadows) and disrupts the group's social behaviour. Removal of badger groups can reduce the risk of bovine TB transmission to nearby cows (e.g. sett 1). However, new territories become available to neighbouring badgers, potentially increasing the scope of movement of infected badgers (red) and increasing exposure to cows (shaded red; e.g. from sett 6 to sett 3). Badgers, bovine TB, and cattle comprise a complex system and this makes the effects of culling and social disruption difficult to predict with any reliability. Other, less-studied components (e.g. the potential reservoir competence of roe deer) pose additional complications.

This illustration shamelessly indulges one of the author's love of badgers, cows, Anthony Caravaggi's badger icon, and Biorender.com.

What to do then? Vaccinating badgers has emerged as a potentially effective way to reduce transmission from wildlife to livestock. The advantages: no need to senselessly slaughter a British cultural icon and no more exacerbated spread of bovine TB through social perturbation. The disadvantages: how do you vaccinate free-living wild badgers?! As mentioned earlier, finding/trapping enough badgers to influence the population (through culling or vaccination) is hard work, and the current vaccine needs to be administered by hand. Nonetheless, the UK government's Department for Environment, Food and Rural Affairs announced in 2021 that it will phase out badger culls and switch emphasis to vaccination.

It's worth remembering that bovine TB dynamics are occurring in badgers, in cattle, and between the two species. If each host species can maintain infection independently, then control interventions targeting only one of these components could be rendered ineffective when spillover occurs again. Instead, a systems approach that simultaneously controls cattle-to-cattle transmission, badger-to-badger transmission, and badger-to-cattle transmission may be most successful (Brooks-Pollock et al. 2014, Crispell et al. 2019).

## 14.5 Culling and wildlife disease reservoirs, more generally

Culling has frequently been adopted as a disease control intervention. For example, culling was used to create wildlife-free corridors around national parks to protect cattle farms from the spread of trypanosomiasis in southern Africa; to reduce red fox numbers in Europe to control rabies; and to reduce bovine TB spread by targeting brushtail possums (*Trichosurus vulpecula*) in New Zealand (itself invasive), Cape buffalo (*Syncerus caffer caffer*) in South Africa, or badgers in the UK (Miguel et al. 2020). However, the effectiveness and cost-benefit balances of wildlife culling is now intensely debated by politicians, scientists, stakeholders, and the general public, and many factors contribute to whether culling might be successful or not (Morters et al. 2013, Miguel et al. 2020).

Efforts to design effective culling programmes are complicated by uncertainty regarding population dynamics and disease dynamics in wildlife communities. For example, culling programmes aim to reduce badger populations by 70% to control bovine tuberculosis. Therefore, in Dorset (a UK county), where the badger population was estimated to include 879–1547 animals (95% confidence interval), culling licensees were required to kill at least 615 badgers (70% of 879). But given population uncertainty, 615 dead badgers could constitute as little as 40% of the population (615/1547), which would be insufficient to have an impact on bovine TB (Donnelly and Woodroffe 2015). If licensees were to instead cull 1083 badgers, they would achieve 70% of the highest population estimate, but they might also completely extirpate the badger population if it is 1083 badgers! And of course, culling 1083 animals would be more expensive and difficult than culling 615 animals.

There are statistical approaches to deal with some of these uncertainties, such as approaches that can incorporate test accuracy or the inability to capture all animals (e.g. some animals are trap-happy and some are trap-shy). There are also modelling approaches that can help predict which control techniques (e.g. cull indiscriminately, test and cull) will be most successful and how much effort will be required for success. There are even methods for conducting cost-benefit analyses to determine whether increased revenue from reduced disease risks exceeds the (direct and indirect) costs of culling (Miguel et al. 2020). But all these approaches require that we have a good understanding of the system, and that is often not the case. For example, even as programmes work to cull one reservoir species, they might be missing other host species that are reservoir hosts. This was so in the case of vaccinating white-footed mice to control Lyme disease dynamics, which is covered in Chapter 7 (Tsao et al. 2004). And of course, even if we know exactly how many hosts of which species need to be culled, the difficulties inherent to field work (Box 14.1) may not allow us to reach those targets.

Culling can also have unpredictable, unintended consequences because we don't understand animal movement and behaviour. For example, we already saw that culling badgers unexpectedly affected badger movement and therefore appeared to exacerbate bovine TB spread in neighbouring cattle herds. Similarly, experimental investigations of deer mouse (*Peromyscus maniculatus*) removal from ranch buildings in Montana revealed that immigrant mice quickly replaced culled resident deer mice and resulted in a higher abundance of mice than in buildings from which no mice were removed (Douglass et al. 2003). Furthermore, in three instances, mice that were seropositive for Sin Nombre virus replaced mice that had been removed but were seronegative. Therefore, removal of deer mice is only useful if it is carried out concurrent with rodent-proofing of the premises.

There are also other ways that culling can have unexpected consequences. For example, after a professional pest control company *had* rodent-proofed a cabin to reduce the risk of hantavirus exposure, students from a high school outdoor education camp that stayed in the cabin were hospitalized with cases of tick-borne relapsing fever (TBRF) caused by *Borrelia hermsii*. Subsequent investigations revealed that *Ornithodoros hermsii* ticks that were trapped in the cabin had become increasingly hungry (two dead wood rats were found in the cabin, suggesting prior sustenance for soft ticks) and opportunistically fed on the sleeping students (Jones et al. 2016). Acaracides in concert with rodent proofing might prevent this kind of occurrence. In another example, from Peru, extensive culling of common vampire bats (*Desmodus rotundus*) has had limited impacts upon rabies incidence in people and livestock or on rabies virus seroprevalence in bat colonies (Chapter 11). Instead, culling campaigns might be counterproductive because they target adult vampire bats, whereas juvenile and subadult bats might have a more important role in rabies virus circulation (Streicker et al. 2012). And in many places, wildlife populations and free-ranging domestic species (i.e. dogs and cats) can respond to culling by compensatory reproduction—a surge in reproductive output to take advantage of newly available resources and less competition (Morters et al. 2013). All of this is to say that culling may sometimes be useful for disease control, but it will rarely, if ever, be sufficient as the only control intervention, and it may even make things worse.

## 14.6 Unintended consequences—bison, elk, cattle, and brucellosis

It was shortly after 9 a.m. and just under zero degrees [F, −18°C], and the elk were hungry. A thousand or so of them huddled together in a snowy field, watching as two towering horses trudged forth pulling a sleigh piled with two tons of hay.                                —Brulliard (2022)

*Greater Yellowstone Ecosystem, US—Brucella abortus,* a bacterial pathogen, was first detected in bison (*Bison bison*) in the Greater Yellowstone Ecosystem

in the early 1900s and has since increased in prevalence (Chapter 5). Bison and elk (*Cervus canadensis*) were struggling in the early 1900s due to over-hunting. Elk are also susceptible to *Brucella abortus,* but elk populations were too small to maintain transmission ($R_0 < 1$) when *Brucella abortus* first spilled over into wildlife populations from cattle (Kamath et al. 2016). That changed over the past century.

To prevent elk starvation during the harsh winters, people started to provide supplemental food

**Figure 14.16** (Top) American bison (*Bison bison*). (Next) Bison and elk (*Cervus canadensis*) sharing habitat. (Next) Locations of elk feed grounds in Wyoming (dark lines are state boundaries), where elk hunt districts are coloured by the estimated brucellosis seroprevalence in elk and the hatched area shows Yellowstone National Park (YNP) and Grand Teton National Park (GTNP). (Bottom) Phylogenetic tree of *Brucella abortus* showing ancestral nodes for five lineages (L1–L5), with different host species (red = bison; blue = elk; green = livestock (both cattle and farmed bison). Dashed lines represent uncertain lineages, though cattle were the original source of introduction into the Greater Yellowstone area, and at the point of the dashed lines, the *Brucella* isolates are more closely related to cattle strains in other parts of the US.

(Top) Photo by Kristin Sherwood. (Next) Photo by USGS, public domain. (Next) Figure from Brennan et al. (2017) and in the public domain. (Bottom) Kamath et al. (2016), used under a Creative Commons Attribution 4.0 International License.

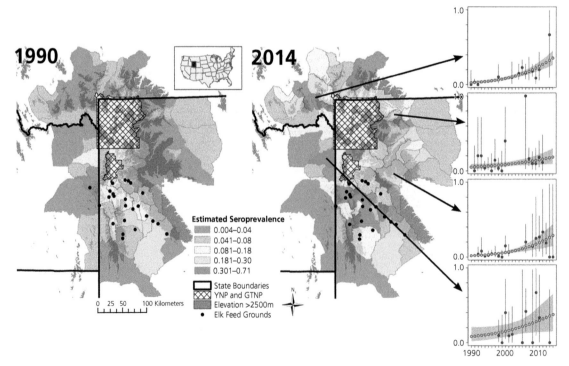

**Figure 14.16** *Continued*

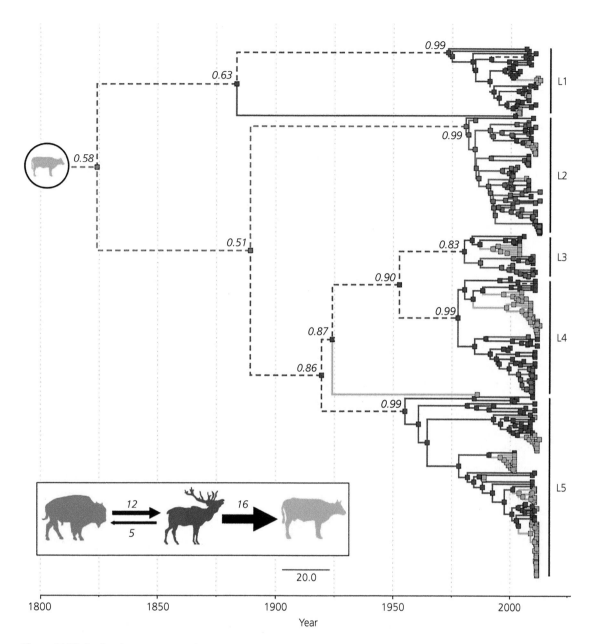

**Figure 14.16** *Continued*

in the early 1990s. These 'feed grounds' were predictable places where elk could find food, so it is
no surprise that they started to congregate there. As
the elk population grew, the elk developed a han-
kering for winter hay, and the feed grounds also
served a second purpose: to keep elk away from
roads where they cause traffic accidents, and from
ranches where elk like to eat hay meant for cattle. In

recent times, there are 22 state-run feed grounds in
Wyoming (Fig. 14.16), and the 20,000 elk that visit
them in the winter will start visiting neighbouring
properties looking for food (Brulliard 2022).

During spring, ranchers don't want wild elk on
their properties, not just because the elk are hungry,
but because as the elk population grew, it became
large enough to sustain *Brucella abortus*, which elk

can then spread to cattle. The risk of abortion and brucellosis transmission is highest in the spring. At first, elk's role in the spread of *Brucella abortus* was small, and bison were blamed for spreading the pathogen to cattle. But over time, as the pathogen increasingly cycled in elk, *Brucella abortus* prevalence became higher in some elk herds (normally 20–40% but sometimes as high as 50%; Anderson et al. 2015), and there is no question that they can move the pathogen around the landscape.

As you might have already guessed, *Brucella abortus* transmission and prevalence is highest in herds that visit the feed grounds (Cotterill et al. 2018), likely the result of density-dependent transmission (Chapter 5). For example, studies of radio-collared elk show that elk made twice as many contacts when they were on feed grounds (Cross et al. 2013). Bigger groups with more contacts likely lead to more within-group *Brucella* transmission and a greater probability that elk are shedding *Brucella* into the environment, leading to transmission to other species. Furthermore, there are concerns that chronic wasting disease, which is caused by an infectious prion, could explode in Wyoming's high-density elk populations.

Closing the feed grounds to reduce pathogen transmission is a controversial subject (Cook et al. 2023). Since supplemental feeding has helped elk survive in winter, people worry that the population will decline if feed grounds are closed, leading to reduced revenue from elk hunting and tourism. Furthermore, without feed grounds, hungry elk and the diseases they carry could become a bigger problem for livestock, at least in the short term. An alternative option may be to continue feeding elk but to spread the feed out to reduce elk aggregation.

The elk feed ground issue reminds us that humans often affect host densities and thereby disease dynamics. In this example and many others, we provide supplemental food or other resources that increase wildlife population sizes and interactions between species (e.g. birds on birdfeeders, raccoons in trash). We also constrain wildlife to smaller and smaller areas, concentrating their remaining populations and pathogen contamination in soil and water reservoirs, while creating ever-larger livestock populations where wildlife used to be (Chapter 11). Though disease control may seem as simple as controlling host population density

(e.g. removing elk feedlots, sterilizing free-ranging dogs, culling rodents), there may be no simple solution after vastly changing the landscapes and ecosystems where infectious diseases circulate.

The elk feed grounds are also a good example of **spatial heterogeneity** (variation across space): some places are elk feed grounds and some places are not, some places are near elk feed grounds and some are not, some places are common elk movement routes and some are not, etc. Spatial heterogeneity can affect disease dynamics in many ways, and thus it cannot be ignored while controlling infectious diseases.

## 14.7 Pathogen invasion and disease control

An ounce of prevention is worth a pound of cure.
—Benjamin Franklin, 1735

There is rarely a single silver bullet that can prevent an emerging infectious disease from spreading to new regions or control an emerging infectious disease in a region where it already occurs. Rather, you will need a hail of silver bullets. And different bullets/methods will be more or less useful depending on the invasion stage of the pathogen, e.g. pre-arrival, invasion front, epidemic, and established (Langwig et al. 2015). Being able to characterize which invasion stage is occurring relies on good surveillance, but frustratingly, the time-course over which management actions must be implemented is often much faster than the pace of scientific research (Langwig et al. 2015).

### Preventing emergence—precautionary measures

It is neither cheap nor easy to prevent pathogens from invading new regions or spilling over into new host species. However, it is *much* cheaper and easier to prevent invasion than it is to control an invasion after it has begun, just as it is much cheaper to fireproof your house than to put out a fire and restore fire damage (e.g. Ben Franklin's famous quote). Given the importance of prevention, there are many examples of ongoing efforts to prevent hosts, vectors, or pathogens from arriving in new areas. For example, there are efforts to prevent illegal smuggling of wildlife across geopolitical borders to reduce the likelihood of spreading novel pathogens; to halt the use of sled dogs in the Antarctic to protect seal populations from canine distemper virus; to prohibit the international movement of pork products to reduce the risk of African swine fever virus; and to recommend that cavers decontaminate their gear to prevent dispersal of *Pseudogymnoascus destructans* fungal spores (Chapter 6).

Precautionary protective measures require good science combined with an ability to navigate the demands of governments and industry. Do we understand the 'anatomy of disease' (Chapter 2) well enough to create effective barriers to pathogen spread? Using models, can we predict the potential impacts of an emerging infectious disease, or the likelihood that it might thrive or fail to invade in a new region or new host community? Can agencies successfully work together and be prepared for pathogen emergence? Are there valid, cheap, easy-to-use diagnostic tests and/or vaccines available? And can surveillance programmes (dashboards, data portals, surveillance networks, etc.) be adapted to new pathogens or take advantage of new approaches like citizen science or phone networks? These questions remind us that while globalization is an important driver of emerging infectious diseases (Chapter 9), our connectedness and technological advances can also be used to prevent pathogen emergence and spread.

## Controlling and slowing invasion

When pathogens are first observed in new regions or new host populations, timing is critical. Interventions should respond as rapidly as possible, whilst the pathogen is still locally restricted, and cases are still limited. At this stage, dramatic or draconian measures might be considered if preventing pathogen establishment outweighs the short-term damage done in its eradication. For example, recall the efforts to fly tadpoles around in helicopters to receive antifungal baths and completely drain and disinfect ponds to eliminate the chytrid fungus from Mallorca. Early invasion stages are the second-easiest time to control pathogens. Unfortunately, this is also often the time when the least is known about the pathogen's dynamics, so disease control efforts must often proceed with a high degree of uncertainty.

## Eliminating pathogen strongholds

If control interventions are successful, then elimination or eradication (Box 14.3) could become a possibility. Often, the feasibility of doing so is limited by how complex the disease system is. For example, smallpox could be eradicated because it was a human-specific disease, a vaccine was available and logistically feasible, and symptoms preceded infectiousness. Elimination and perhaps even eradication are possible in more complex disease systems, but these goals are more challenging to achieve, as we will see in the Guinea worm case study below. Furthermore, public health interventions sometimes suffer a curse: the more effective an intervention, the rarer the disease becomes, the less threatening the disease seems, the less money and effort people opt to devote to disease control. Along similar lines, the stronger the selection pressure of control, the greater the benefit to pathogen or vector strains that evolve traits that escape control. Thus, just as control efforts are making big progress, the disease system may adapt and render control efforts less effective.

## Managing host–pathogen co-existence

Unfortunately, we often fail to prevent pathogen invasion, and after invasion, we often cannot achieve elimination. Instead, we must often try to find optimal ways to help people, livestock, and wildlife to tolerate infectious diseases. Sometimes this has little to do with reducing transmission and more to do with buying threatened populations more time to co-evolve with their new pathogens. For example, treatment options are limited for bat populations threatened by white nose syndrome, and the pathogen is already widespread (Chapter 6). While vaccines and treatments are developed, efforts can be made to try to conserve as much good foraging and roosting habitat for bats as possible, so that they aren't facing additional threats while also suffering mass mortality due to white nose syndrome. In time, bat populations may become more resistant or tolerant to the pathogen, allowing bat populations to slowly rebound in regions where the pathogen persists.

## 14.8 A final case study: Guinea worm disease

*Chad, 1986*—A man feels excruciating pain in his lower leg. There is a small blister there, and under that, he knows that there is a thin, white, parasitic worm. To ease the burning sensation, the man limps from his home to a nearby pond, where he wades into the soothing water. The worm emerges from his leg just a tiny bit, and the man wraps a small stick around the end of the worm's body. The stick is then secured to his leg with some gauze. Over the next two to three weeks, he will wind the stick a little bit each day, being careful not to rip the worm, and eventually fully remove it from his leg. Unfortunately, there is no drug that can cure the man, nor vaccine that can prevent infections like this. While infected, the man will be in too much pain to walk

and harvest his crops. Like 3.5 million people across 21 countries in Africa and Asia in 1986, the man has dracunculiasis, or Guinea worm disease.

Guinea worm disease is caused by the female adult stage of the nematode *Dracunculus medinensis*. When the adult worm has access to water—like when an infected person sticks their painful blister in a nearby pond—the worm releases free-living larvae (L1—first-stage larvae). These larvae are then consumed by aquatic copepods, also known as 'water fleas', and develop into third-stage larvae (L3). People accidentally consume the copepods infected by the L3 larvae when they drink untreated and unfiltered water. Inside a human, the larvae mature into adults and sexually reproduce, and then the female worms migrate to the leg or foot to release larvae and start the life cycle over (Fig. 14.17).

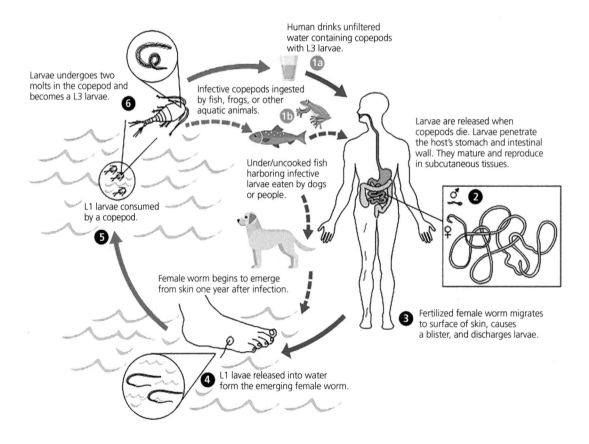

**Figure 14.17** Life cycle of *Dracunculus medinensis*, the cause of Guinea worm disease.
This figure was published in the public domain by the United States Centers for Disease Control and Prevention.

In 1980, the US Centers for Disease Control and Prevention (CDC) launched the Guinea Worm Eradication campaign, which was bolstered by the World Health Assembly and the Carter Center in 1986. Control efforts targeted the environmental reservoir for the pathogen in three ways. First, there were educational campaigns to teach people about the importance of filtering drinking water to avoid accidentally consuming infected copepods. These efforts included providing people with two types of filters to remove copepods from drinking water: filtered drinking pipes/straws and large bucket filters to pour water through. Second, there were active surveillance programmes, where networks of field personnel and village volunteers regularly surveyed for active infections. When cases were found, their blisters were bandaged, and there was case containment to prevent people from exposing the blister to local water sources and thus contaminating the water for everyone. And third, water sources that were known to be contaminated were chemically treated to kill the copepod intermediate hosts. Together, these control methods were remarkably successful! From the starting point of 3.5 million cases in 1986, there were just 15 cases in five African countries in 2021 (Angola, Chad, Ethiopia, Mali, and South Sudan; Hopkins et al. 2022; Fig. 14.18).

Guinea worm disease is a remarkable human disease control story, but the timelines remind us that controlling transmission and surveilling for infectious disease is *difficult*. Coordinating surveillance in several countries is always a huge undertaking, and some of the regions where Guinea worm disease is endemic are especially difficult to monitor due to civil unrest. As a result, some cases may be missed in some years, adding uncertainty to the data or creating lags between when cases are noticed and when health agencies respond to those cases. And then there are the difficulties associated with determining whether a parasite is truly eradicated from a region. For example, after several years with no reported cases in Angola, Guinea worm was found in two people and a dog, with no clear explanation for how Guinea worm arrived in the country. Similarly, after a decade without known cases (and a period of civil unrest), Guinea worm cases again emerged in Chad in 2010. Unexpected events and challenges like this plague any disease control effort.

Furthermore, when control efforts place strong selection pressure on a pathogen species, it is possible that the pathogen will evolve to evade that strong control pressure. In recent years, *Dracunculus medinensis* has been observed infecting non-human animals. It is unclear if this has always occurred but is just now noticeable as human cases decline, or if *Dracunculus medinensis* is infecting more animals now than in the past due to human-centred control efforts. What is clear is that there are hundreds of documented animal infections, especially in domestic dogs; in 2012, a survey of dogs revealed that more than 1000 had active infections along the Chari River in Chad (Eberhard et al. 2014). In Mali, cases were witnessed in a few dogs and cats in 2018, and in Ethiopia, there were infections in dogs, cats, and for the first time, in Anubis baboons (*Papio anubis*) (Hopkins et al. 2022). These observations raised unexpected challenges for disease control efforts.

To potentially control transmission to animals, researchers needed to figure out how animals were becoming infected. Were they accidentally drinking contaminated water, the same way that humans get infected? Perhaps! An alternative hypothesis is trophic transmission: animals may be getting infected when they consume fish or frogs that ate infected copepods. This transmission mode is plausible because it occurs for other species of dracunculoid nematodes, like the ones that are transmitted to snakes and sometimes dogs and other mesomammals through consumption of fish/frog hosts in the United States (Cleveland et al. 2018, 2020, Garrett et al. 2020). These details are important, because along the Chari River in Chad, it may be relatively easy for dogs and other animals to scavenge fish guts left behind by people harvesting fish. By burning or burying fish guts, people may reduce the likelihood of scavenging and transmission to dogs and cats (Cleveland et al. 2017). Additionally, chemical treatment can continue to be used to kill copepods in contaminated waters, disrupting the nematode's life cycle.

Though the transmission pathway to dogs and other animals is still somewhat unclear, there is no question that this new development has impeded efforts to control Guinea worm disease in humans. The World Health Organization pushed back the target date for eradication from 2020 to 2030, hoping

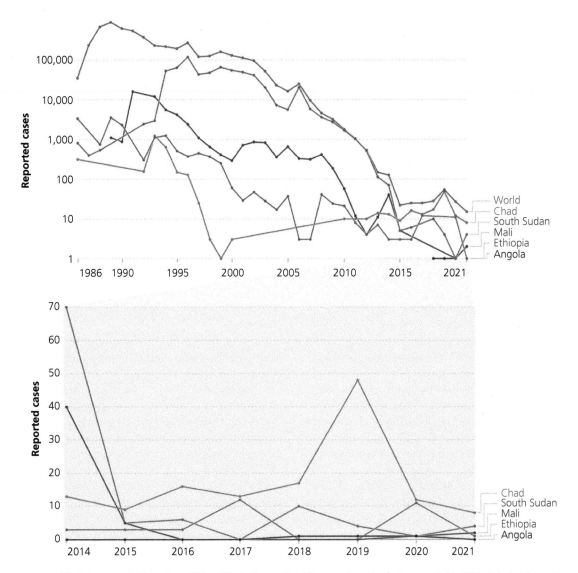

**Figure 14.18** Guinea worm incidence from 1986 to 2022 for the world and five countries with infections reported in 2021. Note that the *y*-axis uses the log scale, so some zeros are missing, and the differences are large. The bottom panel shows just the timeline from 2014 to 2022 for the same five countries using a *y*-axis with a linear scale.

These plots were created using surveillance data from the World Health Organization and published by Our World in Data under a CC BY license.

that the new 2030 deadline would give enough time to both eradicate the disease and officiate the eradication (i.e. spend three years confirming that there were no new cases anywhere). This example illustrates how complex efforts to control and eradicate infectious diseases can be, where both social and ecological complexity create challenges for control.

## 14.9 Summary

For disease control, the best defence is a good offence. There are many ways to prevent pathogens from spreading to new regions or spilling over into new host populations, and though prevention is neither cheap nor easy, it is cheaper and easier than controlling emerging infectious diseases. After pathogen invasion occurs, control interventions usually aim to reduce the effective reproductive rate of the pathogen ($R_e$) to below 1, such that the pathogen will be eliminated over the long term. There may be many ways to reduce $R_e$, depending on the disease system. For example, educational campaigns could be used to make people aware of pathogen risks (e.g. needle sharing is a risk for transmission of several human pathogens) and barriers can be created to reduce the likelihood of transmission (e.g. create sterile syringe exchange programmes). In this chapter, we organized interventions using the BEET-IV mnemonic: Barriers, Education, Elimination/Eradication, Treatment, Isolation, and Vaccination. Since it is difficult to apply single interventions at the necessary spatial and temporal scales (e.g. reaching vaccination targets), disease control programmes are often most successful when they combine multiple control strategies. This may also reduce the likelihood that pathogens evolve mechanisms to escape control efforts. Despite our best efforts, we may not be able to eliminate or eradicate pathogens that threaten the health of humans, domestic species, or threatened plants and wildlife. In those cases, we should not abandon all attempts to reduce transmission, but we should also consider how we might use other interventions to reduce the impacts of infectious diseases on humans, plants, or animals.

## 14.10 References

Abbott, R.C., Osorio, J.E., Bunck, C.M., Rocke, T.E. (2012). Sylvatic plague vaccine: a new tool for conservation of threatened and endangered species? EcoHealth, 9, 243–50.

Anderson, N., Carson, K., Jones, J., et al. (2015). 2015 Targeted Elk Brucellosis Surveillance Post-capture Summary. Montana Fish, Wildlife & Parks.

Arlian, L.G., Morgan, M.S. (2017). A review of *Sarcoptes scabiei*: past, present and future. Parasites & Vectors, 10, 297.

Bains, J., Carver, S., Hua, S. (2022). Pathophysiological and pharmaceutical considerations for enhancing the control of *Sarcoptes scabiei* in wombats through improved transdermal drug delivery. Frontiers in Veterinary Science, 9, 944578.

Barnett, K.M., Civitello, D.J. (2020). Ecological and evolutionary challenges for wildlife vaccination. Trends in Parasitology, 36, 970–78.

BBC. (2013). Badgers 'moved goalposts' says minister Owen Paterson. BBC: https://www.bbc.com/news/uk-england-24459424, accessed 7/9/2022.

Beeton, N.J., Carver, S., Forbes, L.K. (2019). A model for the treatment of environmentally transmitted sarcoptic mange in bare-nosed wombats (*Vombatus ursinus*). Journal of Theoretical Biology, 462, 466–74.

Biggins, D.E., Livieri, T.M., Breck, S.W. (2011). Interface between black-footed ferret research and operational conservation. Journal of Mammalogy, 92, 699–704.

Biggins, D.E., Miller, B.J., Clark, T.W., Reading, R.P. (2006). Restoration of an endangered species: the black-footed ferret. 581–85 in M.J. Groom, G.K. Meffe, C.R. Carroll (Eds.), Principles of Conservation Biology, Sinauer Associates.

Bosch, J., Sanchez-Tomé, E., Fernández-Loras, A., et al. (2015). Successful elimination of a lethal wildlife infectious disease in nature. Biology Letters, 11, 20150874.

Brennan, A., Cross, P.C., Portacci, K., et al. (2017). Shifting brucellosis risk in livestock coincides with spreading seroprevalence in elk. PLoS ONE, 12, e0178780.

Brooks-Pollock, E., Roberts, G., Keeling, M. (2014). A dynamic model of bovine tuberculosis spread and control in Great Britain. Nature, 511, 228–31.

Browne, E., Driessen, M.M., Ross, R., et al. (2021). Environmental suitability of bare-nosed wombat burrows for *Sarcoptes scabiei*. International Journal for Parasitology: Parasites and Wildlife, 16, 37–47.

Brulliard, K. (2022). It began as a tool to save wild elk. A century later, feeding threatens iconic herds. Washington Post, 16 March: https://www.washingtonpost.com/nation/2022/03/16/wyoming-elk-chronic-wasting-disease/, accessed 19/4/2023.

Brumfiel, G. (2021). A 300-year-old tale of one woman's quest to stop a deadly virus. All Things Considered, NPR, 8 March: https://www.npr.org/sections/health-shots/2021/03/08/972978143/a-300-year-old-tale-of-one-womans-quest-to-stop-a-deadly-virus, accessed 6/9/2022.

Carter, S.P., Delahay, R.J., Smith, G.C., et al. (2007). Culling-induced social perturbation in Eurasian badgers *Meles meles* and the management of TB in cattle: an analysis of a critical problem in applied ecology. Proceedings of the Royal Society of London B, 274, 2769–77.

Chang, G.-J.J., Davis, B.S., Stringfield, C., Lutz, C. (2007). Prospective immunization of the endangered California condors (*Gymnogyps californianus*) protects this species from lethal West Nile virus infection. Vaccine, 25, 2325–30.

Cleveland, C.A., Eberhard, M.L., Garrett, K.B., et al. (2020). *Dracunculus* species in meso-mammals from Georgia, United States, and implications for the Guinea Worm

Eradication Program in Chad, Africa. Journal of Parasitology, 106, 616–22.

Cleveland, C.A., Eberhard, M.L., Thompson, A.T., et al. (2017). Possible role of fish as transport hosts for *Dracunculus* spp. larvae. Emerging Infectious Diseases, 23, 1590–92.

Cleveland, C.A., Garrett, K.B., Cozad, R.A., et al. (2018). The wild world of Guinea Worms: a review of the genus *Dracunculus* in wildlife. International Journal for Parasitology: Parasites and Wildlife, 7, 289–300.

Cook, J.D., Cross, P.C., Tomaszewski, E.M., et al. (2023). Evaluating management alternatives for Wyoming elk feedgrounds in consideration of chronic wasting disease (No. 2023-1015). US Geological Survey.

Cotterill, G.G., Cross, P.C., Cole, E.K., et al. (2018). Winter feeding of elk in the Greater Yellowstone Ecosystem and its effects on disease dynamics. Philosophical Transactions of the Royal Society B: Biological Sciences, 373, 20170093.

Crispell, J., Benton, C.H., Balaz, D., et al. (2019). Combining genomics and epidemiology to analyse bidirectional transmission of *Mycobacterium bovis* in a multi-host system. eLife, 8, e45833.

Cross, P.C., Creech, T.G., Ebinger, M.R., et al. (2013). Female elk contacts are neither frequency nor density dependent. Ecology, 94, 2076–86.

Donnelly, C., Woodroffe, R. (2015). Bovine tuberculosis: badger-cull targets unlikely to reduce TB. Nature, 526, 640.

Donnelly, C.A., Woodroffe, R., Cox, D.R., et al. (2005). Positive and negative effects of widespread badger culling on tuberculosis in cattle. Nature, 439, 843–6.

Douglass, R.J., Kuenzi, A.J., Williams, C.Y., et al. (2003). Removing deer mice from buildings and the risk for human exposure to Sin Nombre virus. Emerging Infectious Diseases, 9, 390–92.

Eberhard, M.L., Ruiz-Tiben, E., Hopkins, D.R., et al. (2014). The peculiar epidemiology of dracunculiasis in Chad. American Journal of Tropical Medicine and Hygiene, 90, 61–70.

Escarce, A., Jahan, D.A. (2022). How Rahima came to hold a special place in smallpox history—and help ensure its end. Radio Diaries, NPR, 20 May: https://www.npr.org/2022/05/20/1099830501/smallpox-covid-vaccine-eradication-who, accessed 6/9/2022.

Escobar, L.E., Carver, S., Cross, P.C., et al. (2022). Sarcoptic mange: an emerging panzootic in wildlife. Transboundary and Emerging Diseases, 69, 927–42.

Franklin, B. (1735). Protection of towns from fire. The Pennsylvania Gazette, 4 February 1735.

Fraser, T.A., Holme, R., Martin, A., et al. (2019). Expanded molecular typing of *Sarcoptes scabiei* provides further evidence of disease spillover events in the epidemiology of sarcoptic mange in Australian marsupials. Journal of Wildlife Diseases, 55, 231–7.

Fraser, T.A., Shao, R., Fountain-Jones, N.M., et al. (2017). Mitochondrial genome sequencing reveals potential origins of the scabies mite *Sarcoptes scabiei* infesting two iconic Australian marsupials. BMC Evolutionary Biology, 17, 233.

Garrett, K.B., Box, E.K., Cleveland, C.A., et al. (2020). Dogs and the classic route of Guinea Worm transmission: an evaluation of copepod ingestion. Scientific Reports, 10, 1430.

Goltsman, M., Kruchenkova, E.P., Macdonald, D.W. (1996). The Mednyi Arctic foxes: treating a population imperilled by disease. Oryx, 30, 251–8.

Hopkins, D.R., Weiss, A.J., Yerian, S., et al. (2022). Progress toward global eradication of Dracunculiasis—worldwide, January 2021–June 2022. Morbidity & Mortality Weekly Report (MMWR), 71, 1496–502.

Hopkins, S.R., Lafferty, K.D., Wood, C.L., et al. (2022). Evidence gaps and diversity among potential win–win solutions for conservation and human infectious disease control. The Lancet Planetary Health, 6, e694–e705.

Johnson, C.N. (1998). The evolutionary ecology of wombats. 34–41 in R.T. Wells, P.A. Pridmore, (Eds.), Wombats, Surrey Beatty and Sons.

Jones, J.M., Hranac, C.R., Schumacher, M., et al. (2016). Tick-borne relapsing fever outbreak among a high school football team at an outdoor education camping trip, Arizona, 2014. American Journal of Tropical Medicine & Hygiene, 95, 546–50.

Kamath, P., Foster, J., Drees, K., et al. (2016). Genomics reveals historic and contemporary transmission dynamics of a bacterial disease among wildlife and livestock. Nature Communications, 7, 11,448.

Langwig, K.E., Voyles, J., Wilber, M.Q., et al. (2015). Context-dependent conservation responses to emerging wildlife diseases. Frontiers in Ecology and the Environment, 13, 195–202.

Mariner, J.C., House, J.A., Mebus, C.A., et al. (2012). Rinderpest eradication: appropriate technology and social innovations. Science, 337, 1309–12.

Martin, A.M., Burridge, C.P., Ingram, J., et al. (2018). Invasive pathogen drives host population collapse: effects of a travelling wave of sarcoptic mange on bare-nosed wombats. Journal of Applied Ecology, 55, 331–41.

Martin, A.M., Fraser, T.A., Lesku, J.A., et al. (2018). The cascading pathogenic consequences of *Sarcoptes scabiei* infection that manifest in host disease. Royal Society Open Science, 5, 180018.

Martin, A.M., Richards, S.A., Fraser, T.A., et al. (2019). Population-scale treatment informs solutions for control of environmentally transmitted wildlife disease. Journal of Applied Ecology, 56, 2363–75.

Matchett, M.R., Biggins, D.E., Carlson, V., et al. (2010). Enzootic plague reduces black-footed ferret (*Mustela nigripes*) survival in Montana. Vector-borne & Zoonotic Diseases, 10, 27–35.

Miguel, E., Grosbois, V., Caron, A., et al. (2020). A systemic approach to assess the potential and risks of wildlife culling for infectious disease control. Communications Biology, 3, 353.

Morens, D.M., Holmes, E.C., Davis, A.S., Taubenberger, J.K. (2011). Global rinderpest eradication: lessons learned and why humans should celebrate too. Journal of Infectious Diseases, 204, 502–05.

Morters, M.K., Restif, O., Hampson, K., et al. (2013). Evidence-based control of canine rabies: a critical review of population density reduction. Journal of Animal Ecology, 82, 6–14.

Mounsey, K., Harvey, R.J., Wilkinson, V., et al. (2022). Drug dose and animal welfare: important considerations in the treatment of wildlife. Parasitology Research, 121, 1065–71.

Old, J.M., Sengupta, C., Narayan, E., Wolfenden, J. (2018). Sarcoptic mange in wombats - a review and future research directions. Transboundary and Emerging Diseases, 65, 399–407.

O'Mahony, D.T. (2014). Badger-Cattle Interactions in the Rural Environment: Implications for Bovine Tuberculosis Transmission. Agri-Food and Biosciences Institute, Belfast. Available at: http://www.dardni.gov.uk/badger-cattle-proximity-report.pdf.

Pérez-Morote, R., Pontones-Rosa, C., Gortázar-Schmidt, C., Muñoz-Cardona, Á.I. (2020). Quantifying the economic impact of bovine tuberculosis on livestock farms in south-western Spain. Animals, 10, 2433.

Preston, P. (2016). Black-footed ferrets are hoping to make a historic comeback. http://wyomingpublicmedia.org/post/blackfooted-ferrets-are-hoping-make-historic-comeback, accessed 1/8/2017.

Quammen, D. (2012). Spillover: Animal Infections and the Next Human Pandemic. W.W. Norton.

Robbins, T. (1984). Jitterbug Perfume. Bantam.

Rocke, T.E., Kingstad-Bakke, B., Berlier, W., Osorio, J.E. (2014). A recombinant raccoon poxvirus vaccine expressing both *Yersinia pestis* F1 and truncated V antigens protects animals against lethal plague. Vaccines, 2, 772–84.

Rocke, T.E., Tripp, D.W., Russell, R.E., et al. (2017). Sylvatic plague vaccine partially protects prairie dogs (*Cynomys* spp.) in field trials. EcoHealth, 14, 438–50.

Ross, R., Carver, S., Browne, E., Thai, B.S. (2021). WomBot: An exploratory robot for monitoring wombat burrows. SN Applied Sciences, 3, 1–10.

Rowe, M.L., Whiteley, P.L., Carver, S. (2019). The treatment of sarcoptic mange in wildlife: a systematic review. Parasites & Vectors, 12, 99.

Salkeld, D.J. (2017). Vaccines for conservation: plague, prairie dogs & black-footed ferrets as a case study. EcoHealth, 14, 432–7.

Seery, D.B., Biggins, D.E., Montenieri, J.A., et al. (2003). Treatment of black-tailed prairie dog burrows with deltamethrin to control fleas (Insecta: Siphonaptera) and plague. Journal of Medical Entomology, 40, 718–22.

Senthilingam, M. (2020). Outbreaks and Epidemics. Icon Books.

Simpson, K., Johnson, C.N., Carver, S. (2016). *Sarcoptes scabiei*: the mange mite with mighty effects on the common wombat (*Vombatus ursinus*). PLoS ONE, 11, e0149749.

Skerratt, L.F., Skerratt, J.H.L., Banks, S., et al. (2004). Aspects of the ecology of common wombats (*Vombatus ursinus*) at high density on pastoral land in Victoria. Australian Journal of Zoology, 52, 303–30.

Streicker, D.G., Recuenco, S., Valderrama, W., et al. (2012). Ecological and anthropogenic drivers of rabies exposure in vampire bats: implications for transmission and control. Proceedings of the Royal Society of London: B, 279, 3384–92.

Tripp, D.W., Gage, K.L., Montenieri, J.A., Antolin, M.F. (2009). Flea abundance on black-tailed prairie dogs (*Cynomys ludovicianus*) increases during plague epizootics. Vector-borne & Zoonotic Diseases, 9, 313–21.

Tripp, D.W., Rocke, T.E., Runge, J.P., et al. (2017). Burrow dusting or oral vaccination prevents plague-associated prairie dog colony collapse. EcoHealth, 14, 451–62.

Tripp, D.W., Streich, S.P., Sack, D.A., et al. (2016). Season of deltamethrin application affects flea and plague control in white-tailed prairie dog (*Cynomys leucurus*) colonies, Colorado, USA. Journal of Wildlife Diseases, 52, 553–61.

Tsao, J.I., Wootton, J.T., Bunikis, J., et al. (2004). An ecological approach to preventing human infection: vaccinating wild mouse reservoirs intervenes in the Lyme disease cycle. Proceedings of the National Academy of Sciences, USA, 101, 18,159–64.

Vicente, J., Delahay, R.J., Walker, N.J., Cheeseman, C.L. (2007). Social organization and movement influence the incidence of bovine tuberculosis in an undisturbed high-density badger *Meles meles* population. Journal of Animal Ecology, 76, 348–60.

Weber, N., Carter, S.P., Dall, S.R.X., et al. (2013). Badger social networks correlate with tuberculosis infection. Current Biology, 23, R915–R916.

Wilkinson, V., Takano, K., Nichols, D., et al. (2021). Fluralaner as a novel treatment for sarcoptic mange in the bare-nosed wombat (*Vombatus ursinus*): safety, pharmacokinetics, efficacy and practicable use. Parasites & Vectors, 14, 18.

Williams, E.S., Mills, K., Kwiatkowski, D.R., et al. (1994). Plague in a black-footed ferret (*Mustela nigripes*). Journal of Wildlife Diseases, 30, 581–5.

Williams, E.S., Thorne, E.T., Appel, M.J.G., Belitsky, D.W. (1988). Canine distemper in black-footed ferrets (*Mustela nigripes*) from Wyoming. Journal of Wildlife Diseases, 24, 385–98.

Williams Jr, L.L. (1963). Malaria eradication in the United States. American Journal of Public Health and the Nations Health, 53, 17–21.

Woodroffe, R., Donnelly, C.A., Ham, C., et al. (2016). Badgers prefer cattle pasture but avoid cattle: implications for bovine tuberculosis control. Ecology Letters, 19, 1201–08.

Yang, P.J., Lee, A.B., Chan, M., et al. (2021). Intestines of non-uniform stiffness mold the corners of wombat feces. Soft Matter, 17, 475–88.

# CHAPTER 15

# COVID-19, One Health, and pandemic prevention

## 15.1 Rapid emergence of a novel pathogen

It's going to disappear. One day, it's like a miracle, it will disappear.

—US President Donald Trump referring to the SARS-CoV-2 virus responsible for the COVID-19 pandemic, 28 February 2020

It's important to recognize that pandemics are difficult to talk about. Anything said in advance of a pandemic seems alarmist. After a pandemic begins, anything one has said or done is inadequate.

—Michael Leavitt, Secretary of the US Department of Health and Human Services, 2005–2009

*Wuhan City (population approx. 11,000,000), Hubei Province (population approx. 58,000,000), China*—On 26 December 2019, Dr Zhang Jixian treated a senior couple who had fevers and coughs (Li et al. 2020). Dr Zhang, who had been involved with the 2003 SARS outbreak, found that the couple had irregular thoracic x-rays. The couple's son also presented with lung abnormalities, and similar lung pathologies presented in an unrelated, feverish, coughing patient that arrived at Dr Zhang's hospital the following day. Blood tests from all four cases indicated viral infections but ruled out influenza viruses. Nervous about the situation, Dr Zhang set up an isolation ward to monitor the cases, along with new cases that started to arrive at the hospital.

Dr Zhang reported these events to the head of the hospital on 27 December, and the hospital reported it to the Center for Disease Control in the Jianghan district of Wuhan. At 3:10 p.m. on 30 December 2019, Wuhan Municipal Health and Health Commission issued an 'emergency notice on reporting the treatment of pneumonia of unknown causes' (Li et al. 2020). On 31 December 2019, the World Health Organization (WHO) China Country Office was informed of cases of pneumonia of unknown aetiology (unknown cause) detected in Wuhan City, and by 3 January, 44 cases had been recognized.

Before long, a new coronavirus, temporarily named 2019-nCoV, was isolated from the patients' lower respiratory tracts. Identification of a novel coronavirus was instantly alarming, because other recently emerged coronaviruses included severe acute respiratory syndrome coronavirus (SARS-CoV) (Chapter 8) and Middle East respiratory syndrome coronavirus (MERS-CoV), which had case fatality rates of 10% and 37%, respectively (Huang et al. 2020). To accurately identify 2019-nCoV, a new diagnostic test was quickly developed. Soon after, the 2019-nCoV virus was officially named 'severe acute respiratory syndrome coronavirus 2' or SARS-CoV-2; the pathogen responsible for the COVID-19 pandemic.

Case investigations of hospital patients suggested a link between the disease and the Huanan Wholesale Seafood Market (hereafter, 'Huanan market'), because many patients had visited the market (Huang et al. 2020, Fig. 15.1). In response, the market was closed on 1 January for environmental sanitation and disinfection. At that point 'no evidence of significant human-to-human transmission and no health care worker infections' had been reported (WHO 2020). Soon after, retrospective surveys of atypical pneumonia cases found that an additional 41/59 patients had been infected with 2019-nCoV. Of those, 27/41 patients (66%) had direct exposure to Huanan market as dealers and vendors (Huang

*Emerging Zoonotic and Wildlife Pathogens*. Dan Salkeld, Skylar Hopkins, and David Hayman, Oxford University Press.
© Oxford University Press (2023). DOI: 10.1093/oso/9780198825920.003.0015

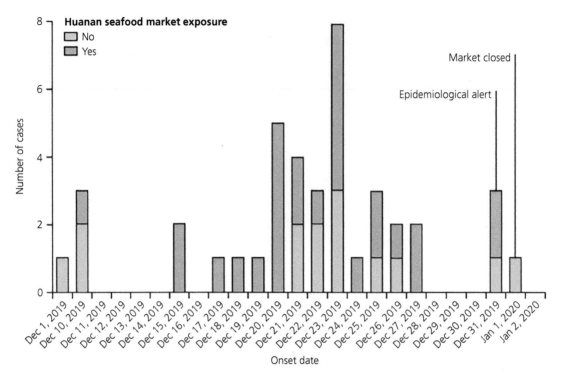

**Figure 15.1** Epidemic curve showing the date of illness onset for cases of laboratory-confirmed 2019-nCoV infection at the early stages of the COVID-19 outbreak, categorized by known exposure to the Huanan market. The Huanan market was shut down on 1 January.
From Huang et al. (2020).

et al. 2020). The first fatal case also had continuous exposure to the Huanan market; however, his wife had no known history of exposure to the market, and she developed pneumonia five days after her husband became sick, indicating human-to-human transmission. All these links made the Huanan market and people who had been there a major focus of epidemiological attention and control efforts.

The most common symptoms at onset of illness were fever, cough, myalgia or fatigue, and difficulty breathing. Some cases deteriorated rapidly from dyspnoea (shortness of breath) to acute respiratory distress syndrome (ARDS) to mechanical ventilation to intensive care (Box 15.1). Early data suggested a terrifying case fatality ratio of 15% (Huang et al. 2020). However, by 24 January 2020, 835 laboratory-confirmed 2019-nCoV infections were reported in China, with 25 fatal cases; still frightening, but a lower case fatality ratio of 3%.

By late January, healthcare workers began to be infected in Wuhan (Huang et al. 2020), demonstrating that human-to-human transmission was occurring. SARS-CoV-2 appears to be well suited for human-to-human transmission because

of its ability to bind to the angiotensin-converting enzyme 2 (ACE2) receptor, which also made other animals like cats, deer mice, bushy-tailed woodrats, and striped skunks susceptible (Anderson et al. 2020, Bosco-Lauth et al. 2020, 2021). (You may remember the incidents of human-mink-human SARS-CoV-2 transmission discussed in Chapter 11; Oreshkova et al. 2020, Oude Munnink et al. 2021.) While many of these details were not discovered until later, it was clear by late January 2020 that despite control efforts, this emerging infectious disease was spreading to other Chinese provinces.

In the subsequent months and years, these early events of the COVID-19 pandemic would be repeatedly re-examined and criticized. There were, of course, new details that came to light and things that could have been done differently. But this outbreak reminds us that even when there are rapid responses to emerging infectious diseases (e.g. reporting early cases to central governments, eliminating potential infection sources, contact tracing), outbreaks can still spread rapidly beyond our control.

---

**Box 15.1  COVID-19 and host heterogeneity**

You may have noticed that hosts are not identical, nor do they move around randomly like gas molecules—the conditions that are easiest to describe with mathematical models (Chapter 5). Instead, there is often substantial variability among exposure, susceptibility, disease, pathogen shedding, and contact rates among individual hosts. For example, you might recall how some badgers were more connected in their social groups than others, and those connections and individual behaviour were important for tuberculosis transmission (Chapter 14). Similar variation among hosts was critical for understanding and predicting SARS-CoV-2 transmission and disease outcomes.

Our abilities to rapidly detect COVID-19 cases, even during the earliest stages of the outbreak, were perhaps better than for any other human infectious disease in history. For example, the Hunan Provincial Center for Disease Control and Prevention was able to compile detailed records for 1178 infected people and their 15,648 contacts, some of whom subsequently became infected and some of whom did not (Sun et al. 2021). These infection chains revealed that most infected people did not infect anyone else. Only a small fraction of infections led to any subsequent transmission events, and a very small fraction of infections led to many subsequent transmission events (Fig. 15.2). As you may

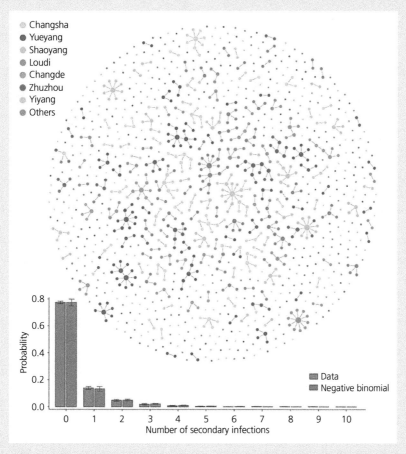

**Figure 15.2** (Top) Possible transmission chains from people infected with SARS-CoV-2 in Hunan province, where the colours indicate different prefectures within the province. Circles or 'nodes' represent infected people and lines or 'links' represent transmission events from one person to another. (Bottom) Density plot showing the number of secondary infections caused by infected individuals, with separate bars showing the 'data' (possible reconstructions of transmission chains) and a negative binomial model fit to the data. Notice that most infected individuals did not spread their infections.

**Box 15.1** *Continued*

recall (Chapter 10), this is consistent with the '80/20 rule', where ~80% of transmission events are caused by ~20% of the host population. (In this case, 80% of secondary infections were caused by 15% of primary infections.)

There was also variation in how sick someone would become when they were infected. (We realize this is a sensitive topic, because many of us lost loved ones during the pandemic, so feel free to skip reading this.) Before and after vaccinations were widely available, COVID-19 mortality rates were highest for people who were physically or socially vulnerable; for example, in the United States, COVID-19 deaths were disproportionately high for people >65 years old and people who were Black or Hispanic (Gold et al. 2020). This variation in disease outcomes was impor-

tant information for targeting control efforts to protect the most vulnerable people (e.g. prioritizing them for vaccination). But even among relatively low-risk groups, such as young white men, severe morbidity and mortality could be caused by COVID-19, which suggests that other sources of variability also contributed to disease outcomes (e.g. initial exposure dose). Though this variation was not surprising to epidemiologists, it did create substantial confusion and misinformation, because there was a widespread belief that only people with pre-existing health conditions would become severely ill due to COVID-19. It's also worth looking up what pre-existing conditions are; you may well be surprised about how mundane they are and how many people you know have them.

## 15.2 From outbreak to pandemic

Initially, transmission chains emanated from Huanan market in late 2019, and the emergence of SARS-CoV-2 seemed like a reprise of the 2002–2003 SARS outbreak: live animal trade, respiratory virus, origins in China. That earlier pandemic was a major scare but was extinguished within months (declared contained in July 2003) after a total reported 8096 recorded cases (Peiris et al. 2004). However, by January and February 2020, the links between new cases of SARS-CoV-2 and the Huanan market were tenuous. Instead, cases were mounting in highly populated areas of the city, particularly those areas with high densities of older people.

By 11 March 2020, 121,564 cases had been confirmed in more than 110 countries, with 4373 deaths (case fatality ratio = 3.6%) (Anderson et al. 2020). On that date, the World Health Organization declared COVID-19 to be a pandemic. By October 2020, more than 36 million people had been infected with SARS-CoV-2 (Fig. 15.3), causing more than 1 million deaths (case fatality ratio = 2.8%) (Oude Munnink et al. 2021). And at the time that we finished writing this book in April 2023, more than 6 million people have died from COVID-19 globally, and the virus is still circulating around the world.

## 15.3 The source of SARS-CoV-2 spillover

Since so many cases were linked to the Huanan market and the market was a place where live animals were sold for food and fur, the Huanan market was suspected to be the site of the original zoonotic spillover. Of course, the market was closed and cleaned on 1 January 2020, in response to the outbreak—a righteously swift control intervention, but one that also made it very hard to go back and describe the animal trade that occurred and possible spillover pathways. Serendipitously, a team had been monitoring live animal trade in multiple markets in Wuhan from May 2017 to November 2019 by talking to vendors and recording which animals were for sale. (The team was investigating the ecology and epidemiology of Severe Fever with Thrombocytopenia Syndrome (SFTS), a tick-borne disease; Xiao et al. 2021).

A broad range of live animals were for sale in the market, often (30%) illegally sourced from the wild (as evidenced by gunshot or snare wounds), often kept in unsanitary conditions, and often kept alive and butchered by vendors on-site (Xiao et al. 2021, Fig. 15.4). In addition to wild animals, there were many captively bred animals. For example, 'wildlife farms' supplying the Huanan market stocked hundreds of thousands of raccoon dogs on farms in Enshi Prefecture in western Hubei

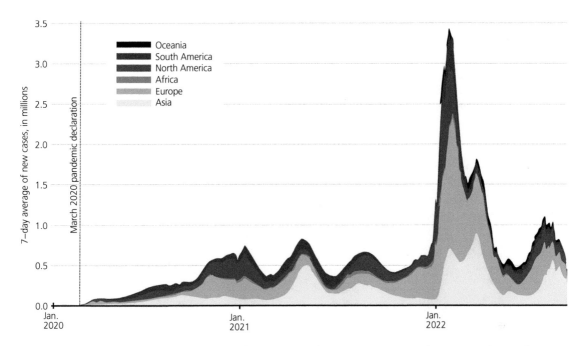

**Figure 15.3** Time series of average daily COVID-19 cases for each continent from January 2020 to September 2022. Note that not all COVID-19 cases were reported and reporting coverage varied by country.

The data were compiled by the COVID-19 Data Repository by the Center for Systems Science and Engineering (CSSE) at Johns Hopkins University (Dong et al. 2020) and the figure was modified from Wood et al. (2022).

Province (Worobey et al. 2022). Some of the species for sale in the Huanan market included animals previously linked to coronavirus infections, such as masked palm civets and raccoon dogs (Xiao et al. 2021, Chapter 8). In short, there were many species that could have introduced a novel coronavirus to the market.

But what was the animal source? Since live, traded animals from the market were not tested, researchers tried testing many wild and farmed species from the surrounding region to look for SARS-CoV-2. Several species harbour viruses closely related to SARS-CoV-2, such as pangolins and bats in China, Thailand, Cambodia, and Japan. For example, SARS-CoV-2-related coronaviruses were isolated in samples from 18 Malayan pangolins (*Manis javanica*) seized during anti-smuggling operations in southern China in August 2017 to January 2018 (Lam et al. 2020). Two sub-lineages of related coronaviruses were identified, exhibiting 85–92% similarity to SARS-CoV-2, and one lineage had similar sequences to SARS-CoV-2 in the ACE2 receptor-binding domain. However, these

sequencing analyses suggest that there is significant evolutionary distance between these progenitors and SARS-CoV-2 (Holmes et al. 2021).

In contrast, a bat coronavirus, RaTG13, collected from a horseshoe bat (*Rhinolophus affinis*) in Yunnan (hundreds of miles away from Wuhan) has a genetic distance of just ~4% (~1150 mutations) from the first SARS-CoV-2 strain (Boni et al. 2020, Zhou et al. 2020, Holmes et al. 2021). While more closely related, this genetic difference still corresponds to decades of evolutionary divergence. If this is the lineage that gave rise to SARS-CoV-2, then the virus has been circulating and evolving unobserved in bats for a long time. More recent analyses have also discovered that there are even closer related viruses to SARS-CoV-2 in bats in Laos than RaTG13 is, but these seem to have mixed ancestries due to recombination (mixing) of viruses, highlighting how hard finding 'a source' might be (Temmam et al. 2022). Further field work is likely to discover other coronaviruses with different and likely closer relationships to SARS-CoV-2, given the numbers of bats and bat species in the region.

**Figure 15.4** Animals on sale in Huanan market: (a) king rat snake (*Elaphe carinata*), (b) Chinese bamboo rat (*Rhizomys sinensis*), (c) Amur hedgehog (*Erinaceus amurensis*) (the finger indicates a tick), (d) raccoon dog (*Nyctereutes procyonoides*), (e) marmot (*Marmota himalayana*) (beneath the marmots is a cage containing hedgehogs), and (f) hog badger (*Arctonyx albogularis*).
From Xiao et al. (2021), used under Creative Commons Attribution 4.0 International License.

Though these data seem to implicate bats as a source of SARS-CoV-2, neither pangolins nor bats were listed as animals available in the live animal markets of Wuhan. (Bats are not typically eaten in Central China. Media footage of bats in wet markets generally depicts markets in Indonesia, though bats could interact with wildlife and farmed animals (Xiao et al. 2021).) Therefore, the reservoir host or hosts that spawned SARS-CoV-2 still have not been identified. This reminds us that identifying reservoir hosts is no mean feat (Chapter 8)! The right animal species and/or populations might not have been sampled yet (Anderson et al. 2020, Holmes et al. 2021); the virus ancestor may be difficult to detect because it has a low prevalence or is transient in the reservoir population. Environmental samples from the Huanan market revealed nucleic acid sequences from a large variety of non-human species, unsurprisingly including poultry, pigs, cattle, and dogs, but also suggesting the presence of hedgehogs, raccoon dogs, ferret badgers, etc., corroborating earlier visual observations (Xiao et al. 2021), and even two bat genera (Liu et al. 2023). Ultimately, the spillover event may in part have been due to the particular alchemy of exposure *and* susceptibility.

Furthermore, the preponderance of early cases linked to the Huanan market does not provide *incontrovertible* evidence that COVID-19 emerged there, though several lines of evidence do support the hypothesis of spillover at the Huanan market. Retrospective spatial analyses showed that SARS-CoV-2-positive environmental samples collected by the Chinese Center for Disease Control and Prevention (CDCC) were in the same section where live animals were sold, including raccoon dogs, hog badgers, and red foxes (Worobey et al. 2022), though environmental samples positive for these species (or at least the genera) were not also positive for SARS-CoV-2 (Liu et al. 2023). Early COVID-19 patients who had connections to the market worked in the same areas of the market with positive environmental samples. And early COVID-19 cases that were not linked to exposure in the market nonetheless lived close to it. So, the simplest and most likely explanation is therefore that spillover occurred at the market, but we cannot rule out the possibility that animals and people at the market were infected by other people.

There have been rumblings about a laboratory origin for SARS-CoV-2, though evidence for this route is much more circumstantial and lacking in detail. Certainly, Wuhan is home to a major virological laboratory that studies coronaviruses. And certainly, laboratory accidents can happen and have included incidents of transmission of Marburg virus, plague, smallpox, and even a small outbreak of the original SARS (Parry, 2004)—incidents described by scientists. But there is no evidence to suggest that the Wuhan Institute of Virology (WIV)—or any other laboratory—was working on SARS-CoV-2, or any virus close enough to have seeded the COVID-19 pandemic. Viral genomic sequencing without cell culture, which was routinely performed at the Wuhan Institute of Virology, uses inactivated non-viable viruses and so is an unlikely spillover route. Extensive contact tracing of early COVID-19 cases has not demonstrated links to laboratory staff at the WIV, and all staff involved in bat virus work were seronegative for SARS-CoV-2 when tested in March 2020 (Holmes et al. 2021).

The precise events that spurred the SARS-CoV-2 pandemic will always be murky, but the most parsimonious explanation for the origin of SARS-CoV-2 remains a zoonotic spillover event, just like that of the 2002–2004 SARS-CoV, and just like so many other pathogens described within this book. Wuhan's large population, and its travel connections to the rest of China and the world, may explain why the spillover could successfully evolve into an outbreak and pandemic.

## 15.4 Controlling the spread of a pandemic virus . . . or not

If you're reading this book, you lived through the COVID-19 pandemic (unless the book somehow remains in circulation for two decades). Consequently, you witnessed how government and public health agencies tried to slow the spread of COVID-19 around the world, and we'll summarize some of those interventions here.

One of the earliest interventions—restricting international travel—did not stop the pandemic from occurring, but it did slow the pathogen's initial spread (Chinazzi et al. 2020, Grépin et al. 2021). Early in the pandemic, most countries completely closed their borders, whereas later in 2020, most countries relaxed to a partial border closure, where only travel from countries with high caseloads was restricted (Bou-Karroum et al. 2021). This included complete bans or requirements for a quarantine period and/or proof of a negative COVID-19 test before travel.

COVID-19 was not the first time that travel restrictions have been used in an attempt to control the spread of infectious diseases. For example, during the 2014 Ebola virus disease outbreak, several high-income countries banned travel to and from the three African countries where Ebola virus was spreading: Guinea, Liberia, and Sierra Leone. While short and early travel restrictions may give nations more time to prepare for epidemics, the economic consequences of travel disruptions, the stigma it creates against 'banned' countries, and people's desire to travel make long-term travel restrictions highly controversial. Furthermore, travel restrictions alone may not be sufficient to stop pandemics in real life or in model simulations; they need to be combined with other methods for reducing transmission.

Restricting human travel can control human-to-human disease spread, but travel and trade restrictions also exist to stop the movement of infected plants and animals across borders. For example, in the United States, there is currently a ban on importing salamanders to prevent the possible importation of the salamander chytrid fungus (*Batrachochytrium salamandrivorans*), which would decimate salamander populations in the world's greatest salamander biodiversity hot spot (Chapter 12; USFWS n.d.). While still potentially controversial for some, animal/plant travel restrictions tend to be better supported by the public, especially because there are historical and present-day examples where imported hosts and pathogens decimated crops, livestock, or wild species (Chapter 9). As with human travel restrictions, some plant and animal travel/trade is allowed following quarantine, vaccination, and/or health screening.

Returning to COVID-19, many countries also restricted travel within the country using 'lockdowns' and social distancing interventions. Businesses and public transportation were closed, people were asked to work from home when they could, large gatherings were prohibited, and travel was only sanctioned for the most urgent situations. These interventions could reduce both density-dependent, disease-relevant contact rates and long-distance spread of COVID (e.g. from one city to another). When implemented rigorously and strictly, models suggested that early lockdowns could have eradicated COVID-19 from some countries, especially if enforced for relatively long periods (i.e. several weeks). However, in many places, lockdowns also incurred economic impacts, and 'voluntary' lockdowns often failed. In countries where lockdowns were successfully implemented and large COVID-19 epidemics were prevented in 2020 (e.g. New Zealand and Australia), there were also benefits from the lockdowns, such as reduced mortality rates from automobile accidents and influenza (Meyerowitz-Katz et al. 2021). Travel restrictions and lockdowns vividly demonstrate that most, if not all, interventions involve trade-offs, where there are benefits *and* costs, and it can be difficult to decide how to balance these trade-offs during rapidly changing, complex situations.

While travel restrictions, lockdowns, and social distancing were implemented to reduce infectious contacts, masking was implemented to reduce the probability of transmission if a contact did occur between an infectious and susceptible person. Face masks serve dual purposes with regard to airborne disease transmission: (1) they reduce the dose of SARS-CoV-2 emitted from an infectious

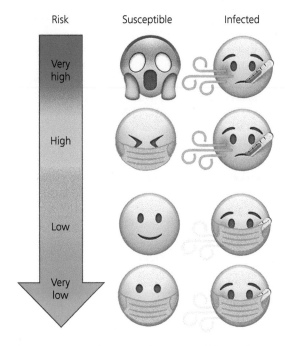

**Figure 15.5** A modified version of an educational diagram depicting how COVID-19 transmission risks could be reduced by masking. Note that N-95 masks have higher efficacy than surgical masks, but the authors of this book have limited artistic talent and spent all of it drawing musk ox and vicuñas in earlier chapters.

person into the air, and (2) masks also reduce the dose of infectious SARS-CoV-2 inhaled from the air by a masked person. When both an infectious and susceptible person are masked, there is maximum transmission reduction (Fig. 15.5). Improving indoor air ventilation and filtering similarly reduces the likelihood that airborne disease agents make it from one host to another; combining improved ventilation with masking is an especially effective way to reduce the probability of transmission parameter (Box 15.2).

In some countries, masks were already popular ways for the public to prevent spreading infectious agents or hide pimples or avoid inhaling air pollutants, so their use was not controversial. In other countries, masks were not already popular, but they were still widely adopted. In still other countries, masks were highly controversial, and seen as an affront to individual freedom. Further complicating the matter, the messaging around masking changed repeatedly. In the United States, people were initially encouraged *not* to wear N95s—advice motivated by the need to preserve the supply of masks for medical professionals. Subsequently, guidance advocated N95s or respirators, and then evolved again, depending on whether people were indoors or outdoors (even though crowded outdoor scenarios could still exhibit a high risk of airborne transmission). Ultimately, masks can be highly effective tools for reducing airborne pathogen transmission (though respirators are more effective than N95s, which are more effective than surgical or cloth masks), and this intervention is more successful for individual and population-level transmission control when it is widely adopted. The controversies that raged around mask use demonstrate both the changing understanding of a system as it unfolds, and the importance of education and science communication.

The creation and rapid mass production of vaccines for COVID-19 is an incredible example of how vaccination and technology saved countless human lives around the world. To provoke human immune systems to rapidly recognize and fight SARS-CoV-2, the most widely distributed vaccinations used one of three methods: two mRNA vaccines (the Pfizer-BioNTech and Moderna vaccines), a protein subunit vaccine (Novavax), and a viral vector vaccine (Johnson & Johnson). Research on mRNA vaccines for other coronaviruses was already underway before COVID-19 (Krammer 2020), kickstarting the fastest vaccine development cycle in history. In the United States, the Food and Drug Administration (FDA) issued emergency approval of the mRNA vaccines in December 2020, just a year after COVID-19 emerged. This is remarkable because vaccine development and safety trials usually take 10–15 years!

For disease control, the goal of vaccination is to shift people from the susceptible class to the resistant class without them ever entering the infectious class (Chapters 5 and 14). COVID-19 vaccines reduced susceptibility to infection (or subsequent infectiousness of infected individuals), though **vaccine efficacy** varied somewhat among vaccines and viral strains, and none of the vaccines completely prevented infection (Eyre et al. 2022). The vaccines also greatly reduced the probability that an exposed person would develop severe disease or die from COVID-19 (Polack et al. 2020, Thomas et al. 2021). The level of protection against severe symptoms was also impacted by the type of vaccine, number of doses, viral strain, and likely the initial exposure dose (Abu-Raddad et al. 2021, Levin et al. 2021).

Though the COVID-19 vaccines were remarkable, vaccine availability also highlighted ongoing disparity issues in public health. High-income countries were better able to purchase vaccines and distribute them than lower-income countries. Even within countries, vaccine access was inequitable, such that people from low socioeconomic backgrounds or otherwise marginalized groups were less likely to have affordable access to vaccines close to home (Wouters et al. 2021). And of course, vaccines work best for people with well-functioning immune systems, so immunocompromised people are often less well protected and still susceptible to infection and disease after vaccination, unless herd immunity is reached, and the pathogen is eventually eliminated from the population (Chapter 5). For COVID-19, we have not reached the herd immunity threshold, which is higher for vaccines that have lower efficacy (Chapter 5).

Though testing for SARS-CoV-2 is not strictly an intervention, it *was* an important surveillance tool and a source of information for governments and individuals making health-related decisions. For example, in many countries with access to diagnostic tests, schools and businesses implemented mandatory weekly testing, and individuals who tested positive for COVID-19 were required to stay home for a designated quarantine period. Of course, individual test results are not 100% accurate (Chapter 3) and sensitivity and specificity of COVID-19 tests varied between the two common test methods: the PCR-based test and the antigen-based test. The PCR test was designed to amplify the virus's RNA (see PCR methods in Chapter 3), and thus it could detect the virus even if there was only a tiny amount present in the test sample. It is difficult to find an exact clinical sensitivity and specificity for the PCR-based test because there is no gold standard to compare it to, but estimates seem to be ~81–92% sensitivity and ~99–100%

specificity (He et al. 2020). In contrast, the antigen tests were designed to detect a viral protein or antigen and were less sensitive: reported sensitivity ranged from 18–96% (depending on viral loads in the patient and other factors) and specificity was ~90–100% (Jegerlehner et al. 2021, Vandenberg et al. 2021). Both the PCR and antigen tests were more sensitive with nasopharyngeal swabs because the swab was more likely to pick up more virus, but that was also the most uncomfortable method for patients, so the less-sensitive anterior nasal swab was often used instead (Callahan et al. 2021, Zhou and O'Leary 2021). While both tests were used by individuals to make decisions about quarantining, they had different pros and cons: the more sensitive PCR-based test usually had to be performed by a medical professional and analysed in a lab, so it took more effort and results were delayed by hours to days, whereas the less sensitive antigen test could be completed by the patient at home, with results in minutes.

Another innovative approach to testing for SARS-CoV-2 was to examine wastewater (Delong and Wilusz 2022). The virus is present in surprisingly large quantities in infected people's faeces (Vaselli et al. 2021), so community-level transmission can be monitored by examining faecal matter from sewage systems. Individual cases aren't identified, but the upside of this tactic is that it can detect pathogen transmission trends without relying on individual-level data, which is influenced by whether cases are symptomatic, the local availability of clinical tests, and people conscientiously reporting their test results. Information at the 'sewershed' level also facilitates more targeted subsequent surveillance, e.g. wastewater that can be attributable to areas such as nursing homes or university residences or aircraft allows subsequent, more targeted screening of individuals (Delong and Wilusz 2022, Morfino et al. 2023). Sampling in this manner allows monitoring of pathogen evolution and strain predominance over time (Fig. 15.6), and this surveillance approach can be adapted to other pathogens that are contained in excrement.

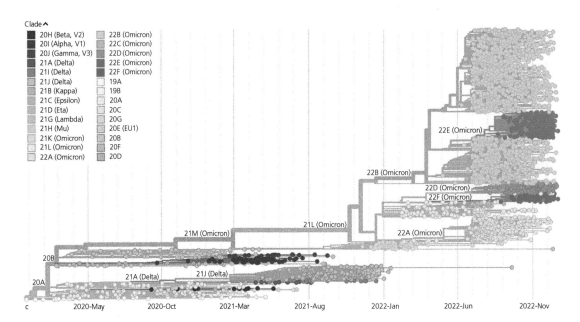

**Figure 15.6** The phylogenetic relationships among the many strains of SARS-CoV-2, the virus that causes COVID-19, and when those strains first emerged.
This open-source figure was created by Nextstrain, a free platform that allows you to map and visualize strain dynamics for multiple pathogens: https://nextstrain.org/.

---

**Box 15.2  A mathematical framework for COVID control**

Stock-and-flow diagrams (see Chapter 5) are useful to visualize the state variables and parameters in compartmental models, like SIR (Susceptible, Infected, and Resistant) models. Below (Fig. 15.7), we have built a model for the purposes of education and understanding; the model contains most of the important components, but we have made several simplifying assumptions that would reduce our accuracy if we tried to use the model to predict disease dynamics. Most importantly, we (incorrectly) assumed that after someone became infected with COVID-19 or was vaccinated, they became resistant to all future infections (no flow from the resistant compartment back to infected or susceptible compartments). Our model also does not include immigration or emigration from other populations (or the effect of travel restrictions), which is not realistic for a pandemic that spread across all human populations via travel. Instead, the population is only impacted by births (b) into the Susceptible compartment and deaths (d) from all compartments. Keep-

ing these simplifying assumptions in mind, we can use our simple model to examine how disease control interventions would affect the various parameters and disease dynamics.

At the start of the pandemic, most people in any given population were susceptible, and there were no vaccines. If COVID were to reach a population (via immigration of infected individuals; not shown), the most critical control interventions were those that reduced the transmission rate, $\beta$. As you may remember from Chapter 5, the transmission rate is the product of two parameters, the disease-relevant contact rate (e.g. how many people's exhalations did you breathe in) and the probability of transmission success given a contact between an infectious and a susceptible individual. To reduce contact rates, which are often a function of host density, there were interventions like social distancing and lockdowns. To reduce the probability of transmission success, there were interventions like wearing face masks or improving indoor ventilation and air filtering.

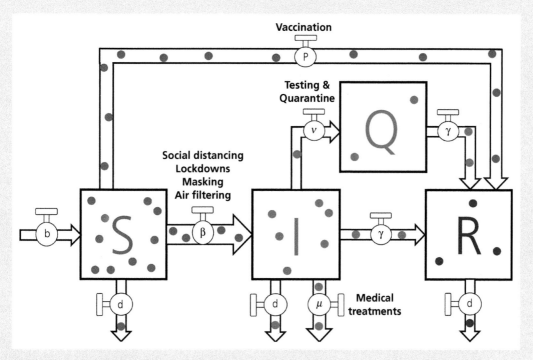

**Figure 15.7** A stock-and-flow diagram showing interventions used to control the spread of COVID-19 and the specific model parameters that those interventions affected.

---

**Box 15.2** *Continued*

When vaccines became available, they could be administered to susceptible individuals to reduce their susceptibility to infection. Here, we show a simplified scenario, where some proportion of people (P) are vaccinated and never become infected. You may remember from Chapter 5 that the vaccine threshold (aka herd immunity threshold), or the proportion of people that need to be vaccinated to start eliminating an infectious disease from a population, is P = $(1 - (1/R_0))*(1/\text{Vaccine Efficacy})$. So, the higher the vaccine efficacy, the lower the required vaccine threshold (P).

For many reasons, people did still become infected. To reduce the probability that people passed on those infections, there were interventions to reduce contact rates and the probability of transmission success. These included interventions that we already discussed, like social distancing, lockdowns, masking, and improved ventilation. It also included quarantining infected people so that they had no interactions with susceptible people; instead of dS/dt = $-\beta*S*I$, we have dS/dt = $-\beta*S*(I - Q)$. The probability that

an infected person went into quarantine ($\nu$) instead of going about their daily routine depended on many factors, including whether a person knew that they were infected. This knowledge depended on whether people were testing regularly and whether symptoms manifested before or after the person became infectious, i.e. the timing of the incubation period and the latent period (Chapter 2).

For some infectious diseases, like influenza, there are public health interventions to reduce the period that a person is infectious ($1/\gamma$), which is the same as increasing the rate that infected people move into the resistant class ($\gamma$). However, this was not a major focus of public health or medical interventions for COVID-19. Of course, there were other medical interventions that aimed to reduce disease severity for people who were infected, such as putting hospitalized people on ventilators. These interventions would not substantially impact transmission dynamics, but when they were available, they did save many lives by reducing the probability of disease-induced mortality ($\mu$).

## 15.5 Drivers of the COVID-19 pandemic

First, animal trade can and does cause disease spillover and amplification from wildlife or domestic species to other animals and people (Chapters 9 and 11). In the case of COVID-19, animals were being brought from the wild, wildlife farms, and captivity to the Huanan market, and enforcement of trade regulations and hygiene was limited (Xiao et al. 2021). This suggests that improved regulations and enforcement could prevent similar outbreaks in the future. And such measures are unlikely to cause food insecurity (Chapter 11), because the extensive trade in Wuhan appears to be driven by demand for a luxury commodity rather than subsistence needs for cheap protein (Xiao et al. 2021); while average retail prices of pork, poultry, and fish in Wuhan were $5.75/kg, $4.25/kg, and $2.32/kg, respectively, more exotic proteins were more expensive: marmots (*Marmota himalayana*) cost >US$25/kg; raccoon dogs (*Nyctereutes procyonoides*) and badgers (*Arctonyx albogularis* and *Meles leucurus*) cost $15–20/kg; hedgehogs (*Erinaceus amurensis*) cost $2–3 each; Indian peafowl (*Pavo cristatus*) cost $56 each;

and sharp-nosed pit vipers (*Deinagkistrodon acutus*) cost $70/kg (Xiao et al. 2021). Demand for exotic wildlife may decline with increasing awareness about pathogen spillover risks.

Second, globalization allows for rapid pathogen spread. Even though this was known in early 2020, and even though some travel restrictions were enforced, COVID-19 still spread around the world. This rapid spread was likely exacerbated by the fact that Wuhan has a large, susceptible urban population that was well-connected (Chapter 14) by local and international travel. At the same time, the spread was likely greatly reduced by available local resources for rapid case identification and contact tracing in Wuhan and other locations that were willing and able to do this. Continuous surveillance and rapid responses are required to control outbreaks, but the best-case scenario is to avoid spillover in the first place.

And third, we cannot always determine which combination of proximate or ultimate drivers caused a given outbreak, which reminds us that our actions may have far-reaching negative consequences that we never realize. For COVID-19, it is impossible to say whether climate change, biodiversity loss, habitat change, or other drivers contributed to disease emergence. They may not have contributed to this pandemic, but they may cause the next one.

## 15.6 Complexity and wicked problems

The COVID-19 pandemic, and so many other topics covered in this book, demonstrate that animal health, human health, and environmental health are inextricably linked. Therefore, trying to understand, predict, or control outbreaks is *difficult*. In the next section, we will discuss how interdisciplinary approaches are needed to tackle these problems and highlight some general ways that problems can be classified and approached.

Traditional scientific approaches often take a **reductionist approach**: they isolate and describe the various parts of a system and then use the knowledge of the parts to try to understand the whole (Waltner-Toews 2017). This works well when processes are **simple**: when you can break them into component parts, put them together, and realize a good approximation of the studied phenomenon. As an analogy, think of cooking a simple meal (cookies, carbonara, or curry). Even without much culinary experience, you can follow step-by-step instructions and have succeeded in your goal by the end of the process. With increased experience in doing each step, the overall quality of the dish likely improves.

Processes can also be **complicated**, involving more intricate steps and requiring more expertise, though often still generating a predictable result. Success, once again, is more likely with increased experience. Instead of a simple meal, imagine cooking several courses, and one of them ought to be blanched asparagus bathed in a coffee-foam with sea-urchin eggs lurking nearby. Or consider launching a rocket to the moon in the 1960s using incredible scientific knowledge.

Infectious disease systems are more likely to be **complex**, with relationships among pathogens, hosts, and the environment that are **non-linear**. In complex systems, it is extraordinarily difficult

to come up with reliable predictions. As an analogy, consider predicting the weather or local-scale impacts of climate change. Faced with similar problems, Edward Lorenz, a meteorologist and mathematician, coined the phrase 'the butterfly effect'—the influence of a seemingly insignificant action to cascade into far-reaching important impacts, like the metaphorical flap of a butterfly's wing provoking tornadoes. This was after he noticed that the simple act of rounding up variables in a computer weather simulation led to significantly different predictions (Dizikes 2011). This is particularly apt for deterministic non-linear systems, where small fluctuations in initial conditions can be exacerbated exponentially. In the world of infectious disease dynamics, downstream effects of land-use change, oil consumption, or wildlife trade may have unimagined and unintended consequences, like vulture declines due to accidental poisoning causing an unanticipated increase in human rabies cases (Markandya et al. 2008). Stochasticity also plays an important role in complex systems.

Finally, there are **wicked** problems, which arise when situations can be defined from a variety of apparently incompatible perspectives (Waltner-Toews 2017), such that people cannot even agree on the variables and their relationships in the infectious disease system. This might be best illustrated by the COVID-19 pandemic, where the media and social media were full of 'experts' who had such different perspectives that they gave completely incompatible advice to the public. Of course, some of those people were not true experts, but wicked problems can exist even when there are no 'bad actors', or people who are purposely misunderstanding or misrepresenting the situation. To an extent, wicked problems can reflect diverse values of stakeholders or can arise simply because the system is so vast and multifaceted that no one person or group can understand it all. Unsurprisingly, emerging infectious diseases are often prime examples of complex, wicked problems, where people have contrasting views on why they arose and spread (e.g. wildlife trade, inadequate surveillance or treatment), how to combat them (e.g. vaccines, lockdowns, face masks, herd immunity), how to measure them, etc.

So, how does one 'solve' a complex or wicked problem given that complex systems are not often fully understood and are open to multiple interpretations? A common approach is to try several complementary interventions (e.g. travel restrictions *and* masking *and* vaccination) and then monitor which work best in different contexts. And by 'work best', we do not mean that one will serve as a miracle cure. Most interventions will have trade-offs (good for some people but not others) and most will only have small to moderate impacts when used alone.

In fact, when wicked problems exist, they may never be 'fixed', but the whole system may be managed in such a way as to improve the situation for many people (while hopefully prioritizing the most vulnerable people). For example, when the scale of COVID-19 transmission became apparent in early 2020, some public health strategies emphasized 'flattening the curve'—not reducing total transmission in the long run, but slowing disease spread enough to avoid overwhelming healthcare systems. Approaches like this may make systems more adaptable or resilient in the face of catastrophic change—the system changes in response to the shock, but there are enough failsafes in place to prevent worst-case scenarios (e.g. healthcare collapse *and* rampant disease spread). Understanding what makes systems resistant and resilient to major shocks is a fast-growing, urgent area of research, with applications to the climate crisis, biodiversity loss, emerging infectious diseases, etc. Unlike a reductionist approach, these topics are studied using a **systems approach**, which emphasizes the importance of cohesively examining the entire system at once to understand drivers, relationships, and the positive and negative consequences of potential interventions.

Investigators into complexity do not seek prediction, control, right answers and efficiency. These are not sensible goals under conditions of complexity. Rather, the investigators *seek understanding, adaptability and resilience*. Scientific inquiry, more than ever, becomes an act of collaborative learning and knowledge integration. The role of the expert shifts from problem solving to an exploration of possibilities, from giving correct advice to sharing information about options and trade-offs.
—James Kay (cited in Waltner-Toews 2017)

## 15.7 The One Health approach and interdisciplinary collaboration

Between animal and human medicine there are no dividing lines—nor should there be.
—Rudolf Virchow, 1856

To tackle complex problems in interconnected health systems, several interdisciplinary fields have emerged over the past few decades, including '**One Health**', '**Planetary Health**', and '**Ecohealth**'. Though each of these fields has its own emphasis and roster of researchers and practitioners, they all adopt and celebrate an interdisciplinary, cohesive approach to simultaneously research and address animal (domestic or wild), human, plant, and environmental health. These frameworks promote the importance of 'the environment' or 'ecosystem health' in a way that is different from traditional 'environmental health', which focuses on the impact of the human environments on human health (e.g. exposure to air and water pollutants at home or at work). Here, we only cover the One Health approach in detail, but we encourage you to look up the other approaches, too.

Imagine a well-intentioned disease ecologist working conscientiously to describe zoonotic pathogen prevalence in communities of rodents and other small mammals as a function of land use types—protected, conserved lands; pastures with different degrees of grazing; and small-scale farms with crops. Can this be considered a One Health approach that looks at land use as a potential driver of emerging infectious diseases or zoonotic spillover to humans? Perhaps. But then imagine inviting an epidemiologist to the project and their first question is, 'Which zoonotic pathogens are the local people suffering from?' Or imagine inviting an anthropologist or economist to the project and their first question is, 'How do the people living here make money, and can they afford and access healthcare if they get sick?' The disease ecologist may not have much of a clue about which pathogens are circulating among people! The epidemiologist is similarly clueless about potential spillovers from the local mammal community. Different experts—e.g. the epidemiologist, the anthropologist, the ecologist, the local

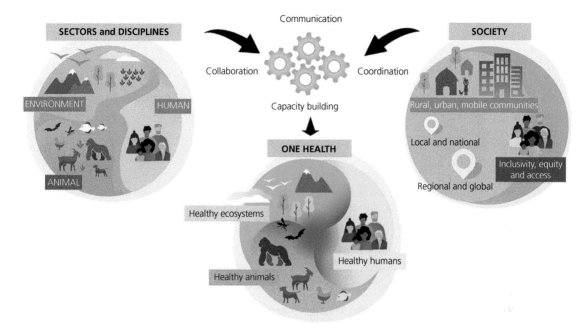

**Figure 15.8** Conceptual approach of One Health.
From OHHLEP et al. 2022, available under the Creative Commons CC0 public domain dedication.

pastoralists, the conservationists—have different tools, experiences, and perspectives. Sharing and translating these intersectoral stances could offer more useful innovative interpretations and solutions than would be yielded by traditional, siloed, single-disciplinary approaches.

The One Health approach makes sense, though a curmudgeon could argue that research on emerging zoonoses that includes aspects of wildlife ecology, livestock, local people, and land use is simply infectious disease research done properly. And perhaps the innovative 'new' One Health approach is simply new clothes for a naked emperor of eco-epidemiology.

Thankfully, a more exciting One Health approach has evolved, which includes interdisciplinary approaches *and* an aim to simultaneously improve the health of people, animals, plants, and the environment—not just research, but also action (Fig. 15.8). A new definition, agreed upon by large intergovernmental organizations—the Food and Agriculture Organization of the United Nations (FAO), the World Organisation for Animal Health (WOAH), the UN Environment Programme

(UNEP), and the World Health Organization (WHO)—defines One Health like this:

**One Health** is an integrated, unifying approach that aims to *sustainably balance and optimize* the health of people, animals, and ecosystems. It recognizes the health of humans, domestic and wild animals, plants, and the wider environment (including ecosystems) are closely linked and interdependent. The approach mobilizes multiple sectors, disciplines, and communities at varying levels of society to work together to foster well-being and tackle threats to health and ecosystems, while addressing the collective need for healthy food, water, energy, and air, *taking action* on climate change and contributing to sustainable development.

(OHHLEP 2022; we added italics)

This One Health approach also echoes another international framework, The United Nations Sustainable Development Goals, which identify 17 interconnected aspirations for improving planetary health, including preventing future pandemics and promoting recovery since the emergence of COVID-19 (Fig. 15.9). These aims, and the processes to achieve them, are explicitly understood to be

**Figure 15.9** The 17 Sustainable Development Goals, which are interconnected.
This figure was used with permission from the United Nations (https://www.un.org/sustainabledevelopment/). The content of this publication has not been approved by the United Nations and does not reflect the views of the United Nations or its officials or Member States.

interconnected—we cannot fully achieve one without achieving the others. Additionally, if we only focus on one (e.g. Goal 8, economic growth), we may take away from the others, due to the existence of **trade-offs**; a simple example is that we cannot keep converting land to grow food, because ultimately ecosystems will collapse.

Of course, there is no simple, effective, one-size-fits-all approach for One Health implementation; each problem must be tackled with an appropriate team and an appropriate approach for the given circumstances (Box 15.3). Solutions that improve situations in one place are not guaranteed to have the same effects elsewhere, because every place is different. For example, remember how reptiles play a role in Lyme disease dynamics in some parts of the United States, but not others? Remember how domestic dogs are the main reservoir of rabies in some places, but not others? Remember how some people were happy to wear masks to prevent COVID transmission, whereas masks were seen by others as an affront to personal freedom? This reminds us that teams should involve *local*

experts, including the people who will live with the 'solutions' applied to their region. One may still take lessons learned from one scenario and try to adapt them to new situations. But one must also realize that flexibility and adaptability are key, and consistent monitoring is needed so that approaches can be changed when they are not working.

Even when a great team is in place, the jargon and disciplinary-specific methods used by different sectors may create barriers or lead to miscommunication. This reminds us that people who specialize in communicating with diverse audiences can be critical for interdisciplinary teams. And even for people who aren't communication experts, making a deliberate attempt to make your work more broadly accessible is invaluable. This can include non-traditional techniques like narrative storytelling or creating conceptual diagrams (Fig. 15.10) that illustrate the issues but avoid technical language. In short, One Health approaches need to meaningfully involve everyone, which is difficult to do but rewarding when it works!

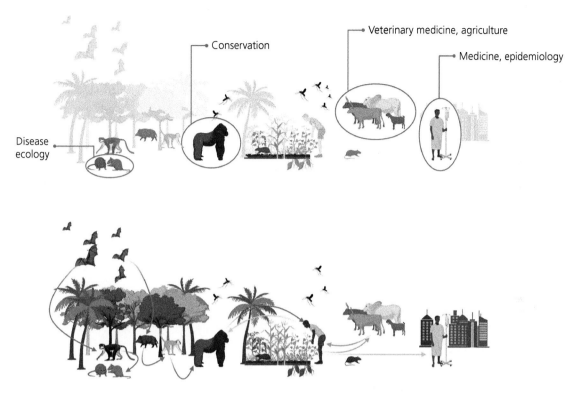

Disease
ecology

Conservation

Veterinary medicine, agriculture

Medicine, epidemiology

**Figure 15.10** Conceptual diagram representing: (top) traditional approaches to health and conservation that are undertaken within single disciplines or expertise; and (bottom) researching the components as part of a whole system may be more informative in understanding and tackling drivers of health and conservation issues.

---

**Box 15.3  A complex problem and one complex solution: African great apes and humans**

Five great ape species occur across Africa: humans (*Homo sapiens*), chimpanzees (*Pan troglodytes*), bonobos (*Pan paniscus*), western gorillas (*Gorilla gorilla*), and eastern gorillas (*Gorilla beringei*) (Fig. 15.11). We, *Homo sapiens*, are the most abundant; consequently, the non-human apes have long been threatened by anthropogenic change, such as deforestation and forest degradation (Strindberg et al. 2018, Chapter 11). At present, gorilla and chimpanzee densities are highest in intact forests with limited human influence. Additionally, as more people encroach into previously intact forests, apes that venture into human-dominated areas and 'raid' crops may be killed because people farming those crops are food- or income-insecure (Hockings 2016). Apes are also hunted for consumption or sale; such practices

are illegal, but the activities may be difficult to prevent and are exacerbated by poverty and food insecurity (Braczkowski et al. 2023, Chapter 11).

Though there is no single solution that will resolve human–ape conflicts across all of Africa, one attempted solution is ecotourism (McKinnie 2016). Ideally, ecotourism allows people from outside an area to travel to that area, observe great apes in a way that does not disturb or otherwise harm those species, and bring income to local communities in an equitable way (Archabald and Naughton-Treves 2001, Blomley et al. 2010). Benefits of community projects have been shown to reduce illegal activities in parks (Bitariho et al. 2022). Additionally, new local livelihoods related to ecotourism can help alleviate food insecurity for

**Box 15.3** *Continued*

local people and increase the perception that great apes have value, thereby improving efforts to conserve apes and reduce human–ape conflicts. Some programmes have come close to achieving this ideal and have truly benefitted both local communities and apes, while other programmes have had negative impacts (Sabuhoro et al. 2021, Bitariho et al. 2022). This brings us back to the topic of this textbook: any time that humans and apes interact closely, be it through hunting, sharing land, or ecotourism programmes, there is the potential for the spread of infectious diseases either from or to apes.

Given the close phylogenetic relationship between humans and apes, it is no surprise that we can and have shared many infectious disease agents, including viruses (e.g. measles virus, polio virus, human respiratory syncytial virus), bacteria (e.g. *Streptococcus pneumoniae*), and parasites (e.g. mange mite, *Giardia* spp.). Humans—whether local or visitors—can spread these disease agents to apes by aerosol transmission or environmental transmission. There are several examples where ape infections and disease have been linked to human transmission in places where apes are habituated to ecotourism (Dunay et al. 2018). This is concerning because some of our shared pathogens can cause high morbidity and mortality for humans and/or great apes, including Ebola viruses.

During Ebola virus disease outbreaks in human populations, it is difficult to track what is happening in nearby gorilla populations, but they are sometimes found dead and test positive for Ebola virus. Among people, Ebola virus outbreaks occur mostly in hot spots of forest fragmentation (Rulli et al. 2017). Whether the same is true for apes is unknown, but gorilla, chimpanzee, and duiker (small antelope) can increase in abundance after forest disturbance, potentially increasing human–wildlife contact in these environments. There is no current evidence that Ebola virus has been spread from humans to gorillas (i.e. anthroponotic transmission), but there is documentation of likely transmission from gorillas to humans in cases when gorilla carcasses were found and consumed. Ebola virus has also been spread from an unconfirmed reservoir species to both gorillas and humans (Rouquet et al. 2005). And several studies have compared gorilla densities in periods before and after Ebola virus outbreaks in human populations and have found catastrophic declines in the gorilla populations (Huijbregts et al. 2003, Bermejo et al. 2006). The situation is the same for chimpanzees (Walsh et al. 2003, Leroy et al. 2004, Rouquet et al. 2005).

Given the risks of anthroponotic disease, what should be done? Ban ecotourism? That may be the best option for apes and people in some places, but in others, it may be a case of throwing the baby out with the bathwater. Ecotourism can benefit ape conservation and local communities in some situations, so many programmes have worked to make human and ape ecotourism interactions safer, e.g. Macfie and Williamson's (2010) 'Best Practice Guidelines for Great Ape Tourism'. These guidelines cover many topics besides disease control, but here we summarize only those pertinent to disease control.

The first disease control guideline will sound familiar to all of us who lived through 'social distancing': people wearing N95 face masks are required to stay at least 7 metres away from apes and unmasked people are required to stay at least 10 metres away. These distances are important even when humans are masked because the endangered apes are not masked. In an ideal world, masks would just be an extra precaution, because no infectious human should come into close contact with apes in the first place. To that end, reputable ecotourism programmes: prohibit children (<15 years old) due to their high likelihood of carrying infectious diseases; require proof of vaccination from tourists for diseases like polio and measles, which have caused mortality in great ape populations; and may have tourists quarantine for several days before tours. Ecotourism programmes are also suggested to have employee health programmes that provide vaccinations, tuberculosis screening, and health education. These same programmes are sometimes also extended to local communities because local communities are also in close contact with habituated apes.

Despite all this, there could be cases where an infected person ends up on a tour, and thus there are also guidelines in addition to masking and distancing to prevent environmental transmission; for example, tourists are asked to wash their hands and disinfect their shoes before tours, and one cannot spit, smoke, eat, dispose of food, defaecate, or otherwise contaminate environments near great apes. Though it is difficult to prove that all these measures have prevented anthroponotic or zoonotic transmission between humans and apes, these efforts should benefit the health of everyone (i.e. One Health) if followed and enforced (Weber et al. 2020).

Of course, COVID-19 presented a crisis to ape ecotourism programmes as global travel restrictions stopped the flow of tourists, creating economic problems for local communities and compromising ape populations. During the first

**Box 15.3** *Continued*

few months of the COVID-19 pandemic, forest patrols in protected areas in Uganda, including Bwindi Impenetrable National Park, found twice as many illegal wildlife snares (Kalema-Zikusoka et al. 2021). As travel restrictions were lifted, programmes needed to decide whether it was safe to re-open tourism programmes, given that COVID-19 could be transmitted from humans to apes (Nerpel et al. 2022). In response, some ecotourism programmes increased their disease control efforts, such as by requiring greater distances between people and apes or negative COVID tests—though

once again, these measures must constantly and consistently be enforced (Weber et al. 2020).

When we discuss 'complex problems' and 'wicked problems', we're talking about dynamic systems with many changing and interacting parts. There are many people involved in these systems, and they have different perspectives. A solution that benefits some people, or apes, or forest integrity, or whatever else, will not benefit everyone everywhere. And a solution that has some benefits today may not be viable tomorrow.

**Figure 15.11** Mountain gorilla, Rwanda.
Photo by Peter Hudson.

## 15.8 Preventing pandemics

*Only a fool learns from his own mistakes. The wise man learns from the mistakes of others.*
—Otto von Bismarck

After HIV, Zika, H1N1, SARS, MERS, and SARS-CoV-2 (to name a few), there is obviously an impetus and a rationale to try to prevent pandemics (Chapter 14) and, given an understanding of the processes involved between spillover and pandemic (Chapters 1–14), there are strategic interventions available. There are numerous new recommendations (e.g. Frieden et al. 2021), but these have been summarized nicely as: 'See the signals. Speed the response. Stop outbreaks.' (Rockefeller Foundation 2023).

**See the signals:** Surveillance can quickly recognize burgeoning outbreaks and facilitate control efforts early on to reduce disease incidence and spread. This requires global training and resources for operations. The rapid release of the SARS-CoV-2 genome exemplifies this philosophy—rapid information collection and sharing are critical for disease control (e.g. Frieden et al. 2021).

**Speed the response:** Multiple control interventions are always available. Given the speed with which interventions must be started, resources should always be ready to control new epidemics. For example, personal protective equipment (PPE) should always be well stocked and we should invest in vaccine technology that can be rapidly adapted to novel pathogens (Keusch et al. 2022, Nuismer and Bull 2020). The mRNA vaccine technologies that were game-changers for COVID-19 were available due to prior long-term investment and development. (And though we are here focusing on preventing pandemics, vaccines and other treatments can also be used to combat existing neglected diseases.)

**Stop outbreaks:** Smart surveillance, vaccine development, and pathogen treatment are all responsive measures. A third approach is to intervene upstream, to control the drivers of infectious disease emergence before emergence ever occurs. For example, the number of spillover events that occur each year could be reduced by reducing habitat loss/degradation, reducing human–wildlife contacts, improving biosecurity in animal husbandry, improving wildlife trade regulations or enforcement, slowing the rate of climate change, or taking other actions to prevent the ultimate or proximate causes of disease emergence.

Just like you can combat high blood cholesterol by increasing exercise, changing your diet, *and* taking lipid-lowering medicines like statins, *we* can reduce pandemic risks by improving surveillance, improving rapid responses, *and* targeting the upstream causes of disease emergence (Fig. 15.12).

## 15.9 Conclusion

This book has taken a very long time to write. During that time, COVID-19 ravaged the world, and our publishers quietly gnashed their teeth and rent their clothes in lockdown because we had missed the chance to be topical. But as we organized this last chapter, we read news about how (1) the first monkeypox pandemic is sputtering out; (2) an unprecedented outbreak of Marburg virus, the haemorrhagic fever virus that afflicted visitors to Python Cave in Uganda (Chapter 9), is happening in Equatorial Guinea (Kritz 2023); and (3) highly pathogenic avian influenza (H5N1 strain) is decimating global seabird populations, wrecking the domestic chicken industry (where culling is a control strategy), killing South American sea lions (*Otaria flavescens*) in Peru, and circulating in mink fur farms in Spain (Agüero et al. 2023, Chappell 2023, Valdez and Aquino 2023). In late February 2023, an 11-year-old girl in Cambodia died from H5N1—the first known human infection in Cambodia in nearly a decade, and additional people are suspected to be sick (Khmer Times 2023). So, unfortunately, our book about emerging infectious diseases is still relevant, and it will continue to be for the foreseeable future. We hope that having read it, you feel better informed.

## 15.10 Notes on sources

Quote from President Donald Trump reported in 'Six months of Trump's Covid denials: "It'll go away ... It's fading"' by Ed Pilkington, The Guardian, 29 July 2020: https://www.theguardian.com/world/2020/jul/29/trump-coronavirus-science-denial-timeline-what-has-he-said, accessed 16/9/2022; and in '"It's going to disappear": a timeline of Trump's claims that Covid-19 will vanish' by

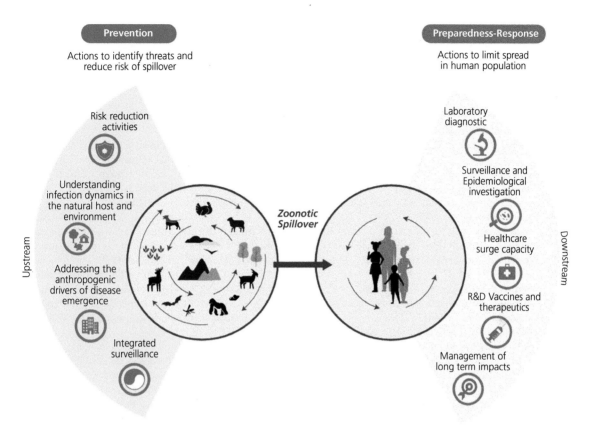

**Figure 15.12** Prevention of pandemics can adopt measures to prevent or reduce incidents of zoonotic spillover *and* to improve preparedness to respond to emerging infectious diseases.

From OHHLEP 2023: https://www.who.int/publications/m/item/prevention-of-zoonotic-spillover.

Daniel Wolfe and Daniel Dale, cnn.com, 31 October 2020: https://edition.cnn.com/interactive/2020/10/politics/covid-disappearing-trump-comment-tracker/, accessed 16/9/2022.

## 15.11 References

Abu-Raddad, L.J., Chemaitelly, H., Butt, A.A. (2021). Effectiveness of the BNT162b2 Covid-19 vaccine against the B. 1.1.7 and B. 1.351 variants. New England Journal of Medicine, 385, 187–9.

Agüero, M., Monne, I., Sánchez, A., et al. (2023). Highly pathogenic avian influenza A (H5N1) virus infection in farmed minks, Spain, October 2022. Eurosurveillance, 28, 2300001.

Andersen, K.G., Rambaut, A., Lipkin, W.I., et al. (2020). The proximal origin of SARS-CoV-2. Nature Medicine, 26, 450–52.

Archabald, K., Naughton-Treves, L. (2001). Tourism revenue-sharing around national parks in Western Uganda: early efforts to identify and reward local communities. Environmental Conservation, 28, 135–49.

Bermejo, M., Rodríguez-Teijeiro, J.D., Illera, G., et al. (2006). Ebola outbreak killed 5000 gorillas. Science, 314, 1564.

Bitariho, R., Akampurira, E., Mugerwa, B. (2022). Long-term funding of community projects has contributed to mitigation of illegal activities within a premier African protected area, Bwindi Impenetrable National Park, Uganda. Conservation Science and Practice, 4, e12761.

Blomley, T., Namara, A., McNeilage, A., et al. (2010). Development AND Gorillas? Assessing Fifteen Years of Integrated Conservation and Development in Southwestern Uganda. Natural Resource Series No. 23. International Institute for Environment and Development (IIED), London and Edinburgh, UK.

Boni, M.F., Lemey, P., Jiang, X., et al. (2020). Evolutionary origins of the SARS-CoV-2 sarbecovirus lineage responsible for the COVID-19 pandemic. Nature Microbiology, 5, 1408–17.

Bosco-Lauth, A.M., Hartwig, A.E., Porter, S.M., et al. (2020). Experimental infection of domestic dogs and cats with SARS-CoV-2: pathogenesis, transmission, and response to reexposure in cats. Proceedings of the National Academy of Sciences, USA, 117, 26, 382–8.

Bosco-Lauth, A.M., Root, J.J., Porter, S.M., et al. (2021). Peridomestic mammal susceptibility to Severe Acute Respiratory Syndrome coronavirus 2 infection. Emerging Infectious Diseases, 27, 2073–80.

Bou-Karroum, L., Khabsa, J., Jabbour, M., et al. (2021). Public health effects of travel-related policies on the COVID-19 pandemic: a mixed-methods systematic review. Journal of Infection, 83, 413–23.

Braczkowski, A.R., O'Bryan, C.J., Lessmann, C., et al. (2023). The unequal burden of human-wildlife conflict. Communications Biology, 6, 182.

Callahan, C., Lee, R.A., Lee, G.R., et al. (2021). Nasal swab performance by collection timing, procedure, and method of transport for patients with SARS-CoV-2. Journal of Clinical Microbiology, 59, e00569–21.

Chappell, B. (2023). What we know about the deadliest U.S. bird flu outbreak in history. All Things Considered, National Public Radio: https://www.npr.org/2022/12/02/1140076426/what-we-know-about-the-deadliest-u-s-bird-flu-outbreak-in-history, accessed 22/2/2023.

Chinazzi, M., Davis, J.T., Ajelli, M., et al. (2020). The effect of travel restrictions on the spread of the 2019 novel coronavirus (COVID-19) outbreak. Science, 368, 395–400.

Delong, S., Wilusz, C. (2022). Wastewater monitoring took off during the COVID-19 pandemic – and here's how it could help head off future outbreaks. The Conversation: https://theconversation.com/wastewater-monitoring-took-off-during-the-covid-19-pandemic-and-heres-how-it-could-help-head-off-future-outbreaks-180775, accessed 17/2/2023.

Dizikes, P. (2011). When the butterfly effect took flight. MIT Technology Review: https://www.technologyreview.com/2011/02/22/196987/when-the-butterfly-effect-took-flight/, accessed 2/11/2022.

Dong, E., Du, H., Gardner, L. (2020). An interactive web-based dashboard to track COVID-19 in real time. Lancet Infectious Diseases, 20, 533–4.

Dunay, E., Apakupakul, K., Leard, S., et al. (2018). Pathogen transmission from humans to great apes is a growing threat to primate conservation. EcoHealth, 15, 148–62.

Eyre, D.W., Taylor, D., Purver, M., et al. (2022). Effect of Covid-19 vaccination on transmission of alpha and delta variants. New England Journal of Medicine, 386, 744–56.

Frieden, T.R., Lee, C.T., Bochner, A.F., et al. (2021). 7-1-7: an organising principle, target, and accountability metric to make the world safer from pandemics. The Lancet, 398, 638–40.

Gold, J.A., Rossen, L.M., Ahmad, F.B., et al. (2020). Race, ethnicity, and age trends in persons who died from COVID-19—United States, May–August 2020. Morbidity and Mortality Weekly Report (MMWR), 69, 1517–21. DOI: 10.15585/mmwr.mm6942e1

Grépin, K.A., Ho, T.L., Liu, Z., et al. (2021). Evidence of the effectiveness of travel-related measures during the early phase of the COVID-19 pandemic: a rapid systematic review. BMJ Global Health, 6, e004537.

He, J.L., Luo, L., Luo, Z.D., et al. (2020). Diagnostic performance between CT and initial real-time RT-PCR for clinically suspected 2019 coronavirus disease (COVID-19) patients outside Wuhan, China. Respiratory Medicine, 168, 105980.

Hockings, K.J. (2016). Mitigating human–nonhuman primate conflict. In M. Bezanson, K.C. MacKinnon, E. Riley, et al. (Eds.), The International Encyclopedia of Primatology, John Wiley & Sons.

Holmes, E.C., Goldstein, S.A., Rasmussen, A.L., et al. (2021). The origins of SARS-CoV-2: a critical review. Cell, 184, 4848–56.

Huang, C., Wang, Y., Li, X., et al. (2020). Clinical features of patients infected with 2019 novel coronavirus in Wuhan, China. Lancet, 395, 497–506.

Huijbregts, B., De Wachter, P., Obiang, L.S.N., Akou, M.E. (2003). Ebola and the decline of gorilla Gorilla gorilla and chimpanzee Pan troglodytes populations in Minkebe Forest, north-eastern Gabon. Oryx, 37, 437–43.

Jegerlehner, S., Suter-Riniker, F., Jent, P., et al. (2021). Diagnostic accuracy of a SARS-CoV-2 rapid antigen test in real-life clinical settings. International Journal of Infectious Diseases, 109, 118–22.

Kalema-Zikusoka, G., Rubanga, S., Ngabirano, A., Zikusoka, L. (2021). Mitigating impacts of the COVID-19 pandemic on gorilla conservation: lessons from Bwindi Impenetrable Forest, Uganda. Frontiers in Public Health, 9, 655175.

Keusch, G.T., Amuasi, J.H., Anderson, D.E., et al. (2022). Pandemic origins and a One Health approach to preparedness and prevention: solutions based on SARS-CoV-2 and other RNA viruses. Proceedings of the National Academy of Sciences, USA, 119, e2202871119.

Khmer Times. (2023). After death of girl, 12 more possibly detected with H5N1 bird flu in Cambodia. Khmer Times: https://www.khmertimeskh.com/501244375/

after-death-of-girl-yesterday-12-more-detected-with-h5n1-bird-flu/, accessed 24/2/2023.

Krammer, F. (2020). SARS-CoV-2 vaccines in development. Nature, 586, 516–27.

Kritz, F. (2023). The Marburg outbreak in Equatorial Guinea is a concern—and a chance for progress. Goats and Soda, National Public Radio: https://www.npr.org/sections/goatsandsoda/2023/02/17/1157810407/the-marburg-outbreak-in-equatorial-guinea-is-a-concern-and-a-chance-for-progress, accessed 22/2/2023.

Lam, T.T.-Y., Jia, N., Zhang, Y.W., et al. (2020). Identifying SARS-CoV-2-related coronaviruses in Malayan pangolins. Nature, 583, 282–5.

Leroy, E.M., Rouquet, P., Formenty, P., et al. (2004). Multiple Ebola virus transmission events and rapid decline of Central African wildlife. Science, 303, 387–90.

Levin, E.G., Lustig, Y., Cohen, C., et al. (2021). Waning immune humoral response to BNT162b2 Covid-19 vaccine over 6 months. New England Journal of Medicine, 385, e84.

Li, X., Cui, W., Zhang, F. (2020). Who was the first doctor to report the COVID-19 outbreak in Wuhan, China? Journal of Nuclear Medicine, 61, 782–3.

Liu, W.J., Liu, P., Lei, W., et al. (2023). Surveillance of SARS-CoV-2 at the Huanan Seafood Market. Nature, in press.

Macfie, E.J., Williamson, E.A. (2010). Best practice guidelines for great ape tourism (No. 38). IUCN: https://portals.iucn.org/library/sites/library/files/documents/ssc-op-038.pdf, accessed 14/4/2023.

Markandya, A., Taylor, T., Longo, A., et al. (2008). Counting the cost of vulture decline—an appraisal of the human health and other benefits of vultures in India. Ecological Economics, 67, 194–204.

McKinnie, T. (2016). Ecotourism. In M. Bezanson, K.C. MacKinnon, E. Riley, et al. (Eds.), The International Encyclopedia of Primatology, John Wiley & Sons.

Meyerowitz-Katz, G., Bhatt, S., Ratmann, O., et al. (2021). Is the cure really worse than the disease? The health impacts of lockdowns during COVID-19. BMJ Global Health, 6, e006653.

Morfino, R.C., Bart, S.M., Franklin, A., et al. (2023). Aircraft wastewater surveillance for early detection of SARS-CoV-2 variants—John F. Kennedy International Airport, New York City, August–September 2022. Morbidity & Mortality Weekly Report (MMWR), 72, 210–11.

Nerpel, A., Yang, L., Sorger, J., et al. (2022). SARS-ANI: a global open access dataset of reported SARS-CoV-2 events in animals. Scientific Data, 9, 438.

Nuismer, S.L., Bull, J.J. (2020). Self-disseminating vaccines to suppress zoonoses. Nature Ecology and Evolution, 4, 1168–73.

One Health High-Level Expert Panel (OHHLEP), Adisasmito, W.B., Almuhairi, S., et al. (2022). One Health: a new definition for a sustainable and healthy future. PLoS Pathogens, 18, e1010537.

One Health High-Level Expert Panel (OHHLEP). (2023). From relying on response to reducing the risk at source. OHHLEP whitepaper/Opinion piece: https://www.who.int/publications/m/item/prevention-of-zoonotic-spillover.

Oreshkova, N., Molenaar, R.J., Vreman, S., et al. (2020). SARS-CoV-2 infection in farmed minks, the Netherlands, April and May 2020. Eurosurveillance, 25, 2001005.

Oude Munnink, B.B., Sikkema, R.S., Nieuwenhuijse, D.F., et al. (2021). Transmission of SARS-CoV-2 on mink farms between humans and mink and back to humans. Science, 371, 172–7.

Parry, J. (2004). Breaches of safety regulations are probable cause of recent SARS outbreak, WHO says. BMJ, 328, 1222.

Peiris, J.S.M., Guan, Y., Yuen, K.Y. (2004). Severe acute respiratory syndrome. Nature Medicine, 10, S88–97.

Polack, F.P., Thomas, S.J., Kitchin, N., et al. (2020). Safety and efficacy of the BNT162b2 mRNA Covid-19 vaccine. New England Journal of Medicine, 383, 2603–15.

Rockefeller Foundation. (2023). Pandemic Prevention Initiative: https://www.rockefellerfoundation.org/pandemicpreventioninitiative/, accessed 16/2/2023.

Rouquet, P., Froment, J.M., Bermejo, M., et al. (2005). Wild animal mortality monitoring and human Ebola outbreaks, Gabon and Republic of Congo, 2001-2003. Emerging Infectious Diseases, 11, 283–90.

Rulli, M.C., Santini, M., Hayman, D.T., D'Odorico, P. (2017). The nexus between forest fragmentation in Africa and Ebola virus disease outbreaks. Scientific Reports, 7, 41613.

Sabuhoro, E., Wright, B., Munanura, I.E., et al. (2021). The potential of ecotourism opportunities to generate support for mountain gorilla conservation among local communities neighboring Volcanoes National Park in Rwanda. Journal of Ecotourism, 20, 1–17.

Strindberg, S., Maisels, F., Williamson, E.A., et al. (2018). Guns, germs, and trees determine density and distribution of gorillas and chimpanzees in Western Equatorial Africa. Science Advances, 4, eaar2964.

Sun, K., Wang, W., Gao, L., et al. (2021). Transmission heterogeneities, kinetics, and controllability of SARS-CoV-2. Science, 371(6526), eabe2424. doi: 10.1126/science.abe2424. PMID: 33234698; PMCID: PMC7857413.

Temmam, S., Vongphayloth, K., Baquero, E., et al. (2022). Bat coronaviruses related to SARS-CoV-2 and infectious for human cells. Nature, 604, 330–36.

Thomas, S.J., Moreira Jr., E.D., Kitchin, N., et al. (2021). Safety and efficacy of the BNT162b2 mRNA Covid-19 vaccine through 6 months. New England Journal of Medicine, 385, 1761–73.

US Fish & Wildlife Service (USFWS). (n.d.). Salamanders as injurious wildlife - what it means for owners and scientists: https://fws.gov/node/266100, accessed 14/2/2023.

Valdez, C., Aquino, M. (2023). Bird flu kills sea lions and thousands of pelicans in Peru's protected areas. Reuters: https://www.reuters.com/business/healthcare-pharmaceuticals/bird-flu-kills-sea-lions-thousands-pelicans-perus-protected-areas-2023-02-21/, accessed 22/2/2023.

Vandenberg, O., Martiny, D., Rochas, O., et al. (2021). Considerations for diagnostic COVID-19 tests. Nature Reviews Microbiology, 19, 171–83.

Vaselli, N.M., Setiabudi, W., Subramaniam, K., et al. (2021). Investigation of SARS-CoV-2 faecal shedding in the community: a prospective household cohort study (COVID-LIV) in the UK. BMC Infectious Disease, 21, 784.

Walsh, P., Abernethy, K., Bermejo, M., et al. (2003). Catastrophic ape decline in western equatorial Africa. Nature, 422, 611–14.

Waltner-Toews, D. (2017). Zoonoses, One Health and complexity: wicked problems and constructive conflict. Philosophical Transactions of the Royal Society B, 372, 20160171.

Weber, A., Kalema-Zikusoka, G., Stevens, N.J. (2020). Lack of rule-adherence during mountain gorilla tourism encounters in Bwindi Impenetrable National Park, Uganda, places gorillas at risk from human disease. Frontiers in Public Health, 8, 1.

World Health Organization (WHO). (2020). COVID-19 – China: https://www.who.int/emergencies/disease-outbreak-news/item/2020-DON229, accessed 9/9/2022.

Wood, D., Adeline, S., Talbot, R., Wilburn, T. (2022). Coronavirus World Map: We've Now Passed the 600 million Mark for Infections. NPR.

Worobey, M., Levy, J.I., Malpica Serrano, L., et al. (2022). The Huanan Seafood Wholesale Market in Wuhan was the early epicenter of the COVID-19 pandemic. Science, 377, 951–9.

Wouters, O.J., Shadlen, K.C., Salcher-Konrad, M., et al. (2021). Challenges in ensuring global access to COVID-19 vaccines: production, affordability, allocation, and deployment. The Lancet, 397, 1023–34.

Xiao, X., Newman, C., Buesching, C.D., et al. (2021). Animal sales from Wuhan wet markets immediately prior to the COVID-19 pandemic. Scientific Reports, 11, 11,898.

Zhou, P., Yang, X.-L., Wang, X.G., et al. (2020). A pneumonia outbreak associated with a new coronavirus of probable bat origin. Nature, 579, 270–73.

Zhou, Y., O'Leary, T.J. (2021). Relative sensitivity of anterior nares and nasopharyngeal swabs for initial detection of SARS-CoV-2 in ambulatory patients: rapid review and meta-analysis. PLoS One, 16(7), e0254559.

# Index

80/20 rule 202*b*
active surveillance 67, 319
*Aedes aegypti* 180, 182*f*, 190, 191, 192
*Aedes albopictus* 182, 190, 191, 192
aerosolized diarrhoea 47
aerosols 19, 20, 20*f*, 47
*Agelaius phoeniceus* 123
aggregated data 74
airborne transmission 19, 20, 20*f*, 28*b*,
  47, 84
*Amblyomma* 1, 3*f*
amoeba, brain-eating 28, 29, 121
Amoy Gardens 47
*Anaplasma phagocytophilum* 148
animal market 155–9, 188, 228, 325–6,
  328–31
anthrax 30, 122, 123, 228
anthroponotic spillover 134, 136*f*,
  151*f*, 230
antibody test 61, 62, 156, 160, 306
antigenic drift 162
antigenic shift 162
anus, hippopotamus 29
anus, man's 21
anus, rhino 29
anus, rockfish 29
anus, rooster 137
anus, sea cucumber 29
arbovirus 178–182
Arctic fox 251, 252
armadillo 134, 135, 136, 135*f*
ascertainment bias 163
Asian tiger mosquito 182, 190, 191,
  192
asymptomatic 31*f*, 32
attenuation 136, 137
autochthonous 134

*Babesia* 39, 40*b*, 148
*Bacillus anthracis* 30, 122, 123, 228
bacteria 40
badger 306–12, 307*f*
badger culling 306–12
barriers 301, 332

bat, fruit 4, 5*f*, 223, 226*b*
bat, horseshoe 159, 160, 159*t*, 160*f*,
  329
bat, North American 115–20
bat, vampire 231–4, 233*f*, 234*f*, 312
*Batrachochytrium dendrobatidis* 188,
  256, 260, 161, 162, 272, 273, 301
*Batrachochytrium salamandrivorans* 262
Bayesian updating 14
beaver 156
beech tree, carnal relations with 262
bioindicators 284, 285*b*
bison 313–16, 313*f*, 314*f*
blackbird, red-winged 123
Black Death 6
Blossom the cow 301, 302
*Borrelia burgdorferi* 67, 143–51
borrelicidal blood factors 150
bovine TB 7, 189, 306–11
Bradford-Hill criteria 164
brain-eating amoeba 28, 29, 121
brucellosis 313–16
Burns, Rabbie 299

*Campylobacter* 1, 51–4, 51*f*, 126, 127
cancer 42
canine distemper virus 38*b*, 140, 258,
  259–60
case definition 81
case finding 81
case, confirmed 82
case, presumptive 82
case, probable 82
case, suspected 81
castration, lamb 1
cats, satanic 158
cattle, brucellosis 316
cattle, tuberculosis 306–11
cause finding 82
ceviche (armadillo, liver, and
  onion) 136
Chagas disease 24, 25, 26*f*, 27*b*
chicken, fried 76
chicken, sentinel 68

chimpanzee 2, 3*f*, 6*f*, 7
chytrid fungus 188, 256, 260, 161, 162,
  272, 273, 301
citizen science 71
clinical diagnosis 35
cluster 3
co-infection 286
Colorado tick fever virus 71*b*
common continuous source 50, 50*f*
common point source 49, 50*f*
common source outbreak 49
condor decline 279
confidence intervals 157
confirmed case 82
congenital transmission 25
contact networks 308, 209, 210
contact rate 96, 97, 98*b*, 335*b*
contact tracing 81
contagious period 31
containment 305*b*
core habitat 224
COVID-19 54–6, 60*f*, 325, 326, 337
critical community size 108
crowd diseases 9
culling 158, 306–12

Darwin, Charles 24, 231
*Dasypus novemcinctus* 134, 135, 136,
  135*f*
dead-end host 107, 128*f*, 133, 141*b*
dead-end zoonoses 4, 5*f*, 133, 141*b*,
  178*b*
Death's envoys 123
definitions 4*b*
definitive host 127, 236
density-dependent transmission 96,
  97*f*, 98*b*, 101, 106, 316
*Dermacentor albipictus* 199, 200, 201,
  200*f*, 203, 204, 204*f*
*Dermacentor andersoni* 71*b*, 73, 74, 73*f*
*Dermacentor variabilis* 73, 74, 73*f*
diagnostic tests 57
Diama dam 235–9, 241

diarrhoea 1, 9, 36, 47, 51, 52, 53, 217, 241, 276
diarrhoea, aerosolized 47, 49
*Dicrocoelium dendriticum* 40, 42*f*
dilution effect 76, 77, 146, 147, 148, 150
*Diphyllobothrium* 21, 23*f*, 41*f*
direct transmission 19, 28*b*
disease ecology 13, 34
disease triangle 36, 36*f*, 261, 261*f*, 264
*Dracunculus medinensis* 318–20
droplets 19, 20, 20*f*
dynamic equilibrium 105

Ebola virus 9, 10, 11, 78, 79, 109, 161, 162, 165, 166, 225, 226, 241, 342
ecologic fallacy 77
ecology 13
ecosystem services 284, 285
edge habitat 224
education 301
EID 1, 9, 11
elimination 301, 304, 305
elk 313–16, 314*f*
emerging infectious disease 1, 9, 11
endemic 3
enemy release 189
environmental reservoir 120, 121
environmental sloshing 208
enzootic 3
epidemic 3
epidemic curve 47, 48, 49, 48*b*, 48*f*
epidemiological triangle 36, 36*f*, 261*f*, 264
epidemiology 13
epizootic 3
eradication 276, 301, 304, 305
*erythema migrans* 35
Ethiopian wolf 256, 257, 258, 257*f*
Everglades virus 194
extrinsic incubation period 179*b*

faeces 20, 22, 30, 40, 121, 123, 124, 126, 127, 128, 207*f*, 236, 241, 334
fade out 4, 76, 108
false negatives 59
false positives 59
ferret, black-footed 294–7, 295*f*
fomites 20, 21
food-borne transmission 21, 22*t*, 28*b*
force of infection 68, 79, 82, 98*b*, 144, 263, 308
Fort Collins, Colorado 177
fox, Arctic 251, 252
Foy, Brian and family 177, 178
fragmented populations 256–9

frequency dependent transmission 96, 98*b*, 106
frog, growling grass 254, 255, 256, 254*f*
fungi 41, 263, 264

genomics 161, 162, 163
germ theory 11, 137
Goldberg, Tony 1
Goldilocks 205
Gordian worms 281, 282
gorilla 7, 226, 267, 341, 342, 243, 343*f*
Grand Canyon National Park 83, 84
grasshopper mouse 74
Guinea worm disease 318, 319, 320

habitat degradation 223, 224, 225
habitat destruction 223, 224, 225
habitat fragmentation 223, 224, 225
Hansen's disease 133–6
hantavirus 33, 34, 33*f*, 164–5, 271
heat waves 215, 216, 217
hedgehog 19, 330
Hendra virus 221, 222, 223
herd immunity 103
hibernaculum 115
Himalayan palm civets 155–9, 329
*Histoplasma capsulatum* 35, 123–6
histoplasmosis 35, 122–6
HIV 6*f*, 7
host community model 146, 147, 148, 150
human immunodeficiency virus 6*f*, 7
hydrophobia 19, 136
hyperendemic 4

immunity 36
inbreeding 258
incidence 56*b*
incubation period 31, 31*f*, 34*f*, 53–4, 137
index case 47
infectious period 31
inoculum 37
intensity dependent disease outcome 201, 202, 261
intermediate hosts 127, 128
intermittent common source 50, 50*f*
intrinsic incubation period 179*b*
invasion stage 316
invasion threshold 99, 104
invasive alien species 188
isolation 301, 302, 332
*Ixodes pacificus* 150, 162*f*
*Ixodes scapularis* 68*f*, 143–51, 213*b*

Jenner, Edward 301

Kafle, Pratap 207*f*
kissing bugs 25*f*
*Klebsiella pneumoniae* 287, 288*b*
Koch's postulates 62*b*

lab accident 302, 331
Lady Mary Wortley Montagu 302
ladybird, two-spotted 166, 167, 167*f*
latent infection 165, 166
latent period 31, 31*f*
law of mass action 92
leprosy 133–6
life cycle, complex 22
limerick interludes 126, 203
lion, African 38*b*, 39*f*
lion, mountain 83, 84
*Listeria* 123
lizard, western fence 150
Lyme disease 35, 67–70, 74, 76, 77, 143–51
lymphocytic choriomeningitis virus 24

Maalin, Ali Maow 302
mad cow disease 9
Maine wedding 54, 55, 56
major histocompatibility complex 258
malaria 4*b*, 7, 24, 30, 40, 104, 120, 177, 192, 195*f*, 205, 206*f*, 226, 241, 304
mangabey, sooty 7
mange, otodectic 251, 252
mange, sarcoptic 279, 280, 291–4
Marburg virus 175, 176, 177
marmot, hibernating 166
mathematical model 89, 91*b*, 107, 109, 141*b*, 189, 201, 209*b*, 254*b*, 280, 294
matrix habitat 224
Meister, Joseph 136, 137
meningeal worm 107, 203, 211
mesopredator release 276
metapopulations 254, 255, 256*b*
microbiome 36
microcephaly 181
mine removal 186*b*
mink 230
monkeypox virus 183–6, 185*f*
moose 107, 199, 200, 201, 200*f*, 203, 204
mountain lion 83, 84
mouse, deer 33, 34, 33*f*, 165, 271
mouse, grasshopper 74
mouse, multimammate 96, 97
mouse, white-footed 143
monkeypox virus 183–6, 185*f*
multi-host communities 133, 138, 141–8
musk ox 206–10, 207*f*, 215, 216

myalgia 33
*Mycobacterium bovis* 7, 189, 306–11
*Mycobacterium leprae* 133–6
*Mycobacterium tuberculosis* 7

*Naegleria foweleri* 28, 29
needles 22
negative predictive value 60*b*
neglected tropical diseases 239, 240*b*
Nelmes, Sarah 301, 302
nematomorphs 281, 282
network theory 308, 309, 310
Ngorongoro Crater 38, 39, 40*b*
niche model 210, 213, 214, 215
Nipah virus 4, 5*f*, 89, 226
norovirus 53, 54
nostril tick 2, 3*f*

oligosymptomatic 35
One Health 338–43
oral-faecal route 22
organ transplant 22, 23, 24
otodectic mange 251, 252
outbreak 3
outbreak investigation 79, 81*f*

pandemic 3
pandemic prevention 344, 345
pangolin 329
parameter 91–4
*Parelaphostrongylus tenuis* 203, 212
party, Christmas 51–4
passive surveillance 70
Pasteur, Louis 136, 137, 299
*Pasteurella multocida* 215, 216, 217
pâté, chicken liver 52, 53
pathogenic 34
PCR 57*b*, 58*f*, 155, 156
pentastome 167, 168, 169, 168*f*
phenology 166, 211, 212
Phipps, James 301, 302
plague 6, 83, 84, 85, 95, 192, 294–7
*Plasmodium falciparum* 4*b*, 7, 8*f*, 205, 206
pornography, hard-core 4*b*
positive predictive value 60*b*
poverty-disease 243
prairie dog 183–6, 294–7, 295*f*, 296*f*
precautionary measures 316, 317, 331
presumptive case 82
predictions, qualitative vs quantitative 90, 91, 99*b*
prevalence 56*b*
primary case 47
prion 41
probable case 82
process-based model 209, 210

propagated outbreaks 50, 50*f*, 54, 55, 56, 55*f*
protozoa 40
*Pseudogymnoascus destructans* 116–20, 117*f*, 264, 265, 266
python, Burmese 167, 168, 169, 168*f*, 173*f*, 192, 193, 194, 192*f*

quarantine 301, 332, 333

rabbit spinal cords 136
rabies virus 4, 19, 23, 75, 76*b*, 136–40, 233, 234, 256, 257, 258
rabies, vaccination 140–1
raccoon dog 155–9, 328, 329
*Raillietiella orientalis* 167, 168, 169, 168*f*
rat, Gambian pouched 184–6
rat, greater bandicoot 186, 188
rat, parts 186
rat, *Rattus* 186, 188
rat, Southern pouched 186, 187*b*, 187*f*
reduviid bugs 24, 25*f*, 26*f*
reproductive number, basic 99, 103, 141, 299*b*, 300*b*
reproductive number, effective 101, 110, 300*b*
reservoir competence 143, 147*f*, 148, 148*f*
reservoir hosts 107, 133, 166, 167*b*, 189, 195*f*, 221, 226, 260, 330
resistance 265, 266
respiratory droplets 19, 20, 20*f*
rinderpest virus 276, 277
Rocky Mountain wood tick 71*b*, 73, 73*f*
$R_0$ 99, 103, 141, 299*b*, 300*b*
roe deer, raw 22*t*

saiga antelope 216, 217
sapronoses 121
sapronotic disease agent 121
sarcoptic mange 279, 280, 291–4
SARS 47, 156–61, 163*f*
SARS-CoV-2 54, 55, 56, 162, 163, 163*f*, 230, 325–37
satanic cats 158
schistosomiasis 235–9, 241
sea otter 128, 129, 129*f*
sensitivity 59, 306, 333, 334
sentinel chickens 68
sentinel monkey 178
sentinel testing 68
Serengeti 38, 39, 40*b*, 75, 76*b*, 138, 139, 140, 276, 277, 278
serology 160
seroprevalence 56*b*, 74, 160

sialorrhoea 136
Simpson's Paradox 78
skin, Celtic 74
slimeball 41, 42*f*
smallpox 301, 302, 303, 317
snake decline 272, 273
snakebite envenomation 239, 240*b*
social perturbation 310
species distribution model 210, 213, 214, 215*b*
specificity 59, 333, 334
spillover 2, 2*f*, 5*f*, 133, 331
spontaneous generation 137
sporadic 4
statistical model 91*b*, 213
stochasticity 107, 254
stuttering zoonoses 4, 5*f*
surveillance 67
susceptibility 36
sushi 21
suspected case 81
sustainable development goals 339, 340
sustainable zoonoses 4, 5*f*
symbionts 286
symptomatic period 32
symptoms 35

tapeworm 21, 22*t*, 23*f*, 41*f*
Tasmanian devil facial tumour disease 42, 258, 259, 259*f*, 274, 275, 276
TB testing 186*b*
test validity 59
testicles, lamb's 1
thermal performance curves 205, 206
tick, American dog 73, 73*f*, 74
tick, black-legged 69*f*, 70*f*, 143–51, 213*b*
tick, lion 39*f*
tick, nostril 2
tick, Rocky mountain wood 71*b*, 73, 73*f*, 74
tick, western black-legged 150
tick, winter 199, 200, 201, 200*f*, 203, 204, 204*f*
tolerance 265, 266
torpor 115
*Toxoplasma gondii* 127, 128, 129
transmission, airborne 19, 20*f*, 28*b*
transmission, biological 24
transmission, congenital 25
transmission, direct 19, 28*b*
transmission, environmental 21
transmission, food-borne 21, 22*t*, 28*b*
transmission, functions 96, 97
transmission, horizontal 25

transmission, iatrogenic 22
transmission, indirect 20
transmission, mechanical 24
transmission, oral-faecal 22
transmission, sexual 19, 178
transmission, trans-ovarial 27
transmission, trophic 22
transmission, vector-borne 24, 28*b*
transmission, vehicle-borne 21, 28*b*
transmission, vertical 25, 28*b*
transmission, water-borne 21
transmission, with needles 22
trans-ovarial transmission 27
treatment 301
trophic transmission 22
true negatives 59
true positives 59

*Trypanosoma cruzi* 25, 27*b*
tuberculosis 7

*Umingmakstrongylus pal-
    likuukensis* 206–10,
    207*f*

vaccination 301–4, 333
vaccination threshold 103
vaccination, wildlife 140, 141, 143,
    296, 297
vector control 291–7
vector-borne transmission 21
vertical transmission 25
vicuna 277–80, 278*f*
virus 40

wedding, Maine 54–6

West Nile virus 4, 5*f*, 60*f*, 243
white nose syndrome 115–20, 117*f*
wicked problems 337, 338, 343
wildebeeste 276, 277, 278, 277*f*
wildlife farming 229, 230
wombat 291–4, 292*f*

xenodiagnosis 144, 145*b*
xenosurveillance 145*b*

yellow fever mosquito 180, 182*f*, 190,
    191, 192
*Yersinia pestis* 6, 35, 83, 84, 85, 192
Yosemite National Park 33

Zika virus 4, 178, 180–3
zoonosis 1, 4*b*, 4–9
zooprophylaxis 234